Alexei Kulik

Ergodic Behavior of Markov Processes

De Gruyter Studies in Mathematics

Volume 67

Alexei Kulik

Ergodic Behavior of Markov Processes

With Applications to Limit Theorems

DE GRUYTER

Mathematics Subject Classification 2010
37A25, 37A30, 47D07, 60F05, 60F17, 60G10, 60J05, 60J25

Author
Prof. Dr. Alexei Kulik
Institute of Mathematics
Ukrainian National Academy of Sciences
01601 Tereshchenkivska str. 3
Kyiv, Ukraine
and
Technical University of Berlin
Institute for Mathematics
Str. 17. Juni 136
10623 Berlin, Germany

ISBN 978-3-11-045870-1
e-ISBN (PDF) 978-3-11-045893-0
e-ISBN (EPUB) 978-3-11-045871-8
Set-ISBN 978-3-11-045894-7
ISSN 0179-0986

Library of Congress Cataloging-in-Publication Data
A CIP catalog record for this book has been applied for at the Library of Congress.

Bibliographic information published by the Deutsche Nationalbibliothek
The Deutsche Nationalbibliothek lists this publication in the Deutsche Nationalbibliografie; detailed bibliographic data are available on the Internet at http://dnb.dnb.de.

Typesetting: Integra Software Services Pvt. Ltd.
Printing and binding: CPI books GmbH, Leck
♾ Printed on acid-free paper
Printed in Germany

www.degruyter.com

To my beloved
Luda, Masha, and Katja

Preface

This book is based on the lecture course [85], furthermore revised and substantially extended. The book pursues two major goals, and is divided into two parts. Part I contains a complete and self-consistent exposition of a set of methods, which allow one to establish ergodic rates for Markov systems. We systematically use the probabilistic coupling approach, which makes it possible to treat, in a unified and quite a transparent way, all the variety of Markov systems, from the classical case of Markov chains with finite state spaces, up to Markov processes with intrinsic memory, where only weak ergodic rates are available. Part II contains a discussion of the limit theorems for functionals of Markov processes, namely, the Law of Large Numbers, the Central Limit Theorem, and their nonadditive generalizations: averaging principle and diffusion approximation. These two topics are naturally connected: in particular, the form the ergodic rates are obtained in Part I is strongly motivated by their further applications in limit theorems of Part II.

The book is aimed for a wide auditory. In its preparation, special attention was devoted for making the presentation systematic, and to simplify the reading for an audience, not necessarily expertised in the field. Therefore, we expect that it will be useful for graduate and postgraduate students, specialized in Probability and Statistics, aiming to get a first acquaintance with the ergodic properties of Markov systems and related applications. Minimal pre-requisites for understanding the core of the book are standard courses of Probability and Measure Theory. However, a basic knowledge in Stochastic Processes, Stochastic Calculus, and SDEs is highly desirable, since it would help a better understanding of the particular examples, which strongly motivate the choice of the form in which the theory is presented. We also expect that the book will be useful for specialists in other fields, where Markov models and related limit theorems are typically applied, for example, Statistics, Monte-Carlo Simulation, Statistical Physics. Finally, we believe that the book will be interesting for the experts in the field, as well: most of the general results presented in the book are scattered in the literature, and their systematic exposition, in a sense, had forced the author to develop for them more direct and transparent versions, which made some of them actually stronger than the originals.

The lecture course [85] has its origin in two minicourses, given by the author in the University of Potsdam, TU Berlin, and Humboldt University of Berlin (March 2013), and in Ritsumeikan University (August 2013). It had been prepared partially during the author's stay at TU Dresden (January 2014), Institute Henry Poincaré (July 2014, "Research in Paris" programme), and Ritsumeikan University (January 2015). The author gladly expresses his gratitude to all these institutions, as well as to the Universitätsverlag Potsdam for the kind permission to use Ref. [85] as a basis for this book. The author would like to thank Sylvie Roelly for the encouragement to compose lecture notes [85] and for a persistent support during their preparation.

DOI 10.1515/9783110458930-202

The author owes special thanks to René Schilling and Niels Jacob; this book would not have appeared without their encouragement for publishing a revision of Ref. [85]. For numerous discussions and helpful comments, the author is grateful to Ilya Pavlyukevich, Oleg Butkovsky, and especially Michael Scheutzow: the majority of the examples in Part I was strongly influenced by his valuable comments, and the name for the central object in the entire Part I – the greedy coupling – was introduced by him. The author is also grateful to Alexander Veretennikov; the entire Part II is deeply influenced by numerous discussions with him.

Berlin – Kyiv, October 2016 – May 2017

Contents

Introduction

It is a kind of “common knowledge” nowadays, that Markov chains and Markov processes provide a natural mathematical background for extremely wide variety of models in natural sciences. Because of a high relevance of Markov systems in applications, it is important to have a well-developed set of methods for study of their asymptotic properties; this, in particular, is crucial for statistical analysis of such systems, and for simulation purposes. The level of complexity of Markov models, available in the literature, drastically varies, starting from simple models, based on Markov chains with finite state spaces, and ending up with realistic yet complicated Markov systems with functional state spaces, such as systems with delay (typical e.g. for population dynamics), stochastic partial differential equations (typical, e.g., for hydrodynamical models), systems with fractional noises (appearing, e.g., in financial models). One general aim of this book is to present systematically one set of tools, well applicable for asymptotic study, in a unified way, of Markov systems of all the levels of complexity.

The theory of Markov chains, from its very beginning, is strongly connected with the limit theorems in Probability theory. The notion of the Markov chain (under the name the “sequence of random variables connected to a chain”) was originally introduced by A. Markov in Ref. [100] in order to design an easily tractable framework, where the basic limit theorems of Probability theory can be obtained without the independence assumption. In this very first paper, the Law of Large Numbers (LLN) for a Markov chain was established, together with the principally important stabilization property of the transition probabilities. The second paper [101], devoted to this topic, had already contained the Central Limit Theorem (CLT) for a Markov chain, though in a quite simple case and proved with an overcomplicated method, which did not receive a further extension. Anyway, that was a crucial step forward in the general theory, since at that time the question, whether the LLN and the CLT are specific for independent setting, was far from being clear. Markov’s insight made it clearly visible that, in that concern, independence can be efficiently replaced by weaker “loss of memory” type conditions, like the stabilization property of the transition probabilities of a Markov chain. This had a further strong (though not always explicit) influence to the entire Probability theory.

The design of this book is aimed to present to the reader a complete and self-sufficient view of one possible route across this very important and interesting field. In few words, the asymptotic study of Markov systems, adopted in the first part of the book, is concentrated around the concept of Harris-type theorems, introduced in Ref. [55] in the spirit of Refs. [50, 51]. This concept appears to be a very promising tool for a description of weak ergodic rates, typical for complicated Markov systems listed above. For the limit theorems, studied in the second part of the book, the main preliminary is given by the weak ergodic rates for the underlying Markov systems, which

DOI 10.1515/9783110458930-001

makes both parts of the book well adjusted, and provides a systematic background for further analysis of particular Markov systems of complicated structure, including their statistical analysis, simulation, and so on.

The proofs of ergodic rates in Part I are based on the coupling technique; see Refs. [95] and [128] for the generalities concerning the coupling methods. Systematic use of the coupling approach allows us to present in a simple and unified way the whole variety of available results, starting from the classic Ergodic Theorem for Markov chains with finite state spaces, and finishing with the Harris-type theorem for weakly ergodic Markov chains. The coupling construction used within all the proofs remains actually unchanged, and particular assumptions effect the subsequent estimates, only. We believe that this will substantially simplify the reading; also, the choice of the coupling method for the proof has some specific advantages. For instance, using this method we are able to separate clearly the ergodic rates and the "loss of memory" rates. Such a separation apparently was not made in the literature before, up to the notable exception of Ref. [53], and it may be important to have these types of "stabilization" effect separated in the settings, where the system slowly tends to equilibrium. Another hidden advantage is that the coupling method actually provides ergodic rates in the "path coupling" form, see eq. (5.3.2), which in some cases makes it possible to obtain more detailed limit theorems than just the similar ergodic rate (5.1.1) for the transition probabilities of the chain.

To treat Markov processes with continuous time, we adopt the time discretization, or "skeleton chain", approach. For various continuous-time Markov processes, which are ergodic in total variation, such an approach may appear not so simple as (and sometimes more restrictive than) a direct coupling approach, see discussion in Section 3.5. However, for weakly ergodic Markov processes such a direct coupling approach is hardly available. On the other hand, the skeleton chain approach made it possible for us to present in a quite unified way the exponential, sub-exponential, and polynomial ergodic rates for such seemingly diverse classes of processes, as diffusions, solutions to Lévy-driven SDEs, and solutions to Stochastic Delay Equations.

In order to keep the reader informed about the diversity of the methods and settings, other than explained in all the details in the main exposition, we endow the exposition by a number of comments and side-by remarks. Let us briefly outline some closely related topics, which were not included into the exposition. In the classical finite state space case, a perfect preliminary for the study of ergodic properties of a Markov chain is provided by the classification of states of the chain. Such an approach has a natural extension to more general setting based on the Harris irreducibility concept. However, in complicated situations, Harris irreducibility is not easy to check, or even may fail; the latter happens typically for Markov models with "intrinsic memory" (see Section 4.1). This is the reason for us to adopt the argument, which avoids the "classification" step systematically. In that sense, this argument is complementary to the one widely used in the literature (see Refs. [28, Chapters V, VI], [106, 109]), where the "classification" step is substantial.

Next, we do not make a detailed discussion of the "analytic" tools, suitable for the description of the stabilization effects in the Markov systems. The contraction-type argument, which dates back to the very origins of the theory ([100]; see also [25]) is just outlined in the proofs of Theorems 1.2.1 and 2.3.1, but its further extensions are not engaged into the discussion; we refer to Ref. [54], where this argument is revisited. Another natural possibility is to describe the stabilization of a Markov system in the terms of corresponding semigroups, and in this concern spectral methods and functional inequalities are widely applicable. We do not discuss this very diverse topic here in any details, referring to Refs. [4, 22, 135], and the survey in Ref. [17].

Within the book, the Markov chains and processes are assumed to be time-homogeneous. Mainly, corresponding results can be further extended to a time-inhomogeneous setting. Such an extension has a long history, for example in Ref. [25] the stabilization for the transition probabilities of a time-inhomogeneous Markov chain was developed as a tool for a CLT, with Markov dependence, without an identical distribution assumption on the summands. While being principally similar with those exposed here, the stabilization properties in the time-inhomogeneous case have some specialities, which can be clearly seen from Ref. [62], where a thorough analysis is given for discrete state space Markov chains, non-homogeneous in time.

Part II, devoted to limit theorems, does not include any Large Deviations (LD) type results, though such an extension of LLN is very natural and available for Markov systems in many cases of interest. In that concern we refer to Ref. [42] for the LD theory for random dynamical systems with small random perturbations, to Ref. [40] for the semigroup-based LD theory, and to Ref. [133] for the survey of the LD theory for diffusions, based on their ergodic rates. The main reason for us to exclude this topic from the discussion is that the corresponding theory apparently is not yet developed in the context of weakly ergodic Markov systems, which is one of our main points of interest. This is just one item in the long list of open questions, still available in this research field.

Part I: Ergodic Rates for Markov Chains and Processes

1 Markov Chains with Discrete State Spaces

1.1 Basic Definitions and Facts

Let $\mathbb{X} = \{i, j, \dots\}$ be a finite or countable set. An $\mathbb{X}$-valued sequence

$$X = \{X_n, n \ge 0\}$$

is called a *Markov chain* (MC), if for any time moments $m < n$, $m_1, \dots, m_k < m$ and any states $j, i, i_1, \dots, i_k \in \mathbb{X}$

$$\mathsf{P}(X_n = j | X_m = i, X_{m_1} = i_1, \dots, X_{m_k} = i_k) = \mathsf{P}(X_n = j | X_m = i).$$

The matrix $\Pi^{m,n} = \{p_{ij}^{m,n}\}_{i,j\in\mathbb{X}}$,

$$p_{ij}^{m,n} = \mathsf{P}(X_n = j | X_m = i),$$

is called the *transition probability matrix* of the MC X on the time interval $[m, n]$.

The basic properties of the family $\{\Pi^{m,n}, m \le n\}$ are listed below:

- Each $\Pi^{m,n}$ is a stochastic matrix:

$$p_{ij}^{m,n} \ge 0, \quad i, j \in \mathbb{X}, \quad \sum_j p_{ij}^{m,n} = 1, \quad i \in \mathbb{X};$$

- $p_{ij}^{m,m} = 1_{i=j}$;
- (the Chapman–Kolmogorov equation): for each $m \le r \le n$,

$$\Pi^{m,n} = \Pi^{m,r}\Pi^{r,n};$$

  that is,

$$p_{ij}^{m,n} = \sum_{k\in\mathbb{X}} p_{ik}^{m,r} p_{kj}^{r,n}, \quad i, j \in \mathbb{X}.$$

The family of transition probability matrices and the *initial distribution*

$$\mu = \{\mu_i = \mathsf{P}(X_0 = i)\}_{i\in\mathbb{X}}$$

of an MC completely define the law of the chain:

$$\mathsf{P}(X_0 = i_0, X_1 = i_1, \dots, X_m = i_m) = \mu_{i_0} p_{i_0 i_1}^{0,1} \cdots p_{i_{m-1} i_m}^{m-1,m}. \tag{1.1.1}$$

DOI 10.1515/9783110458930-002

The law of the chain with the initial distribution μ is denoted by P_μ, and respective expectation is denoted by E_μ. If μ is concentrated at one point, $\mu = \delta_i$, the notation is simplified to P_i, E_i.

In what follows, we restrict our consideration by the case of time-homogeneous chains, which by definition means that $\Pi^{m,n}$ depends on $n - m$ (the length of the corresponding time interval) only. In this case

$$\Pi^{m,n} = \Pi^{0,n-m} = \Pi^{n-m}$$

(the latter equality is just the notation), and if we denote $\Pi = \Pi^1$ (the one-step transition probability matrix), then

$$\Pi^n = \underbrace{\Pi \cdot \dots \cdot \Pi}_{n};$$

that is, Π^n equals to the nth (matrix) power of Π. In what follows, p_{ij}^n, $i, j \in \mathbb{X}$, denotes the entries of Π^n.

By eq. (1.1.1), for an MC with an initial distribution μ the distribution

$$\mu^n = \{\mu_i^n = \mathsf{P}(X_n = i)\}_{i\in\mathbb{X}}$$

of the value X_n of the chain at the time moment n equals

$$\mu_j^n = \sum_{i\in\mathbb{X}} \mu_i p_{ij}^n, \quad j \in \mathbb{X},$$

which can be written as

$$\mu^n = \mu\Pi^n$$

if we adopt for μ, μ^n the notation as row vectors. That is, an MC naturally defines a dynamical system

$$\mu \mapsto \mu\Pi \tag{1.1.2}$$

on the set $\mathcal{P}(\mathbb{X})$ of all probability measures on $\mathbb{X}$, which now can be understood as the set of (row) vectors with nonnegative entries and the total sum of entries equal to 1. A measure $\mu \in \mathcal{P}(\mathbb{X})$ which satisfies

$$\mu = \mu\Pi \tag{1.1.3}$$

is called an *invariant probability measure* (IPM) for the MC X. Clearly, an IPM μ is a fixed point for the dynamical system (1.1.2) and

$$\mu^n = \mu, \quad n \geq 0.$$

This identity and eq. (1.1.1) yield that, once X_0 has the law μ which is invariant for X, one has

$$\begin{aligned}\mathsf{P}(X_k = i_0, X_{k+1} = i_1, \dots, X_{k+m} = i_m) &= \mu_{i_0} p_{i_0 i_1} \cdots p_{i_{m-1} i_m}\\ &= \mathsf{P}(X_0 = i_0, X_1 = i_1, \dots, X_m = i_m)\end{aligned}$$

for any $k \geq 1$ and any states $i_0, \dots, i_k$. That is, the sequence $X^{(k)} = \{X_{n+k}, n \geq 0\}$ (the time shift of X by k) has the same finite-dimensional distributions with the original X. This means that X with $\mathrm{Law}(X_0) = \mu$ is a *strictly stationary* random sequence.

A state j is said to be *accessible* from a state i (written $i \to j$) if $p_{ij}^n > 0$ for some n. A state i is said to be *essential* if for any j such that $i \to j$ it is also true that $j \to i$. Otherwise there exists j such that $i \to j, j \nrightarrow i$ and the state i is called *inessential*. Each two essential states are either *connected* ($i \to j, j \to i$) or disconnected ($i \nrightarrow j, j \nrightarrow i$). A chain is called *irreducible* if all its essential states are connected.

Period $d(i)$ of an essential state i is defined as the greatest common divisor of the set

$$\mathbf{N}_i = \{n \geq 1 : p_{ii}^n > 0\}.$$

It is known that for connected states the periods are equal. An irreducible chain is said to be *aperiodic* if one (and hence any) essential state i has period $d(i) = 1$.

1.2 Ergodic Theorem

1.2.1 Ergodic Theorem for Finite MCs

In this section, we recall the classic Ergodic theorem for MCs with finite state spaces.

Theorem 1.2.1. *Let an MC X with a finite state space be irreducible and aperiodic.*

Then there exists unique IPM μ for X. Moreover, there exist positive constants $C, \rho > 0$ such that

$$\max_i \sum_j |p_{ij}^n - \mu_j| \leq C e^{-\rho n}, \quad i \in \mathbb{X}, \quad n \geq 0. \tag{1.2.1}$$

Theorem 1.2.1 actually states that for any initial distribution $\nu \in \mathscr{P}(\mathbb{X})$ the trajectory of the dynamical system (1.1.2) converges exponentially to the unique stationary point of this system. The following "toy" example gives a good benchmark for this "stabilization" effect.

Example 1.2.2. Let $\{X_n\}$ be a sequence of *independent and identically distributed* random variables valued in $\mathbb{X}$. This sequence can be interpreted as an MC with

$$p_{ij}^n = \mathsf{P}(X_n = j|X_0 = i) = \mathsf{P}(X_n = j) = \mu_j, \quad j \in \mathbb{X}, \quad n \geq 1,$$

the last identity is just the notation. The one-step transition probability matrix for the chain has the form

$$\Pi = \begin{pmatrix} \mu \\ \vdots \\ \mu \end{pmatrix}, \tag{1.2.2}$$

and for any initial distribution $\nu \in \mathscr{P}(\mathbb{X})$

$$\nu^n = \nu \Pi^n = \mu.$$

That is, the one-dimensional distributions of the chain become equal the invariant distribution (i.e., are *stabilized*) after just **one** time step. Theorem 1.2.1 states that in the general irreducible and aperiodic case essentially the same stabilization effect appears; the only difference is that such a stabilization in general holds gradually, while the "toy" example exhibits the stabilization after just the first time step.

We give two different proofs of Theorem 1.2.1 in Sections 1.2.2 and 1.2.3. For that, it will be convenient to reformulate the condition of Theorem 1.2.1.

Proposition 1.2.3. *The following statements are equivalent for a finite MC:*

(i) *the chain is irreducible and aperiodic;*

(ii) *for some $m \geq 1$ and $k \in \mathbb{X}$*

$$p_{ik}^m > 0, \quad i \in \mathbb{X}; \tag{1.2.3}$$

(iii) *for some $m \geq 1$*

$$\sum_k \min(p_{ik}^m, p_{jk}^m) > 0, \quad i, j \in \mathbb{X}. \tag{1.2.4}$$

Proof. (i) ⇒ (ii). Fix an essential state k and note that the set $\mathbf{N}_k$ satisfies

$$m, n \in \mathbf{N}_k \Rightarrow m + n \in \mathbf{N}_k.$$

Since the greatest common divisor of $\mathbf{N}_k$ is 1, this yields that this set contains all $n \geq Q$ for Q large enough. For any other essential state i we have $i \to k$ since the chain is irreducible (i.e., all the essential states are connected). That is, for some $M_i \geq 1$

$$p_{ik}^{M_i+Q} \geq p_{ik}^{M_i} p_{kk}^{n} > 0, \quad n \geq Q. \tag{1.2.5}$$

On the other hand, for any nonessential state i there exists a chain of states $i_1, \dots, i_r$ such that $i_{l-1} \to i_l, i_l \not\to i_{l-1}, l = 1, \dots r$ $(i_0 = i)$ and i_r is essential. This easily follows from the definition of the nonessential state and the fact that $\mathbb{X}$ is finite. Hence $i \to i_r$, an essential state, and eq. (1.2.5) holds true for a nonessential i, as well. Taking M being maximal of all $M_i, i \in \mathbb{X}$, we get

$$p_{ik}^{m} > 0, \quad i \in \mathbb{X}$$

for the fixed k and $m = M + Q$, which proves eq. (1.2.3).

(iii) $\Rightarrow$ (i). If i is essential, then

$$p_{ik}^{n} = 0, \quad n \geq 0$$

for any k which is either not essential or does not belong to the class of all the essential states connected with i. Hence if X contains two disjoint classes of essential states, then eq. (1.2.4) fails for two states i, j taken from these classes.

If an essential state i has a nontrivial period $d(i) > 1$, then the corresponding class of connected states splits into $d(i)$ cyclic subclasses such that for a state k from the rth subclass inequality

$$p_{ik}^{n} > 0$$

is possible only if $n = r \bmod d(i)$. For the original state i, $r = 0$, hence taking j from a cyclic subclass with (say) $r = 1$ we get a pair i, j such that eq. (1.2.4) fails. This completes the proof of implication (iii) $\Rightarrow$ (i). The implication (ii) $\Rightarrow$ (iii) is obvious. □

1.2.2 "Analytic" Proof of Theorem 1.2.1: Contraction Argument

The idea of this proof dates back to Markov's seminal paper [100], and is both natural and transparent: one shows that the matrices of transitional probabilities are *contracting* w.r.t. proper norm on a complement to certain one-dimensional subspace. This subspace actually corresponds to the unique IPM for the chain X.

Denote $N = \#\mathbb{X}$, and consider the norm on the space $\mathbb{R}^N$

$$\|v\|_1 = \sum_i |v_i|, \quad v = \{v_i\} \in \mathbb{R}^N.$$

Because Π is a stochastic matrix,

$$\|\Pi^\top v\|_1 = \sum_j \left|\sum_i v_i p_{ij}\right| \le \sum_i |v_i| \underbrace{\sum_j p_{ij}}_{=1} = \|v\|_1. \tag{1.2.6}$$

Consider the linear subspace

$$\mathbb{R}_0^N = \left\{ v \in \mathbb{R}^N : \sum_i v_i = 0 \right\}.$$

Similar to eq. (1.2.6), one can show that

$$\Pi^\top v \in \mathbb{R}_0^N, \quad v \in \mathbb{R}_0^N. \tag{1.2.7}$$

On the other hand, let $m \ge 1, k \in \mathbb{X}$ be such that eq. (1.2.3) holds, then there exists $c > 0$ such that

$$p_{ik}^m > c, \quad i \in \mathbb{X}. \tag{1.2.8}$$

This yields for $v \in \mathbb{R}_0^N$

$$\begin{aligned} \|(\Pi^\top)^m v\|_1 &= \sum_j \left|\sum_i v_i p_{ij}^m\right| = \sum_j \left|\sum_i v_i p_{ij}^m - c 1_{j=k} \sum_i v_i\right| \\ &\le \sum_i |v_i| \sum_j (p_{ij}^m - c 1_{j=k}) \le (1-c)\|v\|_1, \end{aligned}$$

which means that the matrix $(\Pi^\top)^m$ is *contracting* on $\mathbb{R}_0^N$ w.r.t. the norm $\|\cdot\|_1$. Combined with eqs. (1.2.6) and (1.2.7) this yields that for some $C, \rho > 0$

$$\|(\Pi^\top)^n v\|_1 \le C e^{-\rho n}, \quad v \in \mathbb{R}_0^N. \tag{1.2.9}$$

Taking for arbitrary i, j

$$v = \{v_k\}, \quad v_k = \begin{cases} 1, & k = i; \\ -1, & k = j; \\ 0, & \text{otherwise}, \end{cases}$$

we get from eq. (1.2.9)

$$\max_{i,j} \sum_k |p_{ik}^n - p_{jk}^n| \le C e^{-\rho n}. \tag{1.2.10}$$

Now we can finalize the proof. Since Π is a stochastic matrix, the column vector

$$(1, \dots, 1)^\top$$

is an eigenvector for Π with the eigenvalue 1. Hence there exists also an eigenvector for $\Pi^\top$ with the same eigenvalue, we denote it by $w = \{w_j\}$. Note that $w \notin \mathbb{R}_0^N$, because otherwise the identities

$$w = \Pi^\top w = \cdots = (\Pi^\top)^n w = \cdots$$

contradict to eq. (1.2.9). Hence we can assume that w is properly normalized so that

$$\sum_i w_i = 1.$$

Using eq. (1.2.10) and the identity

$$w_k = ((\Pi^\top)^n w)_k = \sum_j w_j p_{jk},$$

we get that

$$\max_i \sum_k |p_{ik}^n - w_k| = \max_i \sum_k \left| \sum_j w_j (p_{ik}^n - p_{jk}) \right| \le Ce^{-\rho n} \sum_j |w_j|. \tag{1.2.11}$$

That is, the sequence of matrices Π^n converges to the matrix with the identical rows equal to $w^\top = (w_1, \dots, w_N)$. This means that $w^\top \in \mathscr{P}(\mathbb{X})$; in addition, $\mu = w^\top$ is an IPM for X because w is an eigenvector for $\Pi^\top$. Inequality (1.2.1) follows directly from eq. (1.2.11). □

1.2.3 "Probabilistic" Proof of Theorem 1.2.1: Coupling Argument

The idea of this proof dates back to W. Döblin [26, 27] and is based on the notion of *coupling*. To make the exposition transparent, we explain the idea assuming that eq. (1.2.3) holds true with $m = 1$; the general case is analogous.

By the definition, a *coupling* for a pair of probability measures μ, ν is a probability measure κ on the product space $\mathbb{X} \times \mathbb{X}$ such that its *marginal distributions* (i.e., projections on the first and second coordinates, respectively) equal to μ and ν. We denote the set of all couplings for μ, ν by $C(\mu, \nu)$. If $\#\mathbb{X} = N$, an element $C(\mu, \nu)$ is naturally understood as an $N \times N$-matrix $\{\kappa_{ij}\}$ with the prescribed sums in rows and columns:

$$\sum_j \kappa_{ij} = \mu_i, \quad i \in \mathbb{X}; \quad \sum_i \kappa_{ij} = \nu_j, \quad j \in \mathbb{X}. \tag{1.2.12}$$

The core of the coupling proof is provided by the following simple observation: for any random variables ξ, η defined on a common probability space $(\Omega, \mathcal{F}, \mathsf{P})$ and having the joint law $\kappa \in C(\mu, \nu)$,

$$\sum_i |\mu_i - \nu_i| \le 2\mathsf{P}(\xi \ne \eta). \tag{1.2.13}$$

Inequality (1.2.13) follows easily by eq. (1.2.12):

$$\begin{aligned}\sum_i |\mu_i - \nu_i| = \sum_i \left| \sum_j \kappa_{ij} - \sum_j \kappa_{ji} \right| &\le \sum_{i,j} |\kappa_{ij} - \kappa_{ji}| \\ &\le \sum_{i \ne j} (\kappa_{ij} + \kappa_{ji}) = 2 \sum_{i \ne j} \kappa_{ij} = 2\mathsf{P}(\xi \ne \eta).\end{aligned}$$

The main idea of the proof now can be explained as follows: the left-hand side in eq. (1.2.13) can be estimated by choosing a proper pair of "representatives" ξ, η for the given laws μ, ν.

Now we proceed with the proof of Theorem 1.2.1. We define an MC $Z = (X, Y)$ on the product space $\mathbb{X} \times \mathbb{X}$ by the following conventions:

- if the current positions of X, Y are different, then X, Y perform independently one step with the transition probability Π;
- if the current positions of X, Y coincide, then X, Y perform the step simultaneously with the transition probability Π.

That is, the transition probabilities for Z are equal:

$$p_{(i_1,i_2)(j_1,j_2)} = \begin{cases} p_{i_1,j_1} p_{i_2,j_2}, & i_1 \ne i_2; \\ p_{i_1,j_1} 1_{j_1 = j_2}, & i_1 = i_2. \end{cases}$$

By the construction,

- for $Z = (X, Y)$ with $Z_0 = (i, j)$ the laws of $X = \{X_n\}$ and $Y = \{Y_n\}$ equal P_i and P_j, respectively;
- if $X_n = Y_n$, then $X_N = Y_N, N \ge n$ (once coupled, the trajectories stay coupled).

Let $Z_0 = (i, j)$ for some i, j arbitrary but fixed. By the first property, for each $n \ge 1$ the variables X_n and Y_n have the laws $\{p^n_{ik}\}_{k \in \mathbb{X}}$ and $\{p^n_{jk}\}_{k \in \mathbb{X}}$, respectively. Then by eq. (1.2.13) and the second property,

$$\sum_k |p_{ik}^n - p_{jk}^n| \le 2\mathsf{P}(X_n \neq Y_n) = 2\mathsf{P}(L > n), \tag{1.2.14}$$

where we denote

$$L = \min\{n : X_n = Y_n\},$$

the "coupling time" for the chain Z. Recall that we assumed eq. (1.2.3) to hold with $m = 1$, then for any n

$$\mathsf{P}(L = n + 1 | L > n) \ge \min_{i' \neq j'} \sum_l p_{i'l} p_{j'l} \ge \min_{i' \neq j'} p_{i'k} p_{j'k} \ge c^2 > 0;$$

see eq. (1.2.8). This yields eq. (1.2.11). The rest of the proof is the same as in the previous section. □

1.2.4 Recurrence and Transience. Ergodic Theorem for Countable MCs

In this section, we briefly discuss the version of Ergodic theorem for MCs with infinite (but countable) state spaces. In this setting, the tools of the renewal theory are very effective. Since these methods are not at the mainstream of the current book, we do not give a detailed discussion and just outline the main concepts and results, referring to [44, Chapter III, Sect. 6] for details.

Denote for $i \in \mathbb{X}$

$$\tau_i = \inf\{n : X_n = i\}$$

with the usual convention that $\inf \varnothing = \infty$. State i is called

- *recurrent* if

$$\mathsf{P}_i(\tau_i < \infty) = 1;$$

- *transient* if

$$\mathsf{P}_i(\tau_i = \infty) > 0.$$

Recurrent state i

- *zero recurrent* if $\mathsf{E}_i \tau_i = \infty$;
- *positive recurrent* if $\mathsf{E}_i \tau_i < \infty$.

Any two connected states i, j are transient, zero recurrent, or positively recurrent simultaneously. Hence for an irreducible chain this classification relates to the chain itself rather than to individual states.

For a finite state space, each essential state is positively recurrent and each inessential state is transient. In the infinite case the situation is drastically different; the following classical example shows that an essential state, in general, can have any of the types listed earlier.

Example 1.2.4. (The Bernoulli random walk on a half-line). Let $\mathbb{X} = \{0, 1, \dots\}$ and X have one-step transition probabilities

$$p_{00} = q, \quad p_{01} = p, \quad p_{i(i-1)} = q, \quad p_{i(i+1)} = p, \quad i = 1, 2, \dots,$$

$p + q = 1$. The chain is
- transient for $p > q$;
- zero recurrent for $p = q = 1/2$;
- positive recurrent for $p < q$.

Theorem 1.2.5. *Let an MC X with a countable state space be irreducible and aperiodic. Then for any essential state $i \in \mathbb{X}$*

$$p_{ij}^n \to \mu_j = \frac{1}{E_j \tau_j}, \quad n \to \infty, \quad j \in \mathbb{X}. \tag{1.2.15}$$

If the chain X is positive recurrent, then $\mu = \{\mu_i\}$ is the unique IPM for X, otherwise all μ_i equal 0 and an IPM for X does not exist.

Though Theorem 1.2.5 states an "individual" convergence (1.2.15) for i, j fixed, this statement has the following important improvement, which follows directly from Scheffé's lemma (e.g., [10, Appendix II]).

Corollary 1.2.6. *Let an MC X with a countable state space be irreducible, aperiodic, and positive recurrence.*

Then for any essential state $i \in \mathbb{X}$

$$\sum_j |p_{ij}^n - \mu_j| \to 0, \quad n \to \infty, \quad i \in \mathbb{X}. \tag{1.2.16}$$

This statement looks more close to Theorem 1.2.1, but still has substantial differences: convergence (1.2.16) is *individual* w.r.t. starting point i, and the rate of convergence in Corollary 1.2.6 is not specified, while Theorem 1.2.1 provides a *uniform* in i convergence at *exponential* rate.

1.3 Stationary MCs: Ergodicity, Mixing, and Limit Theorems

In this section, we discuss a connection between the stabilization property of the transition probabilities of an MC and ergodic properties of its strictly stationary version. Recall that, once X_0 has the law μ which is an IPM for X, the entire sequence $\{X_n\}$ is strictly stationary. In order to adjust the notation with the one adopted in the theory of strictly stationary random sequences, we will assume X to be defined on the entire discrete time axis $\mathbb{Z}$; one can easily show using the Kolmogorov consistency theorem that such an extension $\{X_n, n \in \mathbb{Z}\}$ exists.

For a strictly stationary random sequence $X = \{X_n, n \in \mathbb{Z}\}$ denote

$$\mathscr{F}_{m,n} = \sigma(X_l, l \in [m,n]), \quad -\infty \le m \le n \le \infty.$$

For ξ measurable w.r.t. $\mathscr{F}_{m,n}$ with $m > -\infty, n < \infty$ one has

$$\xi = F(X_m, \dots, X_n)$$

a.s. with some measurable F. For such a random variable, the family of *time shifts* is defined by

$$\theta_k \xi = F(X_{m+k}, \dots, X_{n+k}), \quad k \in \mathbb{Z}.$$

This family is extended to all $\mathscr{F}_{-\infty,\infty}$-measurable random variables by isometry (which follows by the stationarity of $\{X_n\}$). The invariant σ-algebra $\mathscr{J}_X$ for the sequence X consists of all $A \in \mathscr{F}_{-\infty,\infty}$ such that

$$\theta_k 1_A = 1_A \quad \text{a.s.,} \quad k \in \mathbb{Z}.$$

Strictly stationary sequence X is called *ergodic* if its invariant σ-algebra $\mathscr{J}_X$ is degenerate:

$$A \in \mathscr{J}_X \Rightarrow \mathsf{P}(A) \in \{0,1\}.$$

The Birkhoff theorem states that, for any $\mathscr{F}_{-\infty,\infty}$-measurable ξ with $\mathsf{E}|\xi| < \infty$,

$$\frac{1}{n}\sum_{k=1}^{n} \theta_k \xi \to \mathsf{E}[\xi|\mathscr{J}_X], \quad n \to \infty \tag{1.3.1}$$

a.s. and in mean sense. The ergodicity of X means that the right-hand side of eq. (1.3.1) equals to $\mathsf{E}\xi$.

On the other hand, if

$$\frac{1}{n}\sum_{k=1}^{n}\theta_k\xi \to \mathsf{E}\xi, \quad n\to\infty \tag{1.3.2}$$

in probability for any bounded ξ, then for each $A \in \mathcal{J}_X$

$$1_A = \mathsf{E}1_A = \mathsf{P}(A)$$

a.s. and thus X is ergodic. This motivates the following definition: X is called *mixing* if for any bounded ξ, η

$$\mathrm{Cov}(\theta_n\xi,\eta)\to 0, \quad n\to\infty. \tag{1.3.3}$$

If X is mixing, it is ergodic. Indeed, the mixing property yields convergence (1.3.2) in probability. To see this, denote $\xi_k = \theta_k\xi$ and observe that

$$\mathrm{Cov}(\xi_k,\xi_l)\le C, \quad \nu_M = \sup_{|k-l|>M}\mathrm{Cov}(\xi_k,\xi_l)\to 0, \quad M\to\infty, \tag{1.3.4}$$

hence taking $M(n) = [\sqrt{n}]$ we get

$$\begin{aligned}\left(\frac{1}{n}\sum_{m=1}^{n} f(X_m) - \mathsf{E}f(X_0)\right)^2 &= \frac{1}{n^2}\sum_{l,m=1}^{n}\mathrm{Cov}(\xi_l,\xi_m)\\ &= \left(\sum_{l,m\le n,|m-l|\le M(n)} + \sum_{l,m\le n,|m-l|>M(n)}\right)\mathrm{Cov}(\xi_l,\xi_m)\\ &\le C\frac{M^2(n)}{n^2} + \nu_{M(n)} \to 0, \quad n\to\infty,\end{aligned}$$

which proves eq. (1.3.2).

One can say that the mixing property of a strictly stationary sequence yields law of large numbers (LLN) for the sequence $\{\xi_k = \theta_k\xi\}$ for any $\mathcal{F}_{-\infty,\infty}$-measurable ξ with $\mathsf{E}|\xi| < \infty$. For more elaborate limit theorems, stronger versions of the mixing property are typically required, formulated in terms of *mixing coefficients*. The *strong mixing* (or *complete regularity*, or *Rosenblatt's*) *coefficient* is defined by

$$\alpha(n) = \sup_{A\in\mathcal{F}_{n,\infty},\, B\in\mathcal{F}_{-\infty,0}}\Big|\mathsf{P}(A\cap B) - \mathsf{P}(A)\mathsf{P}(B)\Big|, \quad n\ge 0.$$

The *uniform mixing* (or *Ibragimov's*) *coefficient* is defined by

$$\phi(n) = \sup_{A\in\mathcal{F}_{n,\infty},\, B\in\mathcal{F}_{-\infty,0},\, \mathsf{P}(B)>0}\Big|\mathsf{P}(A) - \mathsf{P}(A|B)\Big|, \quad n\ge 0.$$

We will call a strictly stationary sequence *α-mixing* if

$$\alpha(n) \to 0$$

and *ϕ-mixing* if

$$\phi(n) \to 0.$$

Below we formulate two classical versions of central limit theorem (CLT), stated in terms of strong and uniform mixing coefficients, respectively.

Theorem 1.3.1 ([60, Theorem 18.5.3]). *Let ξ be $\mathscr{F}_{-m,m}$-measurable for some $m \geq 0$ and $E\xi = 0$, $E|\xi|^{2+\delta} < \infty$ for some $\delta > 0$. Assume also that*

$$\sum_n \left(\alpha(n)\right)^{\delta/(2+\delta)} < \infty. \tag{1.3.5}$$

Then for $\{\xi_k = \theta_k \xi\}$ the sum

$$\sigma^2 = E\xi_0^2 + 2\sum_{k=1}^{\infty} E\xi_k \xi_0 \tag{1.3.6}$$

converges, and if $\sigma^2 > 0$

$$\frac{1}{\sqrt{n}} \sum_{k=1}^{n} \xi_k \Rightarrow \mathscr{N}(0, \sigma^2), \quad n \to \infty.$$

Theorem 1.3.2 ([60, Theorem 18.5.2]). *Let ξ be $\mathscr{F}_{-m,m}$-measurable for some $m \geq 0$ and $E\xi = 0$, $E\xi^2 < \infty$. Assume also that*

$$\sum_n \left(\phi(n)\right)^{1/2} < \infty. \tag{1.3.7}$$

Then for $\{\xi_k = \theta_k \xi\}$, sum (1.3.6) converges, and if $\sigma^2 > 0$

$$\frac{1}{\sqrt{n}} \sum_{k=1}^{n} \xi_k \Rightarrow \mathscr{N}(0, \sigma^2), \quad n \to \infty.$$

Remark 1.3.3. Here we formulate only the CLT for ξ which depends on finite number of elements of the sequence X. For general $\mathscr{F}_{-\infty,\infty}$-measurable ξ similar results are available, requiring additional technical conditions on the rate of approximation of ξ by $E[\xi|\mathscr{F}_{-m,m}]$ as $m \to \infty$; see [60, Theorems 18.6.1 and 18.6.2].

In two versions of the CLT given earlier, the strong mixing coefficient $\alpha(n)$ and the uniform mixing coefficient $\phi(n)$ play similar roles: depending on the choice of the coefficient, only the moment condition for ξ and the decay rate of the mixing coefficient are changed. However, we will see below that the difference between these two types of coefficients is substantial because, in particularly interesting models, the sequence may fail to perform a uniform mixing while the strong mixing coefficient decays sufficiently fast.

For a stationary MC, the mixing coefficients are closely related to the family

$$\mathbf{v}_i^n = \sum_j |p_{ij}^n - \mu_j|, \quad i \in \mathbb{X},$$

which quantifies the stabilization property of the chain.

Proposition 1.3.4. *Let X be a stationary MC with $Law(X_0) = \mu$. Then*

(i) $\frac{1}{2}\sup_i \mu_i \mathbf{v}_i^n \le \alpha(n) \le \sum_i \mu_i \mathbf{v}_i^n$;

(ii) $\frac{1}{2}\sup_{i:\mu_i>0} \mathbf{v}_i^n \le \phi(n) \le \sup_{i:\mu_i>0} \mathbf{v}_i^n$.

Proof. For given n, take arbitrary $A \in \mathscr{F}_{n,\infty}$, $B \in \mathscr{F}_{-\infty,0}$, and denote by g such a function that

$$\mathsf{E}[1_A|X_n] = g(X_n),$$

then by the Markov property of X

$$\mathsf{P}(A \cap B) = \mathsf{E}g(X_n)1_B.$$

Denote $B_i = B \cap \{X_0 = i\}$. Using the Markov property once more, we get

$$\mathsf{P}(A \cap B) = \sum_i \mathsf{E}g(X_n)1_{B_i} = \sum_i \mathsf{E}_i g(X_n)\mathsf{P}(B_i) = \sum_{i,j} \mathsf{P}(B_i)p_{ij}^n g(j).$$

The same calculation gives

$$\mathsf{P}(A)\mathsf{P}(B) = \sum_{i,j} \mathsf{P}(B_i)\mu_j g(j),$$

which yields

$$|\mathsf{P}(A \cap B) - \mathsf{P}(A)\mathsf{P}(B)| \le \sum_i \mathsf{P}(B_i)|p_{ij}^n - \mu_j|g(j).$$

Since

$$\mathsf{P}(B_i) \le \mu_i, \quad g(j) \le 1, \quad i, j \in \mathbb{X},$$

this inequality yields the required upper bounds for $\alpha(n), \phi(n)$.

To prove the lower bounds, we choose properly the sets A, B in the above estimates. Fix i such that $\mu_i > 0$ and put $B = B_i = \{X_0 = i\}$. Next, put $A = \{X_n \in C_{i,n}\}$ with $C_{i,n} = \{j : p_{ij}^n \geq \mu_j\}$ and observe that

$$\mathbf{v}_i^n = 2 \sum_{j \in C_{i,n}} (p_{ij}^n - \mu_j).$$

Then

$$\alpha(n) \geq |\mathsf{P}(A \cap B) - \mathsf{P}(A)\mathsf{P}(B)| = \frac{1}{2}\mathsf{P}(B)\mathbf{v}_i^n = \frac{1}{2}\mu_i \mathbf{v}_i^n,$$
$$\phi(n) \geq |\mathsf{P}(A|B) - \mathsf{P}(A)| = \frac{1}{2}\mathbf{v}_i^n. \qquad \square$$

As a corollary, we obtain that a stationary MC, which is irreducible and aperiodic, is α-mixing: this follows directly from Theorem 1.2.5. Note that the fact that the chain is positive recurrent follows from the stationarity assumption, which implies that an IPM exists. If the state space is finite, then the chain is ϕ-mixing at exponential rate:

$$\phi(n) \leq Ce^{-\rho n}, \quad n \geq 0.$$

The Birkhoff theorem, Theorem 1.3.1, and Proposition 1.3.4 give the following.

Corollary 1.3.5. *Let X be a stationary MC, which is irreducible and aperiodic. Then the following LLN holds for any $f : \mathbb{X} \to \mathbb{R}$ such that $\mathsf{E}|f(X_0)| < \infty$*

$$\frac{1}{n}\sum_{k=1}^{N} f(X_k) \to \mathsf{E}f(X_0), \quad n \to \infty \tag{1.3.8}$$

a.s. and in mean sense.

If the state space is finite, then in addition the following CLT holds true: for any f such that $\mathsf{E}f(X_0) = 0$, $\mathsf{E}f^2(X_0) < \infty$ the series

$$\sigma^2 = \mathsf{E}f^2(X_0) + 2\sum_{k=1}^{\infty} \mathsf{E}f(X_k)f(X_0) \tag{1.3.9}$$

converges, and if $\sigma^2 > 0$

$$\frac{1}{\sqrt{n}}\sum_{k=1}^{n} f(X_k) \Rightarrow \mathcal{N}(0, \sigma^2), \quad n \to \infty.$$

In this general statement, the cases of finite and infinite state spaces differ substantially. The reason is that, in the infinite case, the mixing properties of the chain are typically much weaker than the ϕ-mixing property at exponential rate obeyed by a finite chain. One simple, but very typical example is given below.

Example 1.3.6. ("Residual waiting time" model, see [39, Chapter XV.2], Example (k)). Let $\mathbb{X} = \{0, 1, \dots\}$ and

$$p_{i(i-1)} = 1, \quad i = 1, 2, \dots, \quad p_{0i} = p_i, \quad i \in \mathbb{X},$$

where $p_i > 0, i \in \mathbb{X}$,

$$\sum_i p_i = 1.$$

Then the chain is irreducible and aperiodic. The IPM $\mu = \{\mu_i\}$ is determined by the relations

$$\mu_i = \mu_{i+1} + \mu_0 p_i, \quad i \geq 0,$$

and this system has unique solution

$$\mu_i = \frac{\sum_{j=i}^{\infty} p_j}{1 + \sum_{j=1}^{\infty} j p_j}$$

if

$$\sum_{j=1}^{\infty} j p_j < \infty.$$

Otherwise the chain does not have an IPM and is zero recurrent (it is easy to see that the chain is recurrent in any case).

We can give lower bounds for the mixing coefficients using Proposition 1.3.4. For $i \geq n$ we have

$$p_{ij}^n = \begin{cases} 1, & j = i - n; \\ 0, & \text{otherwise}, \end{cases}$$

and thus

$$\mathbf{v}_i^n = \sum_j |p_{ij}^n - \mu_j| \geq 1 - \mu_{i-n}, \quad i > n. \tag{1.3.10}$$

In addition, all μ_i, $i \in \mathbb{X}$, are positive. Hence

$$\phi(n) \geq \frac{1}{2} \sup_i \mathbf{v}_i^n = \frac{1}{2};$$

that is, the chain is not ϕ-mixing. On the other hand, $\mu_0 < 1$ hence

$$\alpha(n) \geq \frac{1}{2} \sup_i \mu_i \mathbf{v}_i^n \geq c\mu_n, \quad c = \frac{1}{2}(1-\mu_0) > 0.$$

That is, the decay rate for the α-mixing coefficient is not faster than the one for the sequence $\{\mu_n\}$, and depending on the choice of $\{p_i\}$ the latter rate can be quite slow. For instance, taking

$$p_i = \Big[(1+i)\log(2+i)\Big]^{-2}, \quad i \geq 0$$

we get

$$\alpha(n) \geq c\mu_i \sim c(1+i)^{-1}\Big[\log(2+i)\Big]^{-2}, \quad i \to \infty,$$

which means that for any $\kappa \in (0,1)$

$$\sum_n \big(\alpha(n)\big)^\kappa = \infty.$$

This (in a sense) negative example should not discourage the reader: in various models of particular interest, it is still possible to show that the α-mixing coefficient decays sufficiently fast, and then Theorem 1.3.1 combined with Proposition 1.3.4 can be applied to prove CLT. This explains a separate attention which we pay to *ergodic rates* in the sequel.

1.4 Comments

1. The aim of this short introductory section is to provide a background for the study of general Markov systems in the rest of the book. Hence we do not provide too much details here, referring the reader, if necessary, either to the classical introduction into discrete MCs at Ref. [39, Chapters XV and XVI], or to the perfect exposition in Ref. [44, Chapter III] (these are just two samples from the vast list of references devoted to this topic).

2. To emphasize the genealogy of the ideas, we note once again – and more explicitly – that the crucial ingredient for the LLN and the CLT stated in Corollary 1.3.5 is the stabilization property for the transition probabilities of the chain; in our exposition these two topics are linked by Proposition 1.3.4. The LLN and CLT for Markov chains (as we call them now) were obtained by A. Markov in the very first and the second papers [100, 101] devoted to this object. Apparently, the entire class of "random variables connected to a chain" was introduced by Markov, mainly, to provide a framework, where the basic limit theorems of probability theory can be obtained without the independence assumption.

3. Markov's method of the proof of LLN is remarkably simple and efficient, and has numerous extensions. It is actually explained in Section 1.3: once the chain exhibits the stabilization property, one can prove the decay of the mutual covariances, see eq. (1.3.4), and then the Chebyshev's proof of LLN becomes applicable. Markov's CLT in Ref. [101] was obtained by the method of moments in a quite restrictive setting. CLTs in Section 1.3 are obtained within the famous *blocks method*, which dates back to S. Bernstein, cf. Ref. [7]. For a perfectly detailed exposition of Bernstein's blocks method we refer to Ref. [60, Chapter XVIII]. An alternative method, based on a martingale structure of the random sequence, was initiated by P. Lévy [92–94]. This method will be discussed in more detail in Chapter 5.

2 General Markov Chains: Ergodicity in Total Variation

2.1 Basic Definitions and Facts

The structural theory of general Markov chains is remarkably similar to the one for countable spaces, discussed in Section 1.1. The only technical difference is that probability measures on $\mathbb{X}$ in general can not be treated as vectors, and the transition probabilities should be specified as *probability kernels*, instead of stochastic matrices.

By definition, a *probability kernel* on a given measurable space $(\mathbb{X}, \mathscr{X})$ is a function $P(x, A), x \in \mathbb{X}, A \in \mathscr{X}$, which is a measurable function in x and a probability measure in A. A general MC with the state space $\mathbb{X}$ is a random sequence $X = \{X_n, n \geq 0\}$ defined on a *filtered probability space* $(\Omega, \mathscr{F}, \{\mathscr{F}_n\}, \mathsf{P})$ such that, for a certain family of probability kernels $\{P^{m,n}, 0 \leq m \leq n\}$

$$\mathsf{P}(X_n \in A | \mathscr{F}_m) = P^{m,n}(X_m, A), \quad 0 \leq m \leq n, \quad A \in \mathscr{X},$$

and the family $\{P^{m,n}\}$ satisfies

- $P^{m,m}(x, A) = 1_A(x) = \delta_x(A)$;
- (the Chapman–Kolmogorov equation): for $m \leq r \leq n$

$$P^{m,n}(x, A) = \int_{\mathbb{X}} P^{m,r}(x, dy) P^{r,n}(y, A).$$

The elements of family $\{P^{m,n}, 0 \leq m \leq n\}$ are called *transition probability kernels*, or simply *transition probabilities* of the chain X. This family and the *initial distribution*

$$\mu(A) = \mathsf{P}(X_0 \in A), \quad A \in \mathscr{X}$$

completely define the law of the chain: for any $m \geq 0, A_0, \dots, A_m \in \mathscr{X}$

$$\begin{aligned} &\mathsf{P}(X_0 \in A_0, X_1 \in A_1, \dots, X_m \in A_m) \\ &\quad = \int_{A_0} \dots \int_{A_{m-1}} \mu(dx_0) P^{0,1}(x_0, dx_1) \dots, P^{m-1,m}(x_{m-1}, A_m). \end{aligned} \tag{2.1.1}$$

The law of the chain with the initial distribution μ is denoted by P_μ, respective expectation is denoted by E_μ. If μ is concentrated at one point, $\mu = \delta_x$, the notation is simplified to $\mathsf{P}_x, \mathsf{E}_x$.

In what follows, we restrict our consideration to the case of time-homogeneous chains, which by definition means that $P^{m,n}$ depends on $n - m$ (the length of the corresponding time interval), only. We denote by $P^n(x, A) = P^{0,n}(x, A)$ and $P(x, A) = P^1(x, A)$ the n-step and the one-step transition probabilities, respectively. Clearly, the kernel

DOI 10.1515/9783110458930-003

$P(x, A)$ completely defines the entire family $\{P^n(x, A), n \geq 0\}$: for $n = 0, 1$ $P^n(x, A)$ are already defined, and for $n > 1$

$$\mathrm{P}^n(x, A) = \int_{\mathbb{X}} \dots \int_{\mathbb{X}} P(x, \mathrm{d}x_1) \dots P(x_{n-1}, A).$$

For an MC with an initial distribution μ, the distribution μ^n of the value X_n of the chain at the time moment n equals

$$\mu^n(A) = \int_{\mathbb{X}} \mu(\mathrm{d}x) P^n(x, A), \quad A \in \mathscr{X}.$$

In other words, MC naturally defines a dynamical system

$$\mu(\mathrm{d}y) \mapsto \int_{\mathbb{X}} \mu(\mathrm{d}x) P(x, \mathrm{d}y) \tag{2.1.2}$$

on the set $\mathscr{P}(\mathbb{X})$ of all probability measures on $\mathbb{X}$. A measure $\mu \in \mathscr{P}(\mathbb{X})$ which satisfies

$$\mu(\mathrm{d}y) = \int_{\mathbb{X}} \mu(\mathrm{d}x) P(x, \mathrm{d}y) \tag{2.1.3}$$

is called an *invariant probability measure* (IPM) for the MC X. Clearly, an IPM μ is a fixed point for the dynamical system (2.1.2) and, for such μ,

$$\mu^n = \mu, \quad n \geq 0.$$

This identity and eq. (2.1.1) yield that, if X_0 has the law μ which is invariant for X, the random sequence is *strictly stationary*.

Denote by $\mathbb{M}(\mathbb{X})$ the family of finite signed measures on $\mathbb{X}$. It is well known that $\mathbb{M}(\mathbb{X})$ is a Banach space w.r.t. the *total variation norm*

$$\|\mu\|_{TV} = \mu_+(\mathbb{X}) + \mu_-(\mathbb{X}),$$

where $\mu = \mu_+ - \mu_-$ is the *Hahn decomposition* for μ. The Banach space $\mathbb{M}(\mathbb{X})$ and the Banach space $\mathbb{B}(\mathbb{X})$ of bounded measurable real-valued functions (with the sup-norm $\|\cdot\|$) are naturally related by the duality

$$\langle f, \mu \rangle = \int_{\mathbb{X}} f(x) \mu(\mathrm{d}x);$$

that is,

$$\|f\| = \sup_{\|\mu\|_{TV}=1} |\langle f, \mu \rangle|, \quad \|\mu\|_{TV} = \sup_{\|f\|=1} |\langle f, \mu \rangle|.$$

Given an MC X with one-step transition probability $P(x, A)$, a pair of mutually adjoint linear operators $P : \mathscr{B}(\mathbb{X}) \to \mathscr{B}(\mathbb{X})$, $P^* : \mathbb{M}(\mathbb{X}) \to \mathbb{M}(\mathbb{X})$ is naturally defined:

$$Pf(x) = \int_{\mathbb{X}} f(y)\,P(x, \mathrm{d}y), \quad f \in \mathbb{B}(\mathbb{X}), \quad P^*\mu(A) = \int_{\mathbb{X}} P(x, A)\,\mu(\mathrm{d}x), \quad \mu \in \mathbb{M}(\mathbb{X}). \tag{2.1.4}$$

Clearly, $\mathscr{P}(\mathbb{X}) \subset \mathbb{M}(\mathbb{X})$ is invariant w.r.t. operator P^* and the dynamical system (2.1.2) is just the restriction of P^* to $\mathscr{P}(\mathbb{X})$. In particular, the n-step transition probability $P^n(x, \cdot)$ for X is just the image of the delta measure δ_x under $(P^*)^n$. Our general aim in this chapter is to establish convergence

$$P^n(x, \mathrm{d}y) \to \mu(\mathrm{d}y), \quad n \to \infty$$

in the total variation norm.

2.2 Total Variation Distance and the Coupling Lemma

In this section, we take a closer look at the properties of total variation distance

$$\|\mu - \nu\|_{TV}$$

with μ, ν taken from the set $\mathscr{P}(\mathbb{X})$ of probability measures on $\mathbb{X}$. Recall that the Hahn decomposition

$$\mu - \nu = (\mu - \nu)_+ - (\mu - \nu)_-$$

has the form

$$(\mu - \nu)_\pm(A) = (\mu - \nu)(A \cap A_\pm), \quad A \in \mathscr{X},$$

where $A_+ \sqcup A_- = \mathbb{X}$ and the choice of the sets A_+, A_- is unique up to a set of zero measure

$$|\mu - \nu| = (\mu - \nu)_+ + (\mu - \nu)_-.$$

This immediately gives

$$\|\mu - \nu\|_{TV} = 2 \max_{A \in \mathscr{X}} (\mu(A) - \nu(A))$$

with the maximum attained at the set A_+.

Next, let λ be a σ-finite measure on $(\mathbb{X}, \mathscr{X})$ such that $\mu \ll \lambda$, $\nu \ll \lambda$. Note that such λ exists, for instance one can take $\lambda = \mu + \nu$. One has

$$A_+ = \left\{x : \frac{d\mu}{d\lambda} \geq \frac{d\nu}{d\lambda}\right\}, \quad A_- = \left\{x : \frac{d\mu}{d\lambda} < \frac{d\nu}{d\lambda}\right\},$$

and therefore

$$\|\mu - \nu\|_{TV} = \int_{\mathbb{X}} \left|\frac{d\mu}{d\lambda} - \frac{d\nu}{d\lambda}\right| d\lambda.$$

Hence the total variation distance between μ, ν is just the $L_1(\mathbb{X}, \lambda)$-distance between their Radon–Nikodym derivatives w.r.t. a common measure λ, which should be chosen in such a way that respective Radon–Nikodym derivatives exist.

The following example shows that we have already used – without naming it explicitly – total variation distance to quantify the stabilization effect of a discrete MC in Theorem 1.2.1 and Corollary 1.2.6.

Example 2.2.1. Let $\mathbb{X}$ be finite, then any $\mu \in \mathcal{P}(\mathbb{X})$ corresponds to a set $\{\mu_i\}_{i\in\mathbb{X}}$ of nonnegative numbers such that

$$\sum_i \mu_i = 1.$$

Take λ equal to the counting measure on $\mathbb{X}$; that is, $\lambda_i = 1, i \in \mathbb{X}$. Then for any $\mu \in \mathcal{P}(\mathbb{X})$ the corresponding Radon–Nikodym derivatives at a point i just equal μ_i, ν_i, hence

$$\|\mu - \nu\|_{TV} = \sum_i |\mu_i - \nu_i|.$$

The following statement, frequently called *the Coupling lemma*, gives a probabilistic characterization for the total variation distance.

Theorem 2.2.2.

$$\|\mu - \nu\|_{TV} = 2 \min_{\kappa\in\mathcal{C}(\mu,\nu)} \kappa\left(\{(x, y) : x = y\}\right). \tag{2.2.1}$$

Remark 2.2.3. One part of the theorem is just inequality (1.2.13) formulated in the general setting: for any ξ, η with the joint law $\kappa \in \mathcal{C}(\mu, \nu)$

$$\|\mu - \nu\|_{TV} \leq 2\mathsf{P}(\xi \neq \eta). \tag{2.2.2}$$

We have seen that such inequality gives a convenient "probabilistic" tool to bound the total variation distance $\|\mu - \nu\|_{TV}$: in order to estimate such a difference it is sufficient to construct a pair of "representatives" ξ, η and bound $\mathsf{P}(\xi \neq \eta)$. The second part of the theorem reveals an important fact that this procedure is exact, in the sense that a proper choice of (ξ, η) turns inequality (2.2.2) into an identity.

Proof. For any $\xi \sim \mu$ and $\eta \sim \nu$, we have

$$\begin{aligned}\|\mu - \nu\|_{TV} &= 2\max_{A\in\mathscr{X}}(\mu(A) - \nu(A)) = 2\max_{A\in\mathscr{X}} \mathsf{E}(\mathbf{I}_{\xi\in A} - \mathbf{I}_{\eta\in A}) \\ &\le 2\max_{A\in\mathscr{X}} \mathsf{P}(\xi \in A, \eta \notin A) \le 2\mathsf{P}(\xi \ne \eta),\end{aligned}$$

which proves eq. (2.2.2).

Next, we construct explicitly the joint law $\kappa \in \mathscr{C}(\mu, \nu)$ of the pair (ξ, η), which turns inequality (2.2.2) into an equality. Consider a probability measure

$$\lambda = \frac{1}{2}(\mu + \nu),$$

then $\mu \ll \lambda$, $\nu \ll \lambda$, and we can put

$$f = \frac{d\mu}{d\lambda}, \quad g = \frac{d\nu}{d\lambda}, \quad h = f \wedge g.$$

If $p = \int h\, d\lambda = 1$, then $\mu = \nu$ and we put $\kappa(A_1 \times A_2) = \mu(A_1 \cap A_2)$; that is, we set the components ξ, η equal to one another, with the same law $\mu = \nu$. Otherwise, we decompose

$$\mu = p\theta + (1-p)\sigma_1, \quad \nu = p\theta + (1-p)\sigma_2, \quad d\theta = \frac{1}{p} h\, d\lambda, \tag{2.2.3}$$

with the convention that $\theta = \lambda$, if $p = 0$. We define

$$\kappa(A_1 \times A_2) = p\theta(A_1 \cap A_2) + (1-p)\sigma_1(A_1)\sigma_2(A_2),$$

then $\kappa \in \mathscr{C}(\mu, \nu)$, because for any $A \in \mathscr{X}$

$$\kappa(A \times \mathbb{X}) = p\theta(A) + (1-p)\sigma_1(A) = \mu(A), \quad \kappa(\mathbb{X} \times A) = p\theta(A) + (1-p)\sigma_2(A) = \nu(A).$$

On the other hand,

$$\kappa\Big(\{(x,y) : x = y\}\Big) \ge p\theta(\mathbb{X}) = p,$$

which proves that for a pair ξ, η with the joint law κ inequality (2.2.2) is actually an identity. □

We will call $\kappa \in \mathscr{C}(\mu, \nu)$ an *optimal coupling for* μ, ν if

$$\|\mu - \nu\|_{TV} = 2\kappa\Big(\{(x,y) : x = y\}\Big).$$

In what follows we will need to have the coupling construction for probability kernels rather than for individual probability measures. For two probability kernels $P_1(x, dy), P_2(x, dy)$ on $(\mathbb{X}, \mathscr{X})$ we call a *coupling kernel* any probability kernel

$Q((x_1, x_2), dy_1\, dy_2)$ on the product space $(\mathbb{X} \times \mathbb{X}, \mathscr{X} \otimes \mathscr{X})$ such that for every $x_1, x_2 \in \mathbb{X} \times \mathbb{X}$

$$Q\Big((x_1, x_2), \cdot\Big) \in \mathscr{C}\Big(P_1(x_1, \cdot), P_2(x_2, \cdot)\Big).$$

The following statement looks like a one-to-one analogue to Theorem 2.2.2; however, some new technical difficulties arise because of necessity to take care about measurability issues.

Theorem 2.2.4 (The Coupling lemma for probability kernels). *Let the measurable space $(\mathbb{X}, \mathscr{X})$ be countably generated; that is, there exists a countable set $\mathscr{X}_0 \subset \mathscr{X}$ such that $\sigma(\mathscr{X}_0) = \mathscr{X}$.*

Then for any probability kernels $P_1(x, dy), P_2(x, dy)$ on $(\mathbb{X}, \mathscr{X})$, there exists a coupling kernel $Q((x_1, x_2), dy_1\, dy_2)$ such that

(i) *for every $(x_1, x_2) \in \mathbb{X} \times \mathbb{X}$,*

$$\|P_1(x_1, \cdot) - P_2(x_2, \cdot)\|_{TV} = 2Q\Big((x_1, x_2), \{(y_1, y_2) : y_1 \neq y_2\}\Big);$$

(ii) *for every $(x_1, x_2) \in \mathbb{X} \times \mathbb{X}$, the measure $Q((x_1, x_2), dy_1\, dy_2)$, restricted to the set $\{(y_1, y_2) : y_1 \neq y_2\}$, is absolutely continuous w.r.t. the product measure $P_1(x_1, dy_1) \otimes P_2(x_2, dy_2)$ with the Radon–Nikodym density bounded by*

$$\left(\frac{1}{2}\|P_1(x_1, \cdot) - P_2(x_2, \cdot)\|_{TV}\right)^{-1}.$$

Proof. Denote

$$\Lambda(x_1, x_2; dy) = \frac{1}{2}\Big(P_1(x_1, dy) + P_2(x_2, dy)\Big).$$

Then for every fixed $x_1, x_2 \in \mathbb{X}$, we have $P_i(x_i, dy) \ll \Lambda(x_1, x_2; dy), i = 1, 2$; that is, there exist respective Radon–Nikodym derivatives $f_i(x_1, x_2; \cdot), i = 1, 2$, such that

$$P_i(x_i, A) = \int_A f_i(x_1, x_2; y)\Lambda(x_1, x_2; dy), \quad i = 1, 2, \quad A \in \mathscr{X}. \tag{2.2.4}$$

Recall that the functions $f_i(x_1, x_2; \cdot), i = 1, 2$ are not uniquely defined and can be changed on sets of zero $\Lambda(x_1, x_2; \cdot)$-measure. Let us show that these functions can be chosen in a jointly measurable way; that is, there exist measurable functions $f_i : \mathbb{X} \times \mathbb{X} \times \mathbb{X} \to \mathbb{R}^+, i = 1, 2$ such that eq. (2.2.4) holds true for all $x_1, x_2 \in \mathbb{X}$.

Let $\mathscr{X}_0$ be a countable algebra which generates $\mathscr{X}$. Denote by $\mathscr{G}$ the countable class of functions representable as finite sums

$$\sum_k c_k 1_{A_k}, \quad \{c_k\} \subset \mathbb{Q}, \quad \{A_k\} \subset \mathscr{X}_0.$$

Then $\mathscr{G}$ is dense in $L_1(\mathbb{X}, \lambda)$ for any probability measure λ on $(\mathbb{X}, \mathscr{X})$. For $x_1, x_2 \in \mathbb{X}$ fixed, the Radon–Nikodym derivative

$$\rho^1_{x_1,x_2}(y) = \frac{P_1(x_1, dy)}{\Lambda(x_1, x_2; dy)}$$

belongs to $L_1(\mathbb{X}, \Lambda(x_1, x_2; \cdot))$, hence for every $\varepsilon > 0$, there exists $g \in \mathscr{G}$ such that

$$\sup_{A \in \mathscr{X}_0} \int_A (\rho^1_{x_1,x_2}(y) - g(y))\Lambda(x_1, x_2; dy) < \frac{\varepsilon}{2}.$$

Observe that this relation is equivalent to

$$\sup_{A \in \mathscr{X}_0} \left(P_1(x_1, A) - \int_A g(y)\Lambda(x_1, x_2; dy) \right) < \frac{\varepsilon}{2}, \tag{2.2.5}$$

and yields

$$\int_E |\rho_{x_1,x_2}(y) - g(y)|\Lambda(x_1, x_2; dy) < \varepsilon.$$

Fix some enumeration of the class $\mathscr{G} = \{g_m, m \in \mathbb{N}\}$, and denote for $n \geq 1$ by $m(x_1, x_2; n)$ the minimal $m \geq 1$ such that eq. (2.2.5) holds true for $g = g_m$ with $\varepsilon = 2^{-n-1}$. Then $m(\cdot; n) : \mathbb{X} \times \mathbb{X} \to \mathbb{N}$ is measurable, and therefore

$$f^n_1(x_1, x_2; y) = g_{m(x_1,x_2;n)}(y)$$

is measurable as a function $\mathbb{X}\times\mathbb{X}\times\mathbb{X} \to \mathbb{R}$. The sequence $\{f^n_1(x_1, x_2; y), n \geq 1\}$ converges to $\rho^1_{x_1,x_2}(y)$ for $\Lambda(x_1, x_2; \cdot)$-a.a. y: this follows from the Borel–Cantelli lemma because, by the construction,

$$\|f^n_1(x_1, x_2; \cdot) - \rho^1_{x_1,x_2}\|_{L_1(E,\Lambda(x_1,x_2;\cdot))} \leq 2^{-n}.$$

Therefore the function

$$f_1(x_1, x_2; y) = \begin{cases} \lim_{n\to\infty} f^n_1(x_1, x_2; y), & \text{if the limit exists} \\ 0, & \text{otherwise} \end{cases}$$

gives the required measurable version of the Radon–Nikodym derivative for $P_1(x_1, dy)$ (the construction for $P_2(x_2, dy)$ is the same).

Now we repeat the construction given in the proof of Theorem 2.2.2, taking care of the measurability issues. Write

$$h(x_1, x_2; y) = \min\Big(f_1(x_1, x_2; y), f_2(x_1, x_2; y)\Big),$$

$$p(x_1, x_2) = \int_E h(x_1, x_2; y)\Lambda(x_1, x_2; dy),$$

$$\Theta(x_1, x_2; dy) = \frac{h(x_1, x_2; y)}{p(x_1, x_2)}\Lambda(x_1, x_2; dy)$$

with the convention that (anything)/0 = 1. Then we have representations

$$P_i(x_i, dy) = p(x_1, x_2)\Theta(x_1, x_2; dy) + (1 - p(x_1, x_2))\Sigma_i(x_1, x_2; dy), \quad i = 1, 2$$

with probability kernels $\Theta, \Sigma_1, \Sigma_2$ and measurable function $p : E \times E \to [0, 1]$.

We remark that the mapping $D : \mathbb{X} \ni x \mapsto (x, x) \in \mathbb{X}\times\mathbb{X}$ is $\mathscr{X} - \mathscr{X} \otimes \mathscr{X}$ measurable: indeed, the class of "measurable parallelepipeds" $A_1 \times A_2$, $A_i \in \mathscr{X}$, $i = 1, 2$ generates $\mathscr{X} \otimes \mathscr{X}$, and for any such set

$$D^{-1}(A_1 \times A_2) = A_1 \cap A_2 \in \mathscr{X}.$$

We denote by $Q_1((x_1, x_2), dy_1\, dy_2)$ the image of $\Theta(x_1, x_2; dy)$ under this mapping, and define $Q_2((x_1, x_2), dy_1\, dy_2)$ as the product of the measures $\Sigma_i(x_1, x_2; dy_i)$, $i = 1, 2$. Then

$$Q((x_1, x_2), \cdot) = p(x_1, x_2)Q_1((x_1, x_2), \cdot) + (1 - p(x_1, x_2))Q_2((x_1, x_2), \cdot)$$

is the required kernel. Observe that the assertion (ii) now holds true because $Q_2((x_1, x_2), \cdot)$ is chosen as a product measure with the components $\Sigma_i(x_1, x_2; \cdot) \ll P_i(x_i, \cdot)$, $i = 1, 2$ and

$$\frac{\Sigma_i(x_1, x_2; dy)}{P_i(x_i, dy)} \le (1 - p(x_1, x_2))^{-1} = \left(\frac{1}{2}\|P_1(x_1, \cdot) - P_2(x_2, \cdot)\|_{TV}\right)^{-1}. \qquad \square$$

In order to be able to apply Theorem 2.2.4, everywhere below in this chapter we assume the space $(\mathbb{X}, \mathscr{X})$ to be countably generated. Let us introduce some terminology which will be used systematically in the sequel.

Let Q be a coupling kernel for the pair of kernels $P_i(x, dy) = P(x, dy)$, $i = 1, 2$, where $P(x, dy)$ is the transition probability for a given chain X. We call an MC Z,

$$Z_n = (X_n, Y_n), \quad n \ge 0,$$

with the transition probability Q *a Markov coupling* for X. If a Markov coupling Z for a chain X is defined on a filtered probability space $(\Omega, \mathscr{F}, \{\mathscr{F}_n\}, \mathsf{P})$, then for any $A \in \mathscr{X}$, $n \ge 0$

$$\mathsf{P}\big(X_{n+1} \in A \big| \mathscr{F}_n\big) = \mathsf{P}\big(Z_{n+1} \in A \times \mathbb{X} \big| \mathscr{F}_n\big) = Q(Z_n, A \times \mathbb{X}) = P(X_n, A),$$

which means that $X_n, n \geq 0$ is an MC with the transition probability $P(x, dy)$; the same argument applies to $Y_n, n \geq 0$. That is, if in the above construction we put $Z_0 = (x, y)$ with given $x, y \in \mathbb{X}$, the components $X_n, n \geq 0$ and $Y_n, n \geq 0$ have distributions P_x and P_y, respectively. In what follows, we denote by $\mathsf{P}^Z_{(x,y)}$ the law of Z with $Z_0 = (x, y)$ and by $\mathsf{E}^Z_{(x,y)}$ the corresponding expectation. We have for all $n \geq 1$

$$\mathsf{P}^Z_{(x,y)}(X_n \in A) = P^n(x, A), \quad \mathsf{P}^Z_{(x,y)}(Y_n \in A) = P^n(y, A), \quad A \in \mathscr{X}. \tag{2.2.6}$$

Then by eqs. (2.2.6) and (2.2.2)

$$\|P^n(x, \cdot) - P^n(y, \cdot)\|_{\mathrm{TV}} \leq 2\mathsf{P}^Z_{(x,y)}\big(X_n \neq Y_n\big) \tag{2.2.7}$$

for any Markov coupling Z.

The kernel Q constructed in Theorem 2.2.4 will be called *an optimal coupling kernel* for P. For the corresponding Markov coupling Z we adopt the name *greedy coupling*, proposed by M. Scheutzow, by the analogy with "greedy algorithms."

2.3 Uniform Ergodicity: The Dobrushin theorem

In this section, we establish ergodicity which is *uniform* in the sense that the corresponding bound for the total variation distance between the transition probabilities is uniform w.r.t. initial positions of the chain. The following theorem is a version of the result established by R. Dobrushin in 1956 in a more general time-inhomogeneous setting; see Ref. [25].

Theorem 2.3.1 (The Dobrushin theorem). *Let the transition probabilities for an MC X satisfy for some $m \geq 1$*

$$\sup_{x,y} \|P^m(x, \cdot) - P^m(y, \cdot)\|_{\mathrm{TV}} < 2. \tag{2.3.1}$$

Then there exist positive constants $C, \rho > 0$ such that

$$\sup_{x,y} \|P^n(x, \cdot) - P^n(y, \cdot)\|_{\mathrm{TV}} \leq Ce^{-\rho n}, \quad n \geq 1. \tag{2.3.2}$$

In addition, there exists unique IPM μ for X, and

$$\sup_{x} \|P^n(x, \cdot) - \mu\|_{\mathrm{TV}} \leq Ce^{-\rho n}, \quad n \geq 1. \tag{2.3.3}$$

Remark 2.3.2. Theorem 2.3.1 closely relates to Theorem 1.2.1. In the finite state space case, eq. (2.3.1) is equivalent to eq. (1.2.4), which by Proposition 1.2.3 is, in turn, equivalent to irreducibility and aperiodicity of the chain. We will see below that essentially the same two proofs for Theorem 1.2.1, developed in Sections 1.2.2 and 1.2.3, apply in the general framework *assuming* eq. (2.3.1) holds true.

In the sequel we call eq. (2.3.1) *the Dobrushin condition*; the name *Markov–Dobrushin condition* is also used in the literature, see for example, [49]. Condition (1.2.4), which coincides with eq. (2.3.1) for finite chains, was introduced by Markov (in somewhat hidden form) in Ref. [102].

"Probabilistic" proof: coupling argument. Assume first that eq. (2.3.1) holds true with $m = 1$. Let Q be the optimal coupling kernel for $P(x, dy)$ and Z be the greedy coupling for X.

Denote $\mathscr{D} = \{(x, y) : x = y\}$, the "diagonal" in $\mathbb{X} \times \mathbb{X}$. By the choice of the kernel Q,

$$p(x, y) = Q(z, \mathscr{D}) = 1 - \frac{1}{2}\|P(x, \cdot) - P(y, \cdot)\|_{TV}, \quad z = (x, y) \in \mathbb{X}.$$

Hence

$$p(x, y) \begin{cases} = 1, & (x, y) \in \mathscr{D}, \\ \geq p, & \text{otherwise}; \end{cases}$$

the second inequality holds true with some $p > 0$ and follows from eq. (2.3.1). We have

$$\mathsf{P}^Z_{(x,y)}(Z_{n+1} \in \mathscr{D}|Z_n) = p(Z_n),$$

and in particular Z_{n+1} stays on the diagonal $\mathscr{D}$ if $Z_n \in \mathscr{D}$. Then

$$\begin{aligned} \mathsf{P}^Z_{(x,y)}\big(X_n \neq Y_n\big) &= \mathsf{E}^Z_{(x,y)}\Big(1_{X_{n-1}\neq Y_{n-1}} \mathsf{P}^Z_{(x,y)}(Z_n \in \mathscr{D}|Z_{n-1})\Big) \\ &\leq (1-p)\mathsf{P}^Z_{(x,y)}(X_{n-1} \neq Y_{n-1}) \leq \cdots \leq (1-p)^n, \end{aligned} \tag{2.3.4}$$

which gives eq. (2.3.2) with $C = 2$, $e^{-\rho} = 1 - p$.

Inequality (2.3.2) implies that for any given $x \in \mathbb{X}$ and any $m < n$,

$$\begin{aligned} \|P^n(x, \cdot) - P^m(x, \cdot)\|_{TV} &= \left\| \int_{\mathbb{X}} \big(P^m(y, \cdot) - P^m(x, \cdot)\big) P^{n-m}(x, dy) \right\|_{TV} \\ &\leq \int_{\mathbb{X}} \left\|P^m(y, \cdot) - P^m(x, \cdot)\right\|_{TV} P^{n-m}(x, dy) \leq Ce^{-\rho m}. \end{aligned}$$

That is, $P^n(x, \cdot), n \geq 1$ is a Cauchy sequence w.r.t. total variation distance, and therefore has a limit μ in this distance. By eq. (2.3.2) for any $y \in \mathbb{X}$ the sequence $P^n(y, \cdot), n \geq 1$ has the same limit. We have

$$P^{n+1}(x, A) = \int_{\mathbb{X}} P(y, A)P^n(x, dy), \quad A \in \mathscr{X},$$

and passing to the limit as $n \to \infty$, we get

$$\mu(A) = \int_{\mathbb{X}} P(y, A)\mu(dy), A \in \mathscr{X};$$

that is, μ is an IPM for X.

Finally,

$$\begin{aligned}\|P^n(x,\cdot) - \mu\|_{TV} &= \left\| \int_{\mathbb{X}} \left(P^n(x,\cdot) - P^n(y,\cdot)\right)\mu(dy) \right\|_{TV} \\ &\le \int_{\mathbb{X}} \left\|P^n(x,\cdot) - P^n(y,\cdot)\right\|_{TV} \mu(dy) \le Ce^{-\rho n},\end{aligned}$$

which gives eq. (2.3.3).

Now we remove the assumption $m = 1$. Repeating the same argument with $P(x, dy)$ replaced by $P^m(x, dy)$, we obtain eq. (2.3.2) for $n = m, 2m, \ldots$. On the other hand, the same coupling argument as above yields

$$\|P^{n+r}(x,\cdot) - P^{n+r}(y,\cdot)\|_{TV} \le \|P^n(x,\cdot) - P^n(y,\cdot)\|_{TV}$$

for any $x, y \in \mathbb{X}$ and $n, r \ge 0$. This completes the proof of eq. (2.3.2). The rest of the proof remains literally the same. □

"Analytic" proof: contraction argument. We assume that eq. (2.3.1) holds true with $m = 1$; this limitation can be removed in the same way we did that in the previous proof. The key idea of the proof is to show that the dual operator P^* (see (2.1.4)) satisfies the following:

$$\|P^*\mu - P^*\nu\|_{TV} \le (1-p)\|\mu - \nu\|_{TV} \tag{2.3.5}$$

with some $p > 0$. This would immediately imply that eq. (2.3.2) holds true, because then

$$\|P^n(x,\cdot) - P^n(y,\cdot)\|_{TV} \le (1-q)^n \|\delta_x - \delta_y\|_{TV} = \begin{cases} 2(1-p)^n, & x \ne y \\ 0, & x = y. \end{cases}$$

Note that P^* *is not* a contraction when considered on the entire $\mathbb{M}(\mathbb{X})$, and this observation gives a hint that, while proving eq. (2.3.5), we need to use the additional property of μ and ν being probability measures (in fact, we will need only that $\mu(\mathbb{X}) = \nu(\mathbb{X})$).

We have

$$\langle f, P^*\mu\rangle = \int_{\mathbb{X}} \int_{\mathbb{X}} f(y)\, P(x, dy)\, \mu(dx) = \langle Pf, \mu\rangle,$$

hence

$$\|P^*\mu - P^*\nu\| = \sup_{\|f\|=1} \left|\langle Pf, \mu - \nu\rangle\right|.$$

For $g \in \mathbb{B}(\mathbb{X})$, denote $g_s(x) = g(x) - s$, $x \in \mathbb{X}$, $s \in \mathbb{R}$, and observe that

$$\langle g, \mu - \nu\rangle = \langle g_s, \mu - \nu\rangle, \quad s \in \mathbb{R},$$

and therefore

$$|\langle g, \mu - \nu\rangle| \le \|\mu - \nu\| \inf_{s\in\mathbb{R}} \|g_s\|.$$

One can show easily that

$$\inf_{s\in\mathbb{R}} \|g_s\| = \frac{1}{2} \sup_{x,y\in\mathbb{X}} |g(x) - g(y)|.$$

Summarizing the above relations, we get

$$\|P^*\mu - P^*\nu\| = \frac{1}{2}\|\mu - \nu\| \sup_{\|f\|=1} \sup_{x,y\in\mathbb{X}} \left|Pf(x) - Pf(y)\right|.$$

By condition (2.3.1)

$$\left|Pf(y) - Pf(y)\right| = \left|\langle f, P(y,\cdot) - P(y,\cdot)\rangle\right| \le \|f\| \left\|P(y,\cdot) - P(y,\cdot)\right\| \le 2(1-p)\|f\|,$$

which completes the proof of eq. (2.3.5). □

2.4 Preliminaries to Nonuniform Ergodicity: Motivation and Auxiliaries

In many particular cases of interest, Theorem 2.3.1 is not applicable because neither assumption (2.3.1) nor the convergence bound (2.3.3) hold true. Below, we give a simple example which explains a typical difficulty.

Example 2.4.1 (Linear regression). Let $\mathbb{X} = \mathbb{R}$ and for some a, σ^2

$$X_n = aX_{n-1} + \sigma\varepsilon_n, \quad n \ge 1,$$

where $\{\varepsilon_n\}$ are i.i.d. random variables with the law $\mathcal{N}(0,1)$. Then

$$X_n = a^n X_0 + \sum_{k=1}^{n} a^{n-k}\xi_k, \quad n \ge 1,$$

and

$$P^n(x,\cdot) = \mathscr{N}\left(a^n x, \sigma^2 \sum_{k=0}^{n-1} a^{2k}\right).$$

Then for any fixed n

$$\left\|P^n(x,\cdot) - P^n(0,\cdot)\right\|_{\mathrm{TV}} \to 2, \quad x \to \infty,$$

which means that

$$\sup_{x,y}\left\|P^n(x,\cdot) - P^n(y,\cdot)\right\|_{\mathrm{TV}} = 2, \quad n \ge 1 \tag{2.4.1}$$

Clearly, this means that eq. (2.3.1) fails and eq. (2.3.3) cannot hold with any $\mu \in \mathscr{P}(\mathbb{X})$. We note that, nevertheless, the chain is ergodic if $|a| < 1$ in the sense that, for a *given* $x \in \mathbb{X}$,

$$P^n(x,\cdot) \to \mu = \mathscr{N}\left(0, \sigma^2 \sum_{k=0}^{\infty} a^{2k}\right), \quad n \to \infty \tag{2.4.2}$$

in total variation distance.

An informal explanation of the effect observed in the above example is that, for “very distant” starting points, it takes a long time for the chain to come back “close to the origin”; hence any bound for the total variation distance between the transition probabilities, which is *uniform* w.r.t. the initial value, is necessarily trivial and is not informative. On the other hand, an “individual” ergodicity, like eq. (2.4.2) in the above example, is still possible. In a sense, the situation here is similar to the one treated, in the discrete setting, by Theorem 1.2.5.

To prove various “individual” forms of ergodicity, we will repeatedly use the following simple construction. Let Z be the greedy coupling for X, defined on some filtered probability space $(\Omega, \mathscr{F}, \{\mathscr{F}_n\}, \mathsf{P})$. Denote

$$A_n = \{X_n \ne Y_n\}, \quad n \ge 1,$$

then

$$\mathsf{P}(A_n | \mathscr{F}_{n-1}) = 1 - p(Z_{n-1}),$$

where $p(z) = Q(z, \mathscr{D})$. Denote

$$\varrho(z) = \frac{1}{1 - p(z)} 1_{p(z)<1}, \quad z = (x, y),$$

and consider the sequences $\{\mathscr{E}_n\}, \{\widetilde{\mathscr{E}}_n\}$ defined by

$$\mathscr{E}_n = \prod_{k=0}^{n-1}(1 - p(Z_k))^{-1}, \quad \widetilde{\mathscr{E}}_n = \prod_{k=0}^{n-1}\varrho(Z_k), \quad n \geq 1, \quad \mathscr{E}_0 = \widetilde{\mathscr{E}}_0 = 1,$$

with the convention (anything)/0 = ∞.

Observe that, once the random event $\{p(Z_k) = 1\}$ happens, one has $X_n = Y_n$ for all subsequent time moments $n > k$. Hence

$$\mathscr{E}_n = \widetilde{\mathscr{E}}_n \text{ on } A_n.$$

Denote

$$M_n = 1_{A_n}\mathscr{E}_n = 1_{A_n}\widetilde{\mathscr{E}}_n, \quad n \geq 0,$$

with the convention $0 \cdot$ (anything)= 0.

Proposition 2.4.2. *The sequence $M_n, n \geq 0$ is a super-martingale w.r.t. $\{\mathscr{F}_n\}$.*

Proof. We have $\{p(Z_n) < 1\} \subset A_n, n \geq 0$, hence

$$\begin{aligned} \mathsf{E}[M_n|\mathscr{F}_{n-1}] &= \mathsf{E}[1_{A_n}|\mathscr{F}_{n-1}]\widetilde{\mathscr{E}}_n \\ &= (1 - p(Z_{n-1}))\frac{1}{1 - p(Z_{n-1})}1_{p(Z_{n-1})<1}1_{A_{n-1}}\widetilde{\mathscr{E}}_{n-1} \\ &\leq 1_{A_{n-1}}\widetilde{\mathscr{E}}_{n-1} = M_{n-1}. \end{aligned}$$

□

Corollary 2.4.3. *Denote*

$$\mathscr{E}_\infty = \prod_{k=0}^{\infty}(1 - p(Z_k))^{-1}$$

and assume that, for some $x, y \in \mathbb{X}$,

$$P^Z_{(x,y)}(\mathscr{E}_\infty = \infty) = 1. \tag{2.4.3}$$

Then

$$\|P_n(x,\cdot) - P_n(y,\cdot)\|_{TV} \to 0, \quad n \to \infty.$$

If μ is an IPM for X and eq. (2.4.3) holds true for some $x \in \mathbb{X}$ and μ-a.a. y, then

$$\|P_n(x,\cdot) - \mu\|_{TV} \to 0, \quad n \to \infty.$$

Proof. We have

$$\mathsf{E}^Z_{(x,y)} 1_{A_n} \mathcal{E}_n = \mathsf{E}^Z_{(x,y)} M_n \le 1,$$

and

$$1_{A_n} \searrow 1_{A_\infty}, \quad \mathcal{E}_n \nearrow \mathcal{E}_\infty, \quad n \to \infty.$$

Then, for any n

$$\mathsf{E}^Z_{(x,y)} 1_{A_\infty} \mathcal{E}_n \le 1,$$

and by the monotone convergence theorem

$$\mathsf{E}^Z_{(x,y)} 1_{A_\infty} \widetilde{\mathcal{E}}_\infty = 1.$$

That is, condition (2.4.3) yields

$$\mathsf{P}^Z_{(x,y)}(A_\infty) = \lim_{n\to\infty} \mathsf{P}^Z_{(x,y)}(A_n) = 0.$$

Hence

$$\mathsf{P}^Z_{(x,y)}(X_n \ne Y_n) \to 0, \quad n \to \infty,$$

which by eq. (2.2.7) completes the proof of the first assertion. The second assertion follows easily from the first one and the bound

$$\|P^n(x,\cdot) - \mu\|_{TV} \le \int_{\mathbb{X}} \|P^n(x,\cdot) - P^n(y,\cdot)\|_{TV} \mu(dy) \to 0, \quad n \to \infty. \qquad \square$$

This corollary well illustrates the reason why "individual" ergodicity may still hold true in situations where uniform ergodicity clearly fails. Of course, one can apply Corollary 2.4.3 in the framework of Theorem 2.3.1: if eq. (2.3.1) holds true, then $p(z) \ge p > 0$ for all z, which immediately implies eq. (2.4.3). However, condition (2.4.3) is much milder and more flexible: for instance, it is satisfied if Z has an infinite number of visits to the set $\{z : p(z) > p\}$ for some $p > 0$. That is, "individual" ergodicity may be provided under a proper combination of a "local mixing" (or "local irreducibility") condition (like $p(z) > p, z \in B$ for some set B) and a "recurrence" condition (e.g., that Z visits B infinitely often). The same heuristics may also lead to convergence rates for such ergodicity, in this context the following corollary appears to be useful.

Corollary 2.4.4. *For any $x, y \in \mathbb{X}$, $n \ge 1$ and $q, r > 1$ such that $1/q + 1/r = 1$,*

$$\|P_n(x,\cdot) - P_n(y,\cdot)\|_{TV} \le 2\Big(E^Z_{(x,y)} \mathcal{E}_{n-1}^{-r/q}\Big)^{1/r}.$$

Proof. We have

$$\|P_n(x,\cdot) - P_n(y,\cdot)\|_{\mathrm{TV}} \le 2\mathsf{P}^Z_{(x,y)}(A_n) = 2\mathsf{E}^Z_{(x,y)} M_n^{1/q} \mathscr{E}_{n-1}^{-1/q}.$$

Then by the Hölder inequality

$$\|P_n(x,\cdot) - P_n(y,\cdot)\|_{\mathrm{TV}} \le 2\Big(\mathsf{E}^Z_{(x,y)} M_n\Big)^{1/q} \Big(\mathsf{E}^Z_{(x,y)} \mathscr{E}_{n-1}^{r/q}\Big)^{1/r} \le 2\Big(\mathsf{E}^Z_{(x,y)} \mathscr{E}_{n-1}^{-r/q}\Big)^{1/r}.$$

2.5 Stabilization of Transition Probabilities: Doob's Theorem

Recall that two measures $\mu, \nu \in \mathscr{P}(\mathbb{X})$ are called *equivalent* (notation $\mu \sim \nu$) if for any $A \in \mathscr{X}$

$$\mu(A) = 0 \Leftrightarrow \nu(A) = 0.$$

Measures μ, ν are called *singular* (notation $\mu \perp \nu$) if there exists $A \in \mathscr{X}$ such that

$$\mu(A) = 0, \quad \nu(A) = 1,$$

otherwise μ, *nu* are non-singular, which we denote by $\mu \not\perp \nu$.

Theorem 2.5.1 (Doob's theorem). *I. Let*

$$P(x,\cdot) \not\perp P(y,\cdot), \quad x, y \in \mathbb{X}. \tag{2.5.1}$$

Then there exists at most one IPM for X. If an IPM μ exists, then for $\mu\times\mu$-a.a. $(x,y) \in \mathbb{X}\times\mathbb{X}$

$$\left\|P^n(x,\cdot) - P^n(y,\cdot)\right\|_{\mathrm{TV}} \to 0, \quad n \to \infty \tag{2.5.2}$$

and, consequently, for μ-a.a. $x \in \mathbb{X}$

$$P^n(x,\cdot) \to \mu, \quad n \to \infty \tag{2.5.3}$$

in total variation.

II. If the stronger condition

$$P(x,\cdot) \sim P(y,\cdot), \quad x, y \in \mathbb{X} \tag{2.5.4}$$

holds true, then eqs. (2.5.2) *and* (2.5.3) *hold true for all $(x,y) \in \mathbb{X}\times\mathbb{X}$ and all $x \in \mathbb{X}$, respectively.*

Remark 2.5.2. Condition (2.5.1) can be written as

$$\left\|P(x,\cdot) - P(y,\cdot)\right\|_{\mathrm{TV}} < 2, \quad x, y \in \mathbb{X}, \tag{2.5.5}$$

which is just a point-wise version of the Dobrushin condition (2.3.1). One can understand Theorem 2.5.1 as an analogue of Theorem 1.2.5 for general MCs, and then eq. (2.5.5) is an analogue of the condition that the chain is irreducible and aperiodic; see also Proposition 1.2.3.

Remark 2.5.3. In its classical form, Doob's theorem assumes eq. (2.5.4) and provides (2.5.3) for every x. Its extension, which gives an a.s. convergence under eq. (2.5.1), is due to Ref. [86]. The difference between these two cases is illustrated by the following example.

Example 2.5.4. Let $\mathbb{X} = \{0, 1, \dots\}$ and

$$p_{00} = 1, \quad p_{i0} = 2^{-i}, \quad p_{i(i+1)} = 1 - 2^{-i}, \quad i = 1, 2, \dots.$$

Then eq. (2.5.1) holds true with $m = 1$, but eq. (2.5.4) fails. For each $i \neq 0$

$$P^n(i, \cdot) = \prod_{k=0}^{n-1}(1 - 2^{-i-n})\delta_{i+n} + \left(1 - \prod_{k=0}^{n-1}(1 - 2^{-i-n})\right)\delta_0,$$

which does not converge to $\mu = \delta_0$ as $n \to \infty$. Of course, $P^n(0, \cdot) = \mu$ converges to μ. Hence assertion II of Theorem 2.5.1 fails, while assertion I is true because all the states $i = 1, 2, \dots$ are exceptional in the sense that the set $\{1, 2, \dots\}$ has zero measure μ. Note that each of these states is nonessential, and hence they are exceptional in the sense of Theorem 1.2.5, as well.

proof of Theorem 2.5.1. I. By Corollary 2.4.3, in order to prove the required statement it is enough to show that eq. (2.4.3) holds true for $\mu \times \mu$-a.a. $(x, y) \in \mathbb{X}$. Clearly, eq. (2.4.3) is equivalent to

$$\mathsf{E}^Z_{(x,y)} \prod_n (1 - p(Z_n))^2 = 0. \tag{2.5.6}$$

Denote by U another MC $U_n = (X_n, Y_n)$, $n \geq 0$ with the transition probability

$$R((x, y), du\, dv) = \mathsf{P}(x, du)\mathsf{P}(y, dv).$$

If $U_0 = (x, y)$, then $X = \{X_n\}$ and $Y = \{Y_n\}$ are independent and have the laws P_x, P_y. We call U an *independent coupling* for the chain X. We note two properties, which make U particularly useful in the proof of eq. (2.5.6).

(i) For any $(x, y) \in \mathbb{X}$, the measure $Q((x, y), \cdot \setminus \mathscr{D})$ is absolutely continuous w.r.t. $R((x, y), \cdot)$ with

$$\frac{dQ((x, y), \cdot)}{dR((x, y), \cdot)}(u, v) \leq (1 - p(x, y))^{-1} \quad (u, v) \notin \mathscr{D}.$$

(ii) The measure $\mu \times \mu$ is an IPM for U.

By the property (i), for any $f : \mathbb{X} \times \mathbb{X} \to [0,1]$ such that $f(z) = 0, z \in \mathscr{D}$,

$$\mathsf{E}^Z_z f(Z_1) \le (1 - p(z))^{-1} \mathsf{E}^U_z f(U_1), \quad z \in \mathbb{X}; \tag{2.5.7}$$

we denote by P^U_z and E^U_z the law of U with $U_0 = z$ and respective expectation.

Now we proceed with the proof of eq. (2.5.6). For $p(x,y) = 1$ this relation is trivial, hence in what follows we assume $p(x,y) < 1$. Take $N \ge 1$ and denote

$$f_1(z) = (1 - p(z))^2.$$

Clearly, $f_1(z) = 0, z \in \mathscr{D}$, hence using eq. (2.5.7), we get

$$\begin{aligned}
\mathsf{E}^Z_{(x,y)} \prod_{n=1}^{N} (1 - p(Z_n))^2 &= \mathsf{E}^Z_{(x,y)} f_1(Z_N) \prod_{n=1}^{N-1} (1 - p(Z_n))^2 \\
&= \mathsf{E}^Z_{(x,y)} \left(\prod_{n=1}^{N-1} (1 - p(Z_n))^2 \mathsf{E}^Z_{Z_n} f_1(Z_1) \right) \\
&\le \mathsf{E}^Z_{(x,y)} f_2(Z_{N-1}) \prod_{n=1}^{N-2} (1 - p(Z_n))^2,
\end{aligned}$$

where we denote

$$f_2(z) = (1 - p(z)) \mathsf{E}^U_z f_1(U_1).$$

Repeating this calculation, we get

$$\begin{aligned}
\mathsf{E}^Z_{(x,y)} \prod_{n=1}^{N} (1 - p(Z_n))^2 &\le \mathsf{E}^Z_{(x,y)} f_3(Z_{N-2}) \prod_{n=1}^{N-3} (1 - p(Z_n))^2 \\
&\le \cdots \le (1 - p(x,y))^{-1} f_N(x,y),
\end{aligned}$$

where the sequence of functions $f_k, k \ge 1$ is defined iteratively,

$$f_k(z) = (1 - p(z)) \mathsf{E}^U_z f_{k-1}(U_1), \quad k > 2.$$

On the other hand,

$$f_N(x,y) = \mathsf{E}^U_{(x,y)} (1 - p(U_1)) \mathsf{E}^U_{U_1} f_{N-1}(U_1) = \mathsf{E}^U_{(x,y)} (1 - p(U_1)) f_{N-1}(U_2),$$

and repeating this calculation, we get

$$f_N(x,y) = \mathsf{E}^U_{(x,y)} f_1(U_N) \prod_{n=1}^{N-1} (1 - p(U_n)).$$

Taking $N \to \infty$, we finally get on the set $\{(x,y) : p(x,y) < 1\}$ the inequality

$$\mathsf{E}^Z_{(x,y)} \prod_{n=1}^{\infty} (1 - p(Z_n))^2 \le (1 - p(x,y))^{-1} \mathsf{E}^U_{(x,y)} \prod_{n=1}^{\infty} (1 - p(U_n)).$$

Consider the chain U with the initial distribution $\mu \times \mu$. This is a strictly stationary sequence, hence the sequence

$$\eta_k = \prod_{n=k}^{\infty} (1 - p(U_n)), \quad k \ge 0,$$

is strictly stationary, as well. On the other hand,

$$\eta_0 = \theta_0 \eta_1$$

with a nonnegative variable

$$\theta_0 = 1 - p(U_0),$$

which by eq. (2.5.1) is strictly less than 1 for every ω. Therefore, $\eta_0 = 0$ a.s., which yields

$$0 = \mathsf{E}^U_{\mu \times \mu} \prod_{n=1}^{\infty} (1 - p(U_n)) = \int_{\mathbb{X} \times \mathbb{X}} \mathsf{E}^U_{(x,y)} \prod_{n=1}^{\infty} (1 - p(U_n)) \mu(dx) \mu(dy).$$

That is, for $\mu \times \mu$-a.a. $(x,y) \in \mathbb{X}$

$$\mathsf{E}^U_{(x,y)} \prod_{n=1}^{\infty} (1 - p(U_n)) = 0,$$

which proves eq. (2.5.6) and completes the proof of eq. (2.5.2) for $\mu \times \mu$-a.a. $(x,y) \in \mathbb{X}$. This yields eq. (2.5.3) for μ-a.a. $x \in \mathbb{X}$ (see Corollary 2.4.3).

II. If eq. (2.5.4) holds true and an IPM μ exists, then for each $x \in \mathbb{X}$

$$P(x, \cdot) \sim \mu = \int_{\mathbb{X}} P(y, \cdot) \mu(dy).$$

Hence for all $(x,y) \in \mathbb{X} \times \mathbb{X}$, the measure

$$Q((x,y), \cdot \setminus \mathscr{D})$$

is absolutely continuous w.r.t. $\mu \times \mu$; denote by $\rho(x,y;u,v)$ the corresponding Radon–Nikodym density. We have already proved eq. (2.5.6) for $\mu \times \mu$-a.a. (x,y), which yields

$$\mathsf{E}^Z_{(x,y)}\prod_{n=1}^{\infty}(1-p(Z_n))^2 \le \mathsf{E}^Z_{(x,y)}1_{Z_1\notin\mathscr{D}}\prod_{n=2}^{\infty}(1-p(Z_n))^2$$
$$= \int_{\mathbb{X}} p(x,y;u,v)\mathsf{E}^Z_{(u,v)}\prod_{n=1}^{\infty}(1-p(Z_n))^2\mu(du)\mu(dv) = 0. \qquad \square$$

We remark that the above proof of Theorem 2.5.1 is essentially based, though in a somewhat hidden form, on the argument outlined after Corollary 2.4.3. Assumption (2.5.1) has the sense of a "local irreducibility" condition, while the "recurrence" in the model is actually guaranteed by the assumption that the chain has an IPM μ: it is exactly this assumption which provides eq. (2.5.6). In the subsequent sections, we will develop this argument more explicitly.

To finish this section, we consider one example which illustrates hidden limitations, which are brought to the entire approach when one type of the coupling, namely, the greedy coupling, is fixed.

Example 2.5.5. Let $\mathbb{X} = [0,1]^2$ and the MC X be defined by the following convention: given $X_n = (X_n^1, X_n^2)$, with probability 1/2 one of the coordinates X_{n+1}^1, X_{n+1}^2 of the next value X_{n+1} of the chain equals to the corresponding coordinate of X_n, while the other coordinate is uniformly distributed on $[0,1]$. That is, the one-step transition probability of the chain equals.

$$P\big((x^1,x^2), dy^1\,dy^2\big) = \frac{1}{2}\Big(\delta_{x^1}(dy^1)\,dy^2 + dy^1\delta_{x^2}(dy^2)\Big).$$

It can be easily seen that, once two initial values $x = (x^1, x^2), y = (y^1, y^2)$ are such that

$$x^1 \neq x^2, \quad y^1 \neq y^2,$$

with probability 1, the same inequalities hold true for all the subsequent values $(X_n, Y_n), n \ge 1$ of the corresponding greedy coupling. On the other hand, one can easily see that the two-step transition probability satisfies

$$P\big((x^1,x^2)\,x, dy^1\,dy^2\big) \ge \frac{1}{4}\,dy^1\,dy^2,$$

hence eq. (2.3.1) holds true and thus the system is uniformly ergodic; see Theorem 2.3.1.

One can say that the reason for the greedy coupling to fail in the previous example is that the joint law of its components outside of the diagonal is not well organized to grant a chance for the components to be coupled in the further attempts. The choice of the law of $Q((x_1,x_2), dy_1\,dy_2)$ outside of the diagonal was in a sense artificial, and the previous example shows that, in some cases, this choice is not optimal. Though, in most cases of interest this choice is completely sufficient (see Sections 3.3 and 3.4). On the other hand, this choice allows one to compare the greedy coupling with the easy-to-analyze independent coupling, like we did that in the above proof of Doob's theorem.

2.6 Nonuniform Ergodicity at Exponential Rate: The Harris Theorem

The strategy to combine local irreducibility with recurrence conditions dates back to T. Harris [57] and gives rise to a wide range of theorems, frequently called *Harris-type theorems*. In order to explain the argument more transparently, we separate the exposition in several parts. In this section, we prove a particularly important theorem of such a type, which provides nonuniform ergodicity at exponential rate; this setting is close to the original one by Harris.

In our exposition, we will systematically separate the "loss of memory"-type results, which control the decay rate for $P^n(x,\cdot) - P^n(y,\cdot)$ with given x, y, and the results which establish "stabilization" of the transition probabilities; that is, their convergence to the IPM. There is a substantial difference between these two types of results; see more discussion after Theorem 2.6.3.

Our first result in this section guarantees the "exponential loss of memory" effect. The recurrence conditions **R**(i),(ii) therein are formulated in terms of the auxiliary coupling chain Z; however, we will see in Section 2.8 that it is easy to provide sufficient conditions for them, formulated in terms of the original chain.

We say that chain X satisfies the local Dobrushin condition on a (measurable) set $B \subset \mathbb{X} \times \mathbb{X}$, if

$$\sup_{(x,y)\in B} \|P(x,\cdot) - P(y,\cdot)\|_{TV} < 2. \tag{2.6.1}$$

Denote by Z the greedy coupling for X, and put

$$\tau_B = \inf\{n \ge 0 : Z_n \in B\}, \quad \theta_B = \inf\{n \ge 1 : Z_n \in B\},$$

the hitting time and the delayed hitting time for the set B, respectively.

Theorem 2.6.1 (On exponential loss of memory). *Let for some set B and function $W : \mathbb{X} \times \mathbb{X} \to [1, \infty)$ the following conditions hold true.*

I. *X satisfies the local Dobrushin condition on B; that is,*

$$\upsilon_B = \frac{1}{2} \sup_{(x,y)\in B} \|P(x,\cdot) - P(y,\cdot)\|_{TV} < 1.$$

R. *There exists $\alpha > 0$ such that*

(i)

$$\mathsf{E}^Z_{(x,y)} e^{\alpha\tau_B} \le W(x,y), \quad (x,y) \in \mathbb{X}\times\mathbb{X};$$

(ii)

$$S_{\alpha,B} = \sup_{(x,y)\in B} \mathsf{E}^Z_{(x,y)} e^{\alpha\theta_B} < \infty.$$

Then for any

$$\beta < \frac{\alpha \log \upsilon_B^{-1}}{\log \upsilon_B^{-1} + \log S_{\alpha,B}} \tag{2.6.2}$$

there exists $C < \infty$ such that

$$\|P^n(x,\cdot) - P^n(y,\cdot)\|_{TV} \le Ce^{-\beta n} W^{\beta/\alpha}(x,y). \tag{2.6.3}$$

Remark 2.6.2. Comparing eq. (2.6.3) with eq. (2.3.2), we see that the only difference is presented by the additional term $W^{\beta/\alpha}(x,y)$. This term well corresponds to the heuristics discussed after Example 2.4.1: for "distant" initial states, it should take more time for the system to stabilize. From this point of view, the term $W^{\beta/\alpha}(x,y)$ plays the role of a "penalty" for the stabilization rate, which corresponds to the "distance" between initial states.

Proof. Denote by N_n the number of consequential visits of the chain Z to B before the time moment n. Then by Corollary 2.4.4 for any $q, r > 1$ with $1/q + 1/r = 1$

$$\|P^n(x,\cdot) - P^n(y,\cdot)\|_{TV} \le 2\left(\mathsf{E}^Z_{(x,y)}\mathscr{E}_n^{-r/q}\right)^{1/r} \le 2\left(\mathsf{E}^Z_{(x,y)}\upsilon_B^{rN_n/q}\right)^{1/r}. \tag{2.6.4}$$

Take

$$r = \frac{\log \upsilon_B^{-1} + \log S_{\alpha,B}}{\log \upsilon_B^{-1}}, \quad q = \frac{\log \upsilon_B^{-1} + \log S_{\alpha,B}}{\log S_{\alpha,B}},$$

then eq. (2.6.2) yields

$$r\beta \le \alpha, \quad q\beta < \alpha\frac{\log \upsilon_B^{-1}}{\log S_{\alpha,B}}. \tag{2.6.5}$$

We have

$$\begin{aligned} \mathsf{E}^Z_{(x,y)}\upsilon_B^{rN_n/q} &= \sum_{k=0}^{\infty} \upsilon_B^{rk/q}\mathsf{P}^Z_{(x,y)}(N_n = k) \\ &= \left(\upsilon_B^{-r/q} - 1\right)\sum_{k=0}^{\infty} \upsilon_B^{rk/q}\mathsf{P}^Z_{(x,y)}(N_n > k). \end{aligned} \tag{2.6.6}$$

Denote

$$\tau_B^1 = \tau_B, \quad \tau_B^{k+1} = \inf\{n > \tau_B^k : Z_n \in B\}, \quad k \geq 1.$$

Then for any $\gamma > 0$

$$\mathsf{P}^Z_{(x,y)}(N_n > k) = \mathsf{P}^Z_{(x,y)}(\tau_B^k > n) \leq e^{-\gamma n}\mathsf{E}^Z_{(x,y)} e^{\gamma\tau_B^k}.$$

Denote

$$\theta_B^k = \tau_B^{k+1} - \tau_B^k, \quad k \geq 1$$

then

$$\tau_B^k = \tau_B + \theta_B^1 + \cdots + \theta_B^{k-1}, \quad k \geq 2. \tag{2.6.7}$$

By the condition **R**(ii) and the strong Markov property of Z,

$$\mathsf{E}^Z_{(x,y)}\left[e^{\alpha\theta_B^j}\Big|\mathscr{F}_{\tau_B^j}\right] = \mathsf{E}^Z_z e^{\alpha\theta_B}\Big|_{z=Z_{\tau_B^j}} \leq S_{\alpha,B},$$

hence

$$\mathsf{E}^Z_{(x,y)} e^{\alpha\tau_B^k} \leq S_{\alpha,B}\mathsf{E}^Z_{(x,y)} e^{\alpha\tau_B^{k-1}} \leq \cdots \leq S_{\alpha,B}^{k-1} W(x,y);$$

in the last inequality we have used condition **R**(i).

Take $\gamma = \beta r$, then by the first inequality in eq. (2.6.5) $\gamma \leq \alpha$ and thus

$$\mathsf{E}^Z_{(x,y)} e^{\gamma\tau_B^k} \leq \left(\mathsf{E}^Z_{(x,y)} e^{\gamma\tau_B^k}\right)^{\gamma/\alpha} \leq S_{\alpha,B}^{\gamma(k-1)/\alpha} W^{\gamma/\alpha}(x,y).$$

The second inequality in eq. (2.6.5) yields

$$S_{\alpha,B}^{\gamma/\alpha} \upsilon_B^{r/q} < 1, \tag{2.6.8}$$

hence we finally obtain

$$\begin{aligned}\mathsf{E}^Z_{(x,y)} \upsilon_B^{rN_n/q} &\leq \left(\upsilon_B^{-r/q} - 1\right) e^{-\gamma n} \sum_{k=0}^{\infty} \upsilon_B^{rk/q} S_{\alpha,B}^{\gamma(k-1)/\alpha} W^{\gamma/\alpha}(x,y) \\ &= Ce^{-\gamma n} W^{\gamma/\alpha}(x,y).\end{aligned}$$

From eq. (2.6.4) and this estimate we get eq. (2.6.3). □

Now we are ready to present the main result of this section, which states exponential convergence of the transition probabilities to (unique) IPM under an additional moment condition for the "penalty" term $W^{\beta/\alpha}(x, y)$.

Theorem 2.6.3 (The Harris theorem). *Let conditions of Theorem 2.6.1 hold true and for each $x \in \mathbb{X}$*

$$\sup_{N\geq 1} \frac{1}{N} \sum_{n=1}^{N} E_x W^{\beta/\alpha}(x, X_n) < \infty. \tag{2.6.9}$$

Then there exists unique IPM μ for X, the function

$$U(x) = \int_{\mathbb{X}} W^{\beta/\alpha}(x, y)\mu(dy), \quad x \in \mathbb{X}$$

takes finite values, and

$$\|P^n(x, \cdot) - \mu\|_{TV} \leq Ce^{-\beta n} U(x). \tag{2.6.10}$$

Proof. The proof follows the lines of the proof of Theorem 2.3.1, with the sequence $P^n(x, \cdot), n \geq 1$ changed to

$$\mu_x^{(N)} = \frac{1}{N} \sum_{n=1}^{N} P^n(x, \cdot), \quad N \geq 1.$$

We have for arbitrary x

$$\begin{aligned} \|P^n(x, \cdot) - P^m(x, \cdot)\|_{TV} &= \left\| \int_{\mathbb{X}} \left(P^{m\wedge n}(y, \cdot) - P^{m\wedge n}(x, \cdot)\right) P^{|n-m|}(x, dy) \right\|_{TV} \\ &\leq \int_{\mathbb{X}} \left\|P^{m\wedge n}(y, \cdot) - P^{m\wedge n}(x, \cdot)\right\|_{TV} P^{|n-m|}(x, dy) \\ &\leq Ce^{-\beta(m\wedge n)} \int_{\mathbb{X}} W^{\beta/\alpha}(x, y) P^{|n-m|}(x, dy). \end{aligned}$$

If $M < N$ and $m \in \{1, \ldots, M\}, n \in \{1, \ldots, N\}$, then $k = m \wedge n \leq M$ and $l = |n - m| \leq N - 1$. Hence

$$\begin{aligned} \|\mu_x^{(N)} - \mu_x^{(M)}\|_{TV} &\leq \frac{1}{MN} \sum_{n=1}^{N} \sum_{m=1}^{M} \|P^n(x, \cdot) - P^m(x, \cdot)\|_{TV} \\ &\leq \frac{2C}{MN} \sum_{k=1}^{M} \sum_{l=0}^{N-1} e^{-\beta k} \int_{\mathbb{X}} W^{\beta/\alpha}(x, y) P^l(x, dy). \end{aligned}$$

By eq. (2.6.9), the sequence

$$\frac{1}{N}\sum_{l=0}^{N-1}\int_{\mathbb{X}} W^{\beta/\alpha}(x,y)\,P^l(x,dy) = \frac{1}{N}\sum_{l=0}^{N-1} \mathsf{E}_x W^{\beta/\alpha}(x,X_l)$$

is bounded, hence

$$\|\mu_x^{(N)} - \mu_x^{(M)}\|_{\text{TV}} \le \frac{C'}{M}\sum_{k=1}^{M} e^{-\beta k} \to 0, \quad M \to \infty.$$

That is, the sequence $\mu_x^{(N)}, N \ge 1$ is fundamental (and thus converges) in total variation distance for any x. Applying eq. (2.6.3) once more we get

$$\|\mu_x^{(N)} - \mu_y^{(N)}\|_{\text{TV}} \le \frac{1}{N}\sum_{n=1}^{N}\|P^n(x,\cdot) - P^n(y,\cdot)\|_{\text{TV}} \le \frac{C}{N} W^{\beta/\alpha}(x,y)\sum_{n=1}^{N} e^{-\beta n} \to 0,$$

hence the limit μ of $\{\mu_x^{(N)}\}$ does not depend on the initial point x. It is easy to show that μ is an invariant measure for X. Since $\mu_x^{(N)} \to \mu$, by the Fatou lemma and condition (2.6.9)

$$U(x) = \int_{\mathbb{X}} W^{\beta/\alpha}(x,y)\mu(dy) \le \liminf_N \frac{1}{N}\sum_{n=1}^{N}\int_{\mathbb{X}} W^{\beta/\alpha}(x,y)P^n(x,dy) < \infty, \quad x \in \mathbb{X}.$$

Finally,

$$\begin{aligned}\|P^n(x,\cdot) - \mu\|_{\text{TV}} &\le \int_{\mathbb{X}} \left\|P^n(x,\cdot) - P^n(y,\cdot)\right\|_{\text{TV}}\mu(dy)\\ &\le Ce^{-\beta n}\int_{\mathbb{X}} W^{\beta/\alpha}(x,y)\mu(dy) = Ce^{-\beta n}U(x).\end{aligned}$$

□

The proof of Theorem 2.6.3 makes clearly visible the main difference between the "loss of memory" type bounds (like eq. (2.6.3)) and the "stability" rates (like eq. (2.6.10)). Since eq. (2.6.3) contains a "penalty term" $W^{\beta/\alpha}(x,y)$, one easily gets eq. (2.6.10) *provided that* the IPM μ exists and

$$U(x) = \int_{\mathbb{X}} W^{\beta/\alpha}(x,y)\mu(dy) < \infty, \quad x \in \mathbb{X}.$$

Then $U(x)$ can be interpreted as a new "penalty term," which now applies to the "stabilization" effect. However, one should be careful: the above integrability and even the existence of an IPM are additional assumptions which may fail even for systems that admit exponential "loss of memory" type bounds.

Example 2.6.4. Let X be a linear autoregressive sequence defined by

$$X_n = aX_{n-1} + \sigma\varepsilon_n + \eta_n, \quad n \ge 1,$$

where $|a| < 1, \sigma^2 > 0$, and $\{\varepsilon_n\}$ and $\{\eta_n\}$ are two independent i.i.d. sequences, with $\varepsilon_n, n \geq 1$ having the standard normal distribution, and the distribution $\eta_n, n \geq 1$ which will be specified later. One can easily prove an analogue of eq. (2.6.3) for X. Indeed, consider first a “truncated” autoregressive sequence, which actually coincides with the one from Example 2.8.4:

$$\widetilde{X}_n = a\widetilde{X}_{n-1} + \sigma\varepsilon_n, \quad n \geq 1.$$

For this sequence, we can easily verify that conditions of Theorem 2.6.1 hold true with $W(x, y) = 2 + x^2 + y^2$; see Example 2.8.4. Then the transition probabilities for $\widetilde{X}$ satisfy

$$\|\widetilde{P}^n(x, \cdot) - \widetilde{P}^n(y, \cdot)\|_{TV} \leq Ce^{-\beta n}(2 + x^2 + y^2)^{\beta/\alpha}$$

with some $C > 0, 0 < \beta < \alpha$. On the other hand, we can represent sequences $X, \widetilde{X}$ with the same initial conditions as

$$X_n = \widetilde{X}_n + \Delta_n, \quad \Delta_n = \sum_{k=1}^{n} a^{n-k}\eta_k.$$

Denote by Ξ_n the law of Δ_n, then

$$\begin{aligned}
\|P^n(x, \cdot) - P^n(y, \cdot)\|_{TV} &= \left\| \int_{\mathbb{R}} \left(\widetilde{P}^n(x, \cdot - z) - \widetilde{P}^n(y, \cdot - z)\right)\Xi_n(dz) \right\|_{TV} \\
&\leq \int_{\mathbb{R}} \left\|\widetilde{P}^n(x, \cdot - z) - \widetilde{P}^n(y, \cdot - z)\right)\right\|_{TV} \Xi_n(dz) \\
&= \|\widetilde{P}^n(x, \cdot) - \widetilde{P}^n(y, \cdot)\|_{TV},
\end{aligned}$$

which gives the required “loss of memory” bound for X. On the other hand, if

$$\mathsf{E}\log(1 + |\eta_n|) = \infty,$$

then

$$\mathsf{E}\log(1 + |\varepsilon_n + \eta_n|) = \infty,$$

and an IPM for X does not exists. The calculation here is similar to the one given in the continuous-time setting in Ref. [122]; we omit the details.

2.7 Nonuniform Ergodicity at Subexponential Rates

In this section, we extend previous results, allowing the system to be subexponentially recurrent, and proving respective ergodicity at subexponential rate. The structure of the argument remains essentially the same, but a new technical issue arise, which makes the formulation of the principal estimate more cumbersome; see Example 2.7.3 and Corollary 2.7.4. This actually was our reason to consider separately the exponential case, where the structure of the argument is more transparent.

Denote by Λ the class of continuous monotone functions $\lambda : [0, \infty) \to [1, \infty)$ with $\lambda(1) = 1$ and $\lambda(\infty) = \infty$ which are *submultiplicative* in the sense that

$$\lambda(t + s) \le \lambda(t)\,\lambda(s), \quad s, t \ge 0.$$

Example 2.7.1. The following functions belong to the class Λ.

(i) Exponential:

$$\lambda(t) = e^{at}, \quad a > 0.$$

(ii) Subexponential:

$$\lambda(t) = e^{at^{\kappa}}, \quad a > 0, \quad \kappa \in (0, 1).$$

(iii) Polynomial:

$$\lambda(t) = (1 + at)^{p}, \quad a > 0, \quad p > 0.$$

Like we did that in previous section, we first state a result about the "loss of memory" effect.

Theorem 2.7.2 (On subexponential loss of memory). *Let for some set B and function $W : \mathbb{X} \times \mathbb{X} \to [1, \infty)$ the following conditions hold true:*

I. *X satisfies the local Dobrushin condition on B.*

R. *There exists function $\lambda \in \Lambda$ such that*

(i)

$$\mathsf{E}^{Z}_{(x,y)}\lambda(\tau_B) \le W(x, y), \quad (x, y) \in \mathbb{X} \times \mathbb{X};$$

(ii)

$$S_{\lambda(\cdot),B} = \sup_{(x,y)\in B} \mathsf{E}^{Z}_{(x,y)}\lambda(\theta_B) < \infty.$$

Then for any

$$\delta < \frac{\log \upsilon_B^{-1}}{\log \upsilon_B^{-1} + \log S_{\lambda(\cdot),B}} \tag{2.7.1}$$

there exists $C < \infty$ such that

$$\|P^n(x,\cdot) - P^n(y,\cdot)\|_{TV} \le C\Big(\lambda(n)\Big)^{-\delta} W^\delta(x,y). \tag{2.7.2}$$

Proof. The proof essentially repeats the proof of Theorem 2.6.1. We have eqs. (2.6.4) and (2.6.6). Take

$$r = \frac{\log \upsilon_B^{-1} + \log S_{\lambda(\cdot),B}}{\log \upsilon_B^{-1}}, \quad q = \frac{\log \upsilon_B^{-1} + \log S_{\lambda(\cdot),B}}{\log S_{\lambda(\cdot),B}},$$

then eq. (2.7.1) yields

$$r\delta \le 1, \quad q\delta < \frac{\log \upsilon_B^{-1}}{\log S_{\lambda(\cdot),B}}. \tag{2.7.3}$$

For any $\gamma > 0$,

$$\mathsf{P}^Z_{(x,y)}(N_n > k) \le \frac{1}{\lambda^\gamma(n)} \mathsf{E}^Z_{(x,y)} \lambda^\gamma(\tau_B^k).$$

By eq. (2.6.7) and submultiplicativity of λ, for any $\gamma > 0$

$$\mathsf{E}^Z_{(x,y)} \lambda^\gamma(\tau_B^k) \le \mathsf{E}^Z_{(x,y)} \lambda^\gamma(\tau_B) \prod_{j=1}^{k-1} \lambda^\gamma(\theta_B^j).$$

Take $\gamma = \delta r$, then by the first inequality in eq. (2.7.3) $\gamma \le 1$. Then the Hölder inequality, condition **R**, and the strong Markov property of Z yield

$$\mathsf{E}^Z_{(x,y)} \lambda^\gamma(\tau_B^k) \le S^{\gamma(k-1)}_{\lambda(\cdot),B} W^\gamma(x,y).$$

The second inequality in eq. (2.7.3) yields

$$S^\gamma_{\lambda(\cdot),B} \upsilon_B^{r/q} < 1, \tag{2.7.4}$$

hence we can finally obtain

$$\mathsf{E}^Z_{(x,y)} \upsilon_B^{rN_n/q} \le \left(\upsilon_B^{-r/q} - 1\right) \frac{1}{\lambda^\gamma(n)} \sum_{k=0}^\infty \upsilon_B^{rk/q} S^{\gamma(k-1)}_{\lambda(\cdot),B} W^\gamma(x,y) = \frac{C}{\lambda^\gamma(n)} W^\gamma(x,y).$$

From eq. (2.6.4) and this estimate we get eq. (2.7.2). □

Clearly, Theorem 2.7.2 is a straightforward extension of Theorem 2.6.1: taking $\lambda(t) = e^{\alpha t}$ and $\beta = \delta\alpha$, we get the statement of Theorem 2.6.1. However, in some situations Theorem 2.7.2 does not give a best possible result.

Example 2.7.3. Let condition **R** hold true with the polynomial function $\lambda(t) = (1 + at)^p$. Then Theorem 2.7.2 provides the stabilization rate (2.7.2) with

$$\Big(\lambda(n)\Big)^{-\delta} = (1 + an)^{-\delta p},$$

where δ satisfies eq. (2.7.1). That is, the order of the polynomial which controls the stabilization rate is substantially smaller than the order of the polynomial which controls the recurrence properties.

This example motivates the following modification of Theorem 2.7.2.

Corollary 2.7.4. *Let conditions of Theorem 2.7.2 hold true. Then for any $\delta \in (0, 1)$ there exist $\gamma > 0$ and $C < \infty$ such that*

$$\|P^n(x, \cdot) - P^n(y, \cdot)\|_{TV} \le C\Big(\lambda(\gamma n)\Big)^{-\delta} W^{\delta}(x, y). \tag{2.7.5}$$

Proof. Fix $\kappa \in (0, 1)$ and denote

$$\lambda_{\gamma,\kappa}(t) = \lambda^{\kappa}(\gamma t), \quad \gamma \le 1.$$

Clearly, condition **R** yields

$$\mathsf{E}^Z_{(x,y)}\lambda_{\gamma,\kappa}(\tau_B) \le \mathsf{E}^Z_{(x,y)}\Big(\lambda(\tau_B)W(x, y)\Big)^{\kappa}, \quad (x, y) \in \mathbb{X} \times \mathbb{X}$$

and

$$S_{\lambda_{\gamma,\kappa}(\cdot),B} \le \sup_{(x,y)\in B} \mathsf{E}^Z_{(x,y)}\Big(\lambda(\theta_B)\Big)^{\kappa} < \infty.$$

Hence eq. (2.7.5) holds true for $\delta = \delta'\kappa$ with

$$\delta' < \frac{\log \upsilon_B^{-1}}{\log \upsilon_B^{-1} + \log S_{\lambda_{\gamma,\kappa}(\cdot),B}}.$$

Observe that

$$S_{\lambda_{\gamma,\kappa}(\cdot),B} \to 1, \quad \gamma \to 0+. \tag{2.7.6}$$

Indeed, for any $N \geq 1$

$$\begin{aligned}\mathsf{E}^Z_{(x,y)}\Big(\lambda(\gamma\theta_B)\Big)^\kappa &\leq \lambda^\kappa(\gamma N) + \mathsf{E}^Z_{(x,y)}\Big(\lambda(\gamma\theta_B)\Big)^\kappa 1_{\theta_B>N}\\ &\leq \lambda^\kappa(\gamma N) + \Big(\mathsf{E}^Z_{(x,y)}\lambda(\theta_B)\Big)\kappa\Big(\mathsf{P}^Z_{(x,y)}1_{\theta_B>N}\Big)^{1-\kappa},\end{aligned}$$

which gives

$$S_{\lambda_{\gamma,\kappa}(\cdot),B} \leq \lambda^\kappa(\gamma N) + \frac{S_{\lambda(\cdot),B}}{\lambda^{1-\kappa}(N)}.$$

That is,

$$\limsup_{\gamma\to 0+} S_{\lambda_{\gamma,\kappa}(\cdot),B} \leq \frac{S_{\lambda(\cdot),B}}{\lambda^{1-\kappa}(N)} \to 0, \quad N\to\infty,$$

which proves eq. (2.7.6).

Now we finalize the proof. For any $\delta \in (0,1)$ we fix $\kappa \in (\delta, 1)$. Using eq. (2.7.6), we can find $\gamma > 0$ small enough, such that

$$\kappa\frac{\log \upsilon_B^{-1}}{\log \upsilon_B^{-1} + \log S_{\lambda_{\gamma,\kappa}(\cdot),B}} > \delta.$$

Then Theorem 2.7.2 applied to $\lambda_{\gamma,\kappa}$ provides the required statement. □

Corollary 2.7.4 improves substantially the polynomial stabilization bound, see Example 2.7.3. Namely, we have

$$(1 + a\gamma n) \geq \gamma(1 + an), \quad \gamma \in (0,1),$$

hence changing the constant C we can derive eq. (2.7.2) with

$$\Big(\lambda(n)\Big)^{-\delta} = (1 + an)^{-\delta p}$$

for any $\delta < 1$; that is, the order of the polynomial which controls the stabilization rate is essentially the same with the one from the recurrence condition.

Now we are ready to present the main result of this section, which states sub-exponential convergence of the transition probabilities to (unique) IPM under an additional moment condition for the "penalty" term $W^{\beta/\alpha}(x,y)$. We call it "Harris-type" to separate the exponential case, which genuinely dates back to Harris.

Theorem 2.7.5 (The Harris-type theorem). *Let conditions of Theorem 2.7.2 hold true, and for given $\delta \in (0,1)$*

$$\sup_{N\geq 1} \frac{1}{N} \sum_{n=1}^{N} \mathsf{E}_x W^{\delta}(x, X_n) < \infty, \quad x \in \mathbb{X}. \tag{2.7.7}$$

Then there exists unique IPM μ for X, the function

$$U(x) = \int_{\mathbb{X}} W^{\delta}(x,y)\mu(dy), \quad x \in \mathbb{X}$$

takes finite values, and

$$\|P^n(x,\cdot) - \mu\|_{TV} \leq C\Big(\lambda(\gamma n)\Big)^{-\delta} U(x), \tag{2.7.8}$$

where $\gamma > 0$ is such that eq. (2.7.5) *holds true.*

With obvious changes, the proof repeats the proof of Theorem 2.6.3; we omit it here.

2.8 Lyapunov-Type Conditions

In this section, we explain a practical method which makes it possible to verify both the recurrence conditions, involved in Theorems 2.6.1 and 2.7.2, and the moment conditions required in Theorems 2.6.3 and 2.7.5. The principal assumption (2.8.1) in this method is very close, both in form and in spirit, to the classical Lyapunov condition in the theory of stability of ordinary differential equations; hence, it is usually called a *Lyapunov-type condition.*

2.8.1 Linear Lyapunov-Type Condition and Exponential Ergodicity

Theorem 2.8.1. *Assume that, for a given MC X, there exists a set $K \in \mathscr{X}$ and a function $V : \mathbb{X} \to [1, +\infty)$ such that*

(i) *V is bounded on K;*
(ii) *for some $a > 0$, $C > 0$*

$$\mathsf{E}_x V(X_1) - V(x) \leq -aV(x) + C, \quad x \in \mathbb{X}, \tag{2.8.1}$$

$$a > \frac{2C}{1 + \inf_{x\notin K} V(x)}. \tag{2.8.2}$$

*Then any Markov coupling Z for the chain X satisfies condition **R** of Theorem 2.6.1 with*

$$B = K \times K, \quad W(x,y) = V(x) + V(y),$$

and

$$\alpha = \log(1 - a + \sigma) > 0, \quad \sigma = \frac{2C}{1 + \inf_{x \notin K} V(x)}.$$

In addition,

$$\mathsf{E}^Z_{x,y} W(Z_n) \le (1-a)^n W(x,y) + \frac{2C}{a}, \quad n \ge 0, \tag{2.8.3}$$

hence eq. (2.6.9) *holds true.*

Remark 2.8.2. We call eq. (2.8.1) a *linear Lyapunov-type condition* because its right hand side is linear w.r.t. V, on the contrary to more general condition (2.8.11) treated below.

Proof. For any Markov coupling $Z = (X, Y)$, we have

$$\mathsf{E}^Z_{(x,y)} W(Z_1) = \mathsf{E}_x V(X_1) + \mathsf{E}_y V(X_1).$$

Hence eq. (2.8.1) implies

$$\mathsf{E}^Z_{(x,y)} W(Z_1) \le (1-a) W(x,y) + 2C. \tag{2.8.4}$$

Iterating this inequality, we easily obtain

$$\begin{aligned} \mathsf{E}^Z_{(x,y)} W(Z_n) &\le (1-a)\mathsf{E}^Z_{(x,y)} W(Z_{n-1}) + 2C \le \dots \\ &\le (1-a)^n W(x,y) + 2C\left(1 + \dots + (1-a)^{n-1}\right), \end{aligned}$$

which proves eq. (2.8.3).

If $(x, y) \notin B$, then at least one of the points x, y does not belong to K. In this case,

$$W(x,y) \ge 1 + \inf_{x' \notin K} V(x'),$$

and

$$(1-a) W(x,y) + 2C \le (1 - a + \sigma) W(x,y).$$

This means that

$$\mathsf{E}^Z_{(x,y)} e^{\alpha} W(Z_1) \le W(x,y), \quad (x,y) \notin B. \tag{2.8.5}$$

From this relation, we easily deduce by induction

$$\mathsf{E}^Z_{(x,y)} e^{\alpha(\tau_B \wedge n)} W(Z_{\tau_B \wedge n}) \le W(x,y), \quad n \ge 0, \quad (x,y) \in \mathbb{X} \times \mathbb{X}. \tag{2.8.6}$$

Since $W \geq 2 > 1$, we get

$$\mathsf{E}^Z_{(x,y)} e^{\alpha(\tau_B \wedge n)} \leq W(x,y), \quad n \geq 0, \quad (x,y) \in \mathbb{X} \times \mathbb{X},$$

and taking $n \to \infty$ we get **R**(i) by the Fatou lemma.

Condition **R**(ii) holds true (almost) trivially: by **R**(i), we have

$$\mathsf{E}^Z_{(x,y)} e^{\alpha\theta_B} = \mathsf{E}^Z_{(x,y)}\left(\mathsf{E}^Z_{Z_1} e^{\alpha\tau_B}\right) \leq \mathsf{E}^Z_{(x,y)} W(Z_1) = \mathsf{E}_x V(X_1) + \mathsf{E}_y V(X_1).$$

Then by eq. (2.8.1)

$$\mathsf{E}^Z_{(x,y)} e^{\alpha\theta_B} \leq (1-a)(V(x)+V(y)) + 2C,$$

which yields **R**(ii) because V is assumed to be bounded on K. □

Combining Theorems 2.6.1, 2.6.3, and 2.8.1, we get the following particularly important corollary, which gives an exponential ergodic bound for an MC.

Corollary 2.8.3 (On exponential ergodic rate). *Let for a given MC X there exists function $V : \mathbb{X} \to [1, +\infty)$ such that X satisfies the local Dobrushin condition on each set $B_c = K_c \times K_c$, $c \geq 1$, where $K_c = \{x : V(x) \leq c\}$ is a* level set *of the function V. Assume also that for some $a,\ C > 0$*

$$\mathsf{E}_x V(X_1) - V(x) \leq -aV(x) + C, \quad x \in \mathbb{X}. \tag{2.8.7}$$

Then there exist $c_1, c_2 > 0$ such that

$$\left\|P_n(x,\cdot) - P_n(y,\cdot)\right\|_{TV} \leq c_1 e^{-c_2 n}\big(V(x) + V(y)\big), \quad x,y \in \mathbb{X}, \quad n \geq 1. \tag{2.8.8}$$

In addition, there exists unique IPM μ for X, measure μ satisfies

$$\int_{\mathbb{X}} V\,d\mu < \infty,$$

and

$$\left\|P_n(x,\cdot) - \mu\right\|_{TV} \leq c_1 e^{-c_2 n}\left(V(x) + \int_{\mathbb{X}} V\,d\mu\right), \quad x \in \mathbb{X}, \quad n \geq 1. \tag{2.8.9}$$

Proof. If eq. (2.8.1) holds true, then for c large enough

$$\frac{2C}{1 + \inf_{x \notin K_c} V(x)} \leq \frac{2C}{1+c} < a.$$

Then Theorems 2.6.1, 2.6.3, and 2.8.1 yield

$$\left\|P_n(x,\cdot) - P_n(y,\cdot)\right\|_{TV} \le C_\beta e^{-\beta n}\left(V(x_1) + V(x_2)\right)^{\beta/\alpha}.$$

Since $V \ge 1$, this implies eq. (2.8.8). Similarly to the proof of Theorem 2.8.1, we have

$$\sup_n \mathsf{E}_x V(X_n) < \infty,$$

which implies

$$\int_{\mathbb{X}} V \, d\mu < \infty$$

and eq. (2.8.9). □

Example 2.8.4. Let

$$X_n = aX_{n-1} + \varepsilon_n, \quad n \ge 1$$

with $|a| < 1$ and i.i.d. $\{\varepsilon_n\}$ which has the standard normal distribution. Then for $V(x) = 1 + x^2$

$$\mathsf{E}_x V(X_1) = 2 + a^2x^2 = a^2V(x) + (2 - a^2).$$

Hence eq. (2.8.7) holds true. Since each level set $K_c \times K_c$ for V is bounded, it is easy to show that the Dobrushin condition is satisfied on $B_c = K_c \times K_c$. Hence, Corollary 2.8.3 can be applied, and eqs. (2.8.8) and (2.8.9) hold true.

2.8.2 Lyapunov-Type Condition for Cesàro Means of Moments

In this short section, we formulate separately a simple but very useful moment bound, formulated in terms of a Lyapunov-type condition.

Proposition 2.8.5. *Let two functions be U, V such that*

$$\mathsf{E}_x U(X_1) - U(x) \le -V(x) + C, \quad x \in \mathbb{X}. \tag{2.8.10}$$

Then

$$\frac{1}{N}\sum_{n=0}^{N-1} \mathsf{E}_x V(X_n) \le C + \frac{1}{N}\Big(U(x) - \mathsf{E}_x U(X_N)\Big), \quad x \in \mathbb{X}, \quad N \ge 1$$

In particular, if $U, V \geq 0$ then for every $x \in \mathbb{X}$ the sequence

$$\frac{1}{N}\sum_{n=1}^{N} \mathsf{E}_x V(X_n), \quad N \geq 1$$

is bounded.

Proof. The required bound follows easily from the inequality

$$\begin{aligned} \mathsf{E}_x U(X_N) &= \mathsf{E}_x \mathsf{E}_{X_{N-1}} U(X_1) \leq \mathsf{E}_x U(X_{N-1}) - \mathsf{E}_x V(X_{N-1}) + C \\ &\leq \cdots \leq U(x) - \sum_{n=0}^{N_1} \mathsf{E}_x V(X_n) + CN. \end{aligned}$$

□

Note that Proposition 2.8.5 gives a bound for Cesáro means of moments but not for the moments themselves. This explains why in Theorems 2.6.3 and 2.7.5 we impose conditions on Cesáro means of moments. In some cases, “individual” moments can be estimated, as well, like we had that in Theorem 2.8.1. However, in general this is no longer the case; that is, the linear Lyapunov-type condition is in a sense “too good” to reveal all possible effects.

2.8.3 Sublinear Lyapunov-Type Conditions and Subexponential Ergodicity

Theorem 2.8.6. *Let for a given MC X, functions $V : \mathbb{X} \to [1, +\infty)$, $\phi : [1, +\infty) \to (0, \infty)$, and constant C, the following holds:*

(i)

$$\mathsf{E}_x V(X_1) - V(x) \leq -\phi(V(x)) + C, \quad x \in \mathbb{X}; \tag{2.8.11}$$

(ii)

$$\phi\left(1 + \inf_{x \notin K} V(x)\right) > 2C; \tag{2.8.12}$$

(iii) *V is bounded on K;*

(iv) *ϕ admits a nonnegative, increasing, and concave extension to $[0, +\infty)$.*

*Then any Markov coupling Z for the chain X satisfies condition **R** of Theorem 2.7.2 with*

$$B = K \times K, \quad W(x, y) = V(x) + V(y),$$

and

$$\lambda(t) = \Phi^{-1}(\kappa t),$$

where

$$\kappa = 1 - \frac{2C}{\phi\left(1 + \inf_{x \notin K} V(x)\right)},$$

and Φ^{-1} *denotes the inverse function to*

$$\Phi(v) = \int_1^v \frac{1}{\phi(w)}\, dw, \quad v \in [1, \infty). \tag{2.8.13}$$

In addition, the function λ *belongs to the class* Λ.

Remark 2.8.7. Since ϕ is concave, it is sublinear in the sense that

$$\phi(v) \le c_1 + c_2 v$$

with some positive c_1, c_2. This is a reason for us to call eq. (2.8.11) a *sublinear Lyapunov-type condition*; this terminology well corresponds to the fact that $\lambda \in \Lambda$ is *sub-exponential* in the sense that

$$\lambda(t) \le c_3 e^{c_4 t}$$

with some positive c_3, c_4.

Proof. Denote

$$H(t, v) = \Phi^{-1}\big(\kappa t + \Phi(v)\big), \quad t \ge 0, \quad v \ge 1. \tag{2.8.14}$$

We will show that for each $n \ge 0$

$$\mathsf{E}^Z_{(x,y)} H\Big(n + 1, W(Z_1)\Big) \le H\Big(n, W(x, y)\Big), \quad (x, y) \notin B. \tag{2.8.15}$$

This relation is a generalization of eq. (2.8.5), and the rest of the proof of the condition **R**(i) will actually repeat respective part of the proof of Theorem 2.8.1. Namely, from eq. (2.8.15), we easily deduce by induction

$$\mathsf{E}^Z_{(x,y)} H\Big(\tau_B \wedge n, W(Z_{\tau_B \wedge n})\Big) \le W(x, y), \quad n \ge 0, \quad (x, y) \in \mathbb{X} \times \mathbb{X}. \tag{2.8.16}$$

Since $W(z) \ge 2 > 1$ and $\Phi(1) = 0$, these inequalities give

$$\mathsf{E}^Z_{(x,y)} \Phi^{-1}\Big(\kappa(\tau_B \wedge n)\Big) \le \mathsf{E}^Z_{(x,y)} H\Big(\tau_B \wedge n, W(Z_{\tau_B \wedge n})\Big) \le W(x, y),$$

and applying the Fatou lemma we complete the proof of **R** (i):

$$\mathrm{E}^Z_{(x,y)}\Phi^{-1}\Big(\kappa\tau_B\Big) \le \liminf_n \mathrm{E}^Z_{(x,y)}\Phi^{-1}\Big(\kappa(\tau_B \wedge n)\Big) \le W(x,y).$$

To prove eq. (2.8.15), we require the following properties of the function $H(t,v)$:

(i) $H'_t(t,v) = \kappa\phi\big(H(t,v)\big)$;

(ii) $H'_v(t,v) = \phi\big(H(t,v)\big)/\phi(v)$;

(iii) $H(t,v)$ is concave w.r.t. the variable v.

The first two properties are verified straightforwardly. To verify (iii), we first additionally assume ϕ to be smooth, and write

$$H''_{vv}(t,v) = \frac{\phi'\big(H(t,v)\big)\phi\big(H(t,v)\big) - \phi\big(H(t,v)\big)\phi'(v)}{\phi^2(v)} \le 0$$

because $H(t,v) \ge v$ and ϕ' is decreasing. For nonsmooth ϕ we approximate it by a sequence of smooth ϕ_n, which shows that $H(t,v)$ is concave w.r.t. the variable v as a point-wise limit of a sequence of concave functions.

We have

$$H(t_2,v) - H(t_1,v) = \int_{t_1}^{t_2} \kappa\phi\big(H(s,v)\big),\mathrm{d}s \le \kappa(t_2-t_1)\phi\big(H(t_2,v)\big), \quad t_1 \le t_2 \tag{2.8.17}$$

by monotonicity of ϕ, and

$$H(t,v_2) - H(t,v_1) \le \frac{\phi\big(H(t,v_1)\big)}{\phi(v_1)}(v_2 - v_1), \quad v_1, \quad v_2 \ge 1 \tag{2.8.18}$$

by concavity of $H(t,\cdot)$.

Now we can proceed with the proof of eq. (2.8.15). We have

$$\begin{aligned} H\Big(n+1, W(Z_1)\Big) - H\Big(n, W(x,y)\Big) &= \Big(H\Big(n+1, W(Z_1)\Big) - H\Big(n+1, W(x,y)\Big)\Big) \\ &\quad + \Big(H\Big(n+1, W(x,y)\Big) - H\Big(n, W(x,y)\Big)\Big) =: \Delta_1 + \Delta_2. \end{aligned}$$

By eq. (2.8.17), we have simply

$$\Delta_2 \le \kappa\phi\Big(H\Big(k+1, W(x,y)\Big)\Big).$$

By eq. (2.8.18),

$$\Delta_1 \le \frac{\phi\Big(H\Big(n+1, W(x,y)\Big)\Big)}{\phi\Big(W(x,y)\Big)}\Big(W(Z_1) - W(x,y)\Big).$$

Then for any Markov coupling Z for X, we have

$$\begin{aligned}\mathsf{E}^Z_{(x,y)}\Delta_1 &\le \frac{\phi\Big(H\Big(n+1, W(x,y)\Big)\Big)}{\phi\Big(W(x,y)\Big)}\mathsf{E}^Z_{(x,y)}\Big(W(Z_1) - W(x,y)\Big)\\ &= \frac{\phi\Big(H\Big(n+1, W(x,y)\Big)\Big)}{\phi\Big(W(x,y)\Big)}\Big(\mathsf{E}_x V(X_1) + \mathsf{E}_y V(X_1) - V(x) - V(y)\Big).\end{aligned}$$

By the Lyapunov-type condition (2.8.11),

$$\mathsf{E}_x V(X_1) + \mathsf{E}_y V(X_1) - V(x) - V(y) \le -\phi(V(x)) - \phi(V(y)) + 2C.$$

Recall that

$$2C = (1-\kappa)\phi\left(1 + \inf_{x\notin K} V(x)\right)$$

and $V(x) \ge 1$, hence for each $(x,y) \notin B$

$$2C \le (1-\kappa)\phi\Big(V(x) + V(y)\Big).$$

Summarizing the above estimates, we get

$$\begin{aligned}\mathsf{E}^Z_{(x,y)}\Big(\Delta_1 + \Delta_2\Big) \le &\frac{\phi\Big(H\Big(n+1, W(x,y)\Big)\Big)}{\phi\Big(V(x) + V(y)\Big)}\\ &\times\Big(-\phi(V(x)) - \phi(V(x)) + \phi(V(x) + V(y))\Big), \quad (x,y) \notin B.\end{aligned}$$

Since ϕ has a concave extension to $[0,\infty)$ with $\phi(0) \ge 0$, we have

$$\phi(v) + \phi(w) \ge \phi(v+w) + \phi(0) \ge \phi(v+w), \quad v, w \ge 1.$$

This completes the proof of eq. (2.8.15) and proves **R**(i).

Condition **R**(ii) holds true (almost) trivially: by **R**(i), we have

$$\mathsf{E}^Z_{(x,y)}\lambda(\theta_B) = \mathsf{E}^Z_{(x,y)}\Big(\mathsf{E}^Z_{Z_1}\lambda(\tau_B)\Big) \le \mathsf{E}^Z_{(x,y)}W(Z_1) = \mathsf{E}_x V(X_1) + \mathsf{E}_y V(X_1),$$

which is bounded on $B = K \times K$ by (2.8.11) and the assumption that V is bounded on K.

The last thing to verify is that λ belongs to Λ. We have $\Phi(1) = 0$, and therefore $\lambda(0) = 1$. Next, ϕ possesses a linear growth bound because ϕ is concave. Hence $\Phi(\infty) = \infty$, which implies $\lambda(\infty) = \infty$. Finally, to prove the submultiplicativity property of λ, it is sufficient to verify that for every fixed $s \geq 0$

$$\frac{\mathrm{d}}{\mathrm{d}t}\left(\frac{\lambda(t+s)}{\lambda(t)}\right) \leq 0. \tag{2.8.19}$$

Because ϕ has a nonnegative concave extension to $[0, \infty)$, one has

$$\phi(b)a \leq \phi(a)b, \quad b \geq a. \tag{2.8.20}$$

Calculating the derivatives straightforwardly and using eq. (2.8.20), one easily verifies eq. (2.8.19). □

It would be natural to proceed further like we did that in Corollary 2.8.3 after proving Theorem 2.8.1: to assume a local Dobrushin condition to hold true, and then use Theorem 2.7.5 in order to obtain ergodic rates. The new difficulty, which arises here and which we did not observe in the setting based on the linear Lyapunov-type condition, is that eq. (2.8.11), in general, does not guarantee the moment condition (2.7.7), and thus Theorem 2.7.5 cannot be used directly. However, applying Proposition 2.8.5 to $U = V$, $V = \phi(V)$, we see that eq. (2.8.11) guarantees a weaker analogue of eq. (2.7.7) with $W(x, y) = V(x) + V(y)$ changed to

$$\widetilde{W}(x, y) = \widetilde{V}(x) + \widetilde{V}(y), \quad \widetilde{V} = \phi(V).$$

This observation leads to the following.

Theorem 2.8.8. *Let for a given MC X there exist function $V : \mathbb{X} \to [1, +\infty)$ such that X satisfies the local Dobrushin condition on each set $B_c = K_c \times K_c, c \geq 1$, where $K_c = \{x : V(x) \leq c\}$ is a* level set *of the function V. Assume also that here exist $C > 0$ and $\phi : [1, \infty) \to (0, \infty)$, which admits a nonnegative, increasing, and concave extension to $[0, +\infty)$, such that*

$$\mathsf{E}_x V(X_1) - V(x) \leq -\phi(V(x)) + C, \quad x \in \mathbb{X};$$

Then for every $\delta \in (0, 1)$ there exist $c, \gamma > 0$ such that

$$\left\|P_n(x, \cdot) - P_n(y, \cdot)\right\|_{\mathrm{TV}} \leq \frac{c}{\left(\Phi^{-1}(\gamma n)\right)^{\delta}}\left(V^{\delta}(x_1) + V^{\delta}(x_2)\right), \quad x, y \in \mathbb{X}, \quad n \geq 1. \tag{2.8.21}$$

In addition, there exists unique IPM μ for X, measure μ satisfies

$$\int_{\mathbb{X}} \phi(V)\, d\mu < \infty,$$

and for every $\delta \in (0,1)$ there exist $c, \gamma > 0$ such that

$$\|P_n(x,\cdot) - \mu\|_{TV} \le \frac{c}{\left(\phi(\Phi^{-1}(\gamma n))\right)^{\delta}} \left(\phi(V(x))^{\delta} + \int_{\mathbb{X}} \phi(V)^{\delta}\, d\mu \right), \quad x \in \mathbb{X}, \quad n \ge 1. \tag{2.8.22}$$

Remark 2.8.9. This theorem exhibits clearly the fact that, under sublinear Lyapunov-type conditions, the “loss of memory” rate and the ergodic rate are substantially different. Such an effect becomes visible if the “penalty term” $V^{\delta}(x) + V^{\delta}(y)$ in the “loss of memory” estimate lacks integrability, and then the system may fail to perform ergodicity at the same rate. We note that the general ergodic rate (2.8.22), in some particular cases, can be slightly improved, see Remark 2.8.11.

Proof. Statement (2.8.21) follows by Corollary 2.7.4 and Theorem 2.8.6; the argument here is the same as in the proof of Corollary 2.8.3, and we omit the details.

Next, observe that Theorem 2.8.6 and the Jensen inequality yield that, for $B_c = K_c \times K_c, K_c = \{V \le c\}$ with c large enough, an analogue of condition **R** of Theorem 2.7.2 holds true with W, λ replaced by

$$\widetilde{W}(x,y) = \phi\big(W(x,y)\big),$$
$$\widetilde{\lambda}(t) = \phi\big(\lambda(t)\big),$$

respectively. The function $\widetilde{\lambda}$ not necessarily belongs to the class Λ because $\widetilde{\lambda}(0) = \phi(1)$ may fail to be equal 1. Taking

$$\widehat{\lambda}(t) = \frac{\widetilde{\lambda}(t)}{\phi(1)}, \quad \widehat{W}(x,y) = \frac{\widetilde{W}(x,y)}{\phi(1)},$$

we get another pair of functions such that an analogue of condition **R** of Theorem 2.7.2 holds true. In addition, $\widehat{\lambda}$ belongs to the class Λ. Indeed, the relations $\widehat{\lambda}(0) = 1, \widehat{\lambda}(\infty) = \infty$ are obvious. To prove the submultiplicativity property, we assume first that ϕ is smooth and observe that for $t, s \ge 0$

$$\frac{d}{dt}\left(\frac{\widehat{\lambda}(t+s)}{\widehat{\lambda}(t)}\right) = \frac{\phi(\lambda(t))\phi(\lambda(t+s))\big(\phi'(\lambda(t+s)) - \phi'(\lambda(s))\big)}{\big(\phi(\lambda(t))\big)^2} \le 0.$$

This proves submultiplicativity of $\widehat{\lambda}$ if ϕ is smooth; the general case can be proved by approximation argument.

We have

$$\widehat{W}(x,y) = \frac{\phi(V(x)+V(y))}{\phi(1)} \le \frac{\phi(V(x)) + \phi(V(y))}{\phi(1)}.$$

Hence the analogue of eq. (2.7.7) with W replaced by $\widehat{W}$ holds true by Proposition 2.8.5. That is, we can apply Theorem 2.7.5 with λ, W replaced by $\widehat{\lambda}$, $\widehat{W}$, which after simple re-arrangements completes the proof of eq. (2.8.22). □

At the end of this section, we give two corollaries which, similarly to Corollary 2.8.3, specify sufficient conditions for polynomial and subexponential ergodic rates, respectively.

Corollary 2.8.10 (On polynomial ergodic rate). *For a given MC X, assume that there exists a function $V : \mathbb{X} \to [1, +\infty)$ such that X satisfies the Dobrushin condition on every level set $\{x : V(x) \le c\}$ of the function V, and for some a, $C > 0$ and $\sigma \in (0,1)$*

$$\mathsf{E}_x V(X_1) - V(x) \le -aV^{\sigma}(x) + C, \quad x \in \mathbb{X}. \tag{2.8.23}$$

Then for every $\delta \in (0,1)$, there exist c_1, $c_2 > 0$ such that

$$\left\|P_n(x,\cdot) - P_n(y,\cdot)\right\|_{TV} \le c_1(1+c_2 n)^{-\delta/(1-\sigma)}\big(V(x)+V(y)\big)^{\delta}, \quad x, y \in \mathbb{X}, \quad n \ge 1. \tag{2.8.24}$$

In addition, there exists unique IPM μ for X, this measure satisfies

$$\int_{\mathbb{X}} V^{\sigma}\, d\mu < \infty, \tag{2.8.25}$$

and

$$\left\|P_n(x,\cdot) - \mu\right\|_{TV} \le c_1(1+c_2 n)^{-\sigma/(1-\sigma)}\left(V^{\sigma}(x) + \int_{\mathbb{X}} V^{\sigma}\, d\mu\right), \quad x \in \mathbb{X},\ n \ge 1. \tag{2.8.26}$$

Remark 2.8.11. Ergodic rate (2.8.26), in this particular case, improves the general bound (2.8.22): applying directly eq. (2.8.22), we will obtain only the main term of the form

$$c_1(1+c_2 n)^{-\delta\sigma/(1-\sigma)}, \quad \delta \in (0,1).$$

Proof. Take $\phi(v) = av^{\sigma}, v \ge 0$, a concave increasing function with $\phi(0) = 0$, $\phi(\infty) = \infty$. Taking $K = \{x : V(x) \le c\}$ with sufficiently large c, we get

$$\phi\left(1 + \inf_{x \notin K} V(x)\right) = a(1+c)^{\sigma} > 2C,$$

and Theorem 2.8.6 can be applied. An explicit calculation gives

$$\Phi(v) = \frac{1}{a(1-\sigma)}\left(v^{1-\sigma} - 1\right), \quad \Phi^{-1}(t) = \left(1 + a(1-\sigma)t\right)^{1/(1-\sigma)},$$

and combining Theorems 2.7.2 and 2.8.6 we get eq. (2.8.24).

Now we specify the δ and put it equal to σ. We have

$$\mathsf{E}_x U(X_1) - U(x) \le -V^{\sigma}(x) + C', \quad x \in \mathbb{X} \tag{2.8.27}$$

with $U(x) = a^{-1}V(x)$. Then Proposition 2.8.5 and elementary inequality

$$(V(x) + V(y))^{\sigma} \le V^{\sigma}(x) + V^{\sigma}(y)$$

yield eq. (2.7.7) with $\delta = \sigma$. Applying Theorem 2.7.5, we complete the proof. □

Corollary 2.8.12 (On subexponential ergodic rate). *For a given MC X, assume that there exists function $V : \mathbb{X} \to [1, +\infty)$ such that X satisfies the Dobrushin condition on every level set $\{x : V(x) \le c\}$ of the function V, and, for some $a, b, \sigma, C > 0$,*

$$\mathsf{E}_x V(X_1) - V(x) \le -aV(x)\log^{-\sigma}\left(V(x) + b\right) + C, \quad x \in \mathbb{X}. \tag{2.8.28}$$

Then for any $\delta \in (0, 1)$, there exist $c_1, c_2 > 0$ such that

$$\left\|P_n(x, \cdot) - P_n(x, \cdot)\right\|_{TV} \le c_1 e^{-c_2 n^{1/(1+\sigma)}}\left(V(x) + V(y)\right)^{\delta}, \quad x, y \in \mathbb{X}, \quad n \ge 1. \tag{2.8.29}$$

In addition, there exists a unique IPM μ for X, this measure satisfies eq. (2.8.25), *and*

$$\left\|P_n(x, \cdot) - \mu\right\|_{TV} \le c_1 e^{-c_2 n^{1/(1+\sigma)}}\left(V^{\delta}(x) + \int_{\mathbb{X}} V^{\delta}\, d\mu\right), \quad x \in \mathbb{X}, \; n \ge 1. \tag{2.8.30}$$

Proof. Now the proof is slightly more cumbersome because the function

$$v \mapsto av\log^{-\sigma}(v + b),$$

although well-defined on $[1, \infty)$, may fail to have a nonnegative concave increasing extension to $[0, \infty)$ (e.g., when $b > 0$ is small). Hence we can not apply Theorem 2.8.6; instead, we tune up the relation (2.8.28) first.

We can check straightforwardly that, for a given σ, there exists $b_\sigma > 1$ such that the function $v \mapsto v\log^{-\sigma} v$ is concave and increasing on $[b_\sigma, \infty)$. Hence rather than using eq. (2.8.28) with Theorem 2.8.6, we can apply a (weaker) inequality (2.8.11) with

$$\phi(v) = \tilde{a}(v + b_\sigma)\log^{-\sigma}(v + b_\sigma),$$

where

$$\tilde{a} = a \inf_{v \geq 1} \frac{v \log^{-\sigma}(v + b)}{(v + b_\sigma) \log^{-\sigma}(v + b_\sigma)} > 0.$$

We have

$$\begin{aligned}
\Phi(v) &= \frac{1}{\tilde{a}(1+\sigma)} \Big(\log^{1+\sigma}(v + b_\sigma) - \log^{1+\sigma}(1 + b_\sigma) \Big), \\
\Phi^{-1}(t) &= \exp \left\{ \Big(\log^{1+\sigma}(1 + b_\sigma) + \tilde{a}(1+\sigma)t \Big)^{1/(1+\sigma)} \right\} - b_\sigma \\
&= (1 + b_\sigma) \exp \left\{ \Big(1 + \frac{\tilde{a}(1+\sigma)t}{\log^{1+\sigma}(1 + b_\sigma)} \Big)^{1/(1+\sigma)} \right\} - b_\sigma \\
&\geq \exp \left\{ \Big(\frac{\tilde{a}(1+\sigma)t}{\log^{1+\sigma}(1 + b_\sigma)} \Big)^{1/(1+\sigma)} \right\}.
\end{aligned}$$

Applying Theorem 2.8.6, we get eq. (2.8.29). We have

$$v^\delta \leq C v \log^{-\sigma}(v + b), \quad v \geq 1,$$

hence using Proposition 2.8.5 we prove that eq. (2.7.7) holds true for each $\delta \in (0, 1)$. Applying Theorem 2.7.5 we complete the proof. □

2.9 Dobrushin Condition: Sufficient Conditions and Alternatives

In this section, we explain one natural way to verify the local Dobrushin condition. Combined with the detailed analysis of the recurrence assumptions, made in the previous sections, this will provide an efficient set of tools for proving ergodic rates in total variation distance. We will also shortly discuss possible alternatives to this condition, used in the literature for similar purposes.

Assume $\mathbb{X}$ to be a metric space with the metric ρ and $\mathscr{X}$ be the corresponding Borel σ-algebra.

Proposition 2.9.1. *Let, for a given measure ν, there exist a decomposition of the transition probability $P(x, dy)$ of the form*

$$P(x, dy) = P_c(x, dy) + P_d(x, dy),$$

where

(i) $P_c(x, dy), P_d(x, dy)$ *are nonnegative kernels;*

(ii)

$$P_c(x, dy) \ll \nu(dy);$$

(iii) *for a given point* $x_* \in \mathbb{X}$, *the mapping*

$$x \mapsto \frac{P_c(x, dy)}{\nu(dy)} \in L_1(\mathbb{X}, \nu) \tag{2.9.1}$$

is continuous at x_* *and*

$$P_c(x_*, \mathbb{X}) > 0. \tag{2.9.2}$$

Then for $\varepsilon > 0$ *sufficiently small, the local Dobrushin condition holds true on the set* $B = B_\varepsilon(x_*) \times B_\varepsilon(x_*)$, *where* $B_\varepsilon(x_*) = \{x : \rho(x, x_*) < \varepsilon\}$ *denotes the open ball in* $\mathbb{X}$ *with center* x_* *and radius* ε.

The proof is straightforward, once we recall the interpretation of the total variation distance as the L_1-distance for the Radon–Nikodym densities:

$$\begin{aligned}
\|P(x,\cdot) - P(y,\cdot)\|_{TV} &\le \|P_c(x,\cdot) - P_c(y,\cdot)\|_{TV} + \|P_d(x,\cdot) - P_d(y,\cdot)\|_{TV} \\
&\le \int_{\mathbb{X}} \left| \frac{P_c(x, dz)}{\nu(dz)} - \frac{P_c(y, dz)}{\nu(dz)} \right| \nu(dz) + 2 - P_c(x, \mathbb{X}) - P_c(y, \mathbb{X}) \\
&\to 2 - P_c(x_*, \mathbb{X}) - P_c(y_*, \mathbb{X}) < 2, \quad x, y \to x_*.
\end{aligned}$$

This result just outlines a convenient way to verify the local Dobrushin condition by means of L_1-continuity of (a nontrivial part of) the transition probability kernel. For further applications, it is useful to extend the set B where local Dobrushin condition is verified. The first extension is straightforward: assume conditions of Proposition 2.9.1 to hold true uniformly w.r.t. $x_* \in K$ for some given set K; that is,

$$\sup_{x_* \in K, \rho(x, x_*) < \delta} \left\| \frac{P_c(x, dy)}{\nu(dy)} - \frac{P_c(x_*, dy)}{\nu(dy)} \right\|_{L_1} \to 0, \quad \delta \to 0,$$

$$\inf_{x_* \in K} P_c(x, \mathbb{X}) > 0.$$

Then the statement of Proposition 2.9.1 holds true with

$$B = \bigcup_{x_* \in K} B_\varepsilon(x_*) \times B_\varepsilon(x_*).$$

Remark 2.9.2. The uniform conditions given above hold true, for instance, if eqs. (2.9.1) and (2.9.2) hold true for all $x_* \in K$ for a compact set K.

The second extension can be made using the Markov property.

Proposition 2.9.3. *Let conditions of Proposition 2.9.1 hold true and, for a given set K, for any $\varepsilon > 0$*

$$\inf_{x\in K} P(x, B_\varepsilon(x_*)) > 0. \tag{2.9.3}$$

Then for the two-step transition probability $P^2(x, dy)$ the local Dobrushin condition holds true on the set $B = K \times K$.

Again, the proof is straightforward: for $x, y \in B$,

$$\begin{aligned}\|P^2(x,\cdot) - P^2(y,\cdot)\|_{TV} &\le \int_{\mathbb{X}} \|P(x',\cdot) - P(y',\cdot)\|_{TV} P(x, dx')P(y, dy') \\ &\le 2 \int_{\mathbb{X}\setminus B_\varepsilon(x_*)\times B_\varepsilon(x_*)} P(x, dx')P(y, dy') \\ &\quad + \upsilon \int_{B_\varepsilon(x_*)\times B_\varepsilon(x_*)} P(x, dx')P(y, dy') \\ &= 2 - (2-\upsilon)P(x, B_\varepsilon(x_*))P(y, B_\varepsilon(x_*)),\end{aligned}$$

where

$$\upsilon = \sup_{x,y\in B_\varepsilon(x_*)} \|P(x,\cdot) - P(y,\cdot)\|_{TV} < 2.$$

Remark 2.9.4. A chain is called *topologically irreducible*, if for every $x, y \in \mathbb{X}$

$$P(x, B_\varepsilon(y)) > 0, \quad \varepsilon > 0.$$

Note that if the chain is *Feller*, that is, the mapping

$$x \mapsto P(x, dy)$$

is continuous w.r.t. weak convergence in $\mathscr{P}(\mathbb{X})$, then for each $y \in \mathbb{X}$, $\varepsilon > 0$ the mapping

$$x \mapsto P(x, B_\varepsilon(y))$$

is lower semicontinuous; that is

$$\liminf_{x'\to x} P(x', B_\varepsilon(y)) \ge P(x, B_\varepsilon(y)).$$

Thus a Feller chain which is topologically irreducible satisfies eq. (2.9.3) for any compact set K.

The choice of a particular form of the irreducibility condition for an MC is far from trivial and may depend both on the structure of the particular chain and the goals of the analysis. A variety of forms of the irreducibility assumption other than the Dobrushin condition used above are available in the literature. Below, we briefly discuss several such assumptions.

The chain X is said to satisfy the *minorization condition* if there exist a probability measure λ and $\rho \in (0, 1)$ such that

$$P(x, \mathrm{d}y) \ge \rho\,\lambda(\mathrm{d}y).$$

Clearly, the minorization condition implies the Dobrushin condition with the same ρ. The inverse implication, in general, is not true.

The *Harris irreducibility condition* requires that there exist a measure λ and a positive function $f \in \mathbb{B}(\mathbb{X})$ such that

$$P(x, \mathrm{d}y) \ge f(x)\,\lambda(\mathrm{d}y).$$

In a sense, the Harris irreducibility condition is a relaxed minorization condition, adapted to the duality structure between the spaces of measures $\mathscr{M}(\mathbb{X})$ and functions $\mathbb{B}(\mathbb{X})$.

The *Döblin condition* requires that there exist a probability measure λ and $\varepsilon > 0$ such that

$$P(x, A) \le 1 - \varepsilon \quad \text{if } \lambda(A) < \varepsilon.$$

In the case of $\mathbb{X}$ being a compact metric space, this condition can be verified in the terms of the *Doob condition* which claims that X is strong Feller; that is, for every bounded and measurable $f : \mathbb{X} \to \mathbb{R}$, the mapping $x \mapsto \mathsf{E}_x f(X_1) = \int_{\mathbb{X}} f(y)\,P(x, \mathrm{d}y)$ is continuous.

All the conditions listed above can also be relaxed by imposing them on some N-step transition probability $P_N(x, \mathrm{d}y)$ instead of $P(x, \mathrm{d}y)$ or, more generally, on a “mixed” (or “sampled”, cf. Ref. [106]) kernel

$$P^{(Q)}(x, \mathrm{d}y) = \sum_{n=1}^{\infty} Q(n)\,P^n(x, \mathrm{d}y)$$

for some weight sequence $\{Q(n),\ n \ge 1\} \subset (0, \infty)$ with $\sum_n Q(n) < \infty$. This typically makes it possible to give conditions for the ergodicity which are sufficient and close to necessary. For instance, the Dobrushin condition for an N-step transition probability $P^N(x, \mathrm{d}y)$ yields that the “N-step skeleton chain” possesses a uniform exponential ergodic rate (2.3.2). As we have explained in the proof of Theorem 2.3.1, this would yield a similar rate (with other constants) for the chain X itself. On the other hand, if X possesses a uniform ergodic bound

$$\sup_{x,y\in\mathbb{X}} \left\|P^n(x, \cdot) - P^n(y, \cdot)\right\|_{TV} \to 0, \quad n \to \infty \tag{2.9.4}$$

with *arbitrary* rate, then for some N, the Dobrushin condition for an N-step transition probability $P^N(x, dy)$ holds.

This simple argument reveals the general feature that, on the level of *uniform* ergodic rates, the only rate which actually occurs is an exponential one. In addition, the N-step version of the Dobrushin condition appears to be necessary and sufficient for X to possess a uniform exponential ergodic rate. This property is not exclusive to the Dobrushin condition: the N-step versions of the Harris irreducibility condition and the Döblin condition are necessary and sufficient, as well. The necessity of the N-step version of the Döblin condition is simple. Indeed, eq. (2.9.4) implies the uniform convergence of $P^n(x, dy)$ to the (unique) IPM μ for X in the total variation distance, and then take the Döblin condition holds true with $\lambda = \mu$. The proof of the necessity of the Harris irreducibility condition is much more involved; see Ref. [109, Chapter 3].

All the conditions listed above are also available in local versions, that is, on a set $K \in \mathscr{X}$, and together with proper recurrence conditions typically would lead to nonuniform ergodic rates. In the terminology explained above, the central notions of a *small set* and a *petite set* from the Meyn–Tweedie approach [106], widely used in the literature, can be formulated as follows: a set $K \in \mathscr{X}$ is *small* if X verifies the N-step version of the minorization condition, and $K \in \mathscr{X}$ is *petite* if a "mixed" (or "sampled") chain verifies the minorization condition.

Note that, although all the irreducibilty conditions discussed above are genuinely equivalent at least in their global versions, there are still good reasons to use them separately for various purposes and classes of Markov models. The Döblin condition and the strong Feller property are the mildest within this list of assumptions, and hence are easiest to verify. However, using these assumption makes it rather difficult to get ergodic bounds with *explicit* constants. The minorization condition leads to much more explicit ergodic bounds, but it is too restrictive for models which have transition probabilities with a complicated local behavior. A good example here is given by a Markov process solution to an SDE with a Lévy noise; see Section 3.4. The Dobrushin condition is more balanced in the sense that it leads to explicit ergodic bounds though it is more flexible and might be less restrictive than the minorization condition. This is the reason for us to ground our exposition mainly on the Dobrushin condition.

2.10 Comments

1. Traditional proofs of the Harris theorem are based on the decomposition of the MC into excursions away from the irreducibility set, combined with an analysis of the length of these excursions, see Refs. [106, 109] for exponential rates and Refs. [3, 29, 30, 65, 129] for subexponential rates. In our exposition the same strategy is used, but in a unified and more simple way, which is based on auxiliary martingale-type construction from Section 2.4. This construction makes it possible to separate within the proofs the contraction phase, which now is enabled

by the local Dobrushin condition, from the estimates of the length of the excursions. Similar martingale-type constructions can be found in Refs. [15, 32].

2. For *strictly concave* ϕ, statement of Theorem 2.8.8 is apparently suboptimal: by Ref. [53, Theorem 4.1], stronger versions of eqs. (2.8.21) and (2.8.22) hold true with $\gamma = 1, \delta = 1$. This difference is not crucial, at least in particular models studied in Sections 3.3 and 3.4. Hence we keep this statement in its current form, which allows us to preserve the structure of the overall presentation: repeating the proof almost literally, we will extend it in Chapter 4 for the weak ergodicity framework. It is an open question whether statement 2 of Theorem 4.5.2 can be similarly improved, that is, is it possible to get eq. (4.5.7) with $\gamma = 1, \delta = 1$ for strictly submultiplicative λ.
3. Under the linear Lyapunov condition (2.8.7) and the local Dobrushin condition, one can show P^* to be contraction on $\mathscr{P}(\mathbb{X})$ w.r.t. a certain weighted total variation distance, see Ref. [54] or [71]. This extends the "analytic" proofs of Theorems 1.2.1 and 2.3.1, and for exponentially ergodic systems provides a convenient point of view on stabilization properties of the system, see the discussion in Ref. [54] or [53]. In the subexponential setting such an approach is still feasible (e.g., Ref. [13]), but it is more complicated and model dependent.

3 Markov Processes with Continuous Time

3.1 Basic Definitions and Facts

In this chapter, we consider continuous-time Markov processes $\{X_t, t \in [0, \infty)\}$, which are time-homogeneous. That is, we assume that, for a certain family of probability kernels $\{P^t, t \in [0, \infty)\}$,

$$\mathsf{P}(X_t \in A|\mathscr{F}_t) = P^{t-s}(X_s, A), \quad s \le t, \quad A \in \mathscr{X},$$

and the family $\{P^t\}$ satisfies
- $P^0(x, A) = 1_A(x) = \delta_x(A)$;
- (the Chapman–Kolmogorov equation): for $s, t \ge 0$

$$P^{t+s}(x, A) = \int_{\mathbb{X}} P^t(x, dy)P^s(y, A).$$

The process $X = \{X_t\}$ is defined on a filtered probability space $(\Omega, \mathscr{F}, \{\mathscr{F}_t\}, \mathsf{P})$.

Likewise to the discrete time setting, the elements of family $\{P^t, t \ge 0\}$ are called *transition probabilities* of the process X. This family and the initial distribution

$$\mu(A) = \mathsf{P}(X_0 \in A), \quad A \in \mathscr{X},$$

completely define the law of the chain: for any $m \ge 0, 0 \le t_1 \le \cdots \le t_m, A_0, \ldots, A_m \in \mathscr{X}$

$$\begin{aligned}\mathsf{P}(X_0 \in A_0, X_{t_1} \in A_1, \ldots X_{t_m} \in A_m)\\ = \int_{A_0} \cdots \int_{A_{m-1}} \mu(dx_0)P^{t_1}(x_0, dx_1)\ldots P^{t_m-t_{m-1}}(x_{m-1}, A_m).\end{aligned} \tag{3.1.1}$$

The same notation P_μ, E_μ is used for the law of the process with the initial distribution μ, and for $\mu = \delta_x$ the notation is simplified to P_x, E_x.

A measure $\mu \in \mathscr{P}(\mathbb{X})$ which satisfies

$$\mu(dy) = \int_{\mathbb{X}} \mu(dx)P^t(x, dy), \quad t \ge 0 \tag{3.1.2}$$

is called an *invariant probability measure* (IPM) for the Markov process X. If μ is an IPM, then the laws

$$\mu^t(dy) = \mathsf{P}_x(X_t \in dy) = \int_{\mathbb{X}} \mu(dx)P^t(x, dy), \quad t \ge 0$$

of $X_t, t \ge 0$ under P_μ are the same, and by eq. (3.1.1) the random process $\{X_t\}$ is strictly stationary.

DOI 10.1515/9783110458930-004

To simplify the overall exposition, we will impose certain mild continuity assumptions on the process X. Within this chapter, we assume $\mathbb{X}$ to be a *Polish space* (i.e., a complete separable metric space) with the metric ρ, and $\mathscr{X} = \mathscr{B}(\mathbb{X})$, the *Borel* σ-algebra. We assume that for each $\varepsilon > 0, x \in \mathbb{X}$

$$P^t\Big(x, \{y : \rho(x, y) > \varepsilon\}\Big) \to 0, \quad t \to 0. \tag{3.1.3}$$

Condition eq. (3.1.3) actually means that the process $\{X_t\}$ is *continuous in probability* at the point $t = 0$ w.r.t. any $\mathsf{P}_x, x \in \mathbb{X}$. We also assume that, w.r.t. any P_x, there exists a *modification* of the process $\{X_t\}$ with *càdlàg* trajectories (the French abbreviation for *right continuous with left limits*). The latter assumption is known to hold true under mild assumptions on the transition probabilities of the process, one possible sufficient condition is that eq. (3.1.3) holds true for any $\varepsilon > 0$ uniformly on each bounded ball in $\mathbb{X}$; we refer for a detailed analysis of this topic to Ref. [44, Chapter IV, §4] and [38, Chapter 3]. In what follows we denote by $\mathbb{D} = \mathbb{D}([0, \infty), \mathbb{X})$ the *Skorokhod space* of càdlàg functions, defined on $[0, \infty)$ and taking values in $\mathbb{X}$, and assume without a further notice that $X_t, t \geq 0$ has trajectories in $\mathbb{D}$.

The process X generates a *semigroup of operators* in the Banach space $\mathbb{B}(\mathbb{X})$:

$$\mathbf{T}_t f(x) = \int_{\mathbb{X}} f(y)\, P^t(x, dy) = \mathsf{E}_x f(X_t), \qquad f \in \mathbb{B}(\mathbb{X}), \quad t \geq 0.$$

Its *generator* $\mathbf{A}$ is an unbounded linear operator, defined by

$$\mathbf{A}f = \lim_{t\to 0} \frac{\mathbf{T}_t f - f}{t}, \quad f \in D(\mathbf{A}) \tag{3.1.4}$$

with the domain $D(\mathbf{A})$ which consist of all $f \in \mathbb{B}(\mathbb{X})$ such that the limit in eq. (3.1.4) exists w.r.t. the sup-norm in $\mathbb{B}(\mathbb{X})$. *The Dynkin formula* states that for any $f \in D(\mathbf{A})$ the process

$$f(X_t) - \int_0^t \mathbf{A}f(X_s)\, ds, \quad t \geq 0$$

is an $\{\mathscr{F}_t\}$-martingale w.r.t. any $\mathsf{P}_x, x \in \mathbb{X}$.

3.2 Ergodicity in Total Variation: The Skeleton Chain Approach

Our main aim in this chapter is to establish for continuous-time Markov processes the analogues of statements obtained in the previous chapter in the discrete time setting.

One natural way to design such statements is based on time discretization. Namely, for a given Markov process $\{X_t\}$, for any $h > 0$, one can consider an associated *skeleton chain* with the discretization step h:

$$X^h = \{X_{nh},\ n \in \mathbb{Z}_+\}.$$

Clearly, for any $s < t$

$$\left\|P^t(x,\cdot) - P^t(y,\cdot)\right\|_{TV} \le \left\|P^s(x,\cdot) - P^s(y,\cdot)\right\|_{TV}, \quad x, y \in \mathbb{X};$$

that is, once we have a "loss of memory" effect for a skeleton chain, we have it also for the process X itself. The same holds true also for the ergodic rates: if μ is an IPM for the process X, then

$$\left\|P^t(x,\cdot) - \mu\right\|_{TV} \le \left\|P^s(x,\cdot) - \mu\right\|_{TV}, \quad x \in \mathbb{X}, \quad s < t.$$

Hence one can expect that the stabilization properties for the transition probabilities of the process X can be derived from the same properties for a skeleton chain. The main difficulty in such an approach is that one should typically require the conditions to be formulated in terms of the process X itself, rather than the auxiliary skeleton chains. Below, we explain one practical way to guarantee Lyapunov-type conditions for a skeleton chain under the assumptions imposed on X.

Let us begin from the key calculation, which would then clarify Definition 3.2.1. Let nonnegative function V belong to the domain $D(\mathbf{A})$ of the generator $\mathbf{A}$ for the process X, and satisfy

$$\mathbf{A}V \le -aV + C \tag{3.2.1}$$

with some $a, C > 0$. Take $v(x,t) = \mathsf{E}_x V(X_t)$, then by the Dynkin formula for any $t_1 \le t_2$ the inequality

$$v(x,t_2) - v(x,t_1) \le \int_{t_1}^{t_2} \Big(-av(x,s) + C\Big)\, ds \tag{3.2.2}$$

holds true. Since $V \in D(\mathbf{A})$, the function $v(x,\cdot)$ is continuous (it is actually even continuously differentiable). Fix $t > 0$ and denote $t_{k,n} = tk/n$, then

$$\begin{aligned}
e^{at}v(x,t) - v(x,0) &= \lim_{n\to\infty} \sum_{k=1}^{n} \Big(e^{at_{k,n}} v(x,t_{k,n}) - e^{at_{k-1,n}} v(x,t_{k-1,n})\Big) \\
&\le \lim_{n\to\infty} \sum_{k=1}^{n} \Big((e^{at_{k,n}} - e^{at_{k-1,n}}) v(x,t_{k-1,n}) \\
&\qquad + e^{at_{k,n}} \int_{t_{k-1,n}}^{t_{k,n}} \Big(-av(x,s) + C\Big)\, ds\Big) \\
&= \int_0^t Ce^{as}\, ds = \frac{C}{a}(e^{at} - 1).
\end{aligned}$$

This gives for any $h > 0$

$$\mathsf{E}_x V(X_1^h) - V(x) \leq -a^h V(x) + C^h \tag{3.2.3}$$

with

$$a^h = 1 - e^{-ah}, \quad C^h = C\frac{1 - e^{-ah}}{a}.$$

This simple calculation outlines a general idea that, in order to verify a Lyapunov-type condition for skeleton chain, one can use a Lyapunov-type condition, similar to eq. (3.2.1), formulated in the terms of the generator of the initial process. Implementation of this idea meets two principal difficulties.

The first difficulty is caused by the fact that, typically, the Lyapunov function V has to be unbounded. Indeed, typical sufficient conditions designed in the previous chapter require that the Dobrushin condition is satisfied on $K_c \times K_c$ for each level set $K_c = \{V \leq c\}$. For a bounded V this would mean that the Dobrushin condition holds true on the entire space; that is, the process is uniformly ergodic, and there is no need to evaluate any recurrence condition. For nonuniformly ergodic processes, this is definitely not true, which brings unbounded V's to consideration. However, any unbounded V does not belong to $D(\mathbf{A}) \subset \mathbb{B}(\mathbb{X})$. Hence, to keep the whole argument operational, we have to extend the domain of the generator in order to include unbounded functions V therein. One very natural way to do this is to remove from the definition of the *extended generator* all technicalities except the main feature required in the argument; that is, the Dynkin formula.

Definition 3.2.1. A measurable function $f : \mathbb{X} \to \mathbb{R}$ belongs to the domain of the *extended generator* $\mathscr{A}$ of the Markov process X if there exists a measurable function $g : \mathbb{X} \to \mathbb{R}$ such that the process

$$M_t^f = f(X_t) - \int_0^t g(X_s)\, ds, \quad t \in \mathbb{R}^+ \tag{3.2.4}$$

is well-defined and is a local $\{\mathscr{F}_t\}$-martingale w.r.t. every measure P_x, $x \in \mathbb{X}$. For any such pair (f, g), we write $f \in Dom(\mathscr{A})$ and $\mathscr{A}f = g$.

Remark 3.2.2. This definition, both very convenient and very useful, apparently dates back to H. Kunita [90]. It is a kind of mathematical "common knowledge," used widely in research papers with various technical modifications, though somehow it is missing from the classical textbooks (with the important exception of Ref. [119, Chapter VII.1]).

We will see below that, using the Itô formula, for particular classes of processes the Lyapunov-type condition

$$\mathscr{A}V \leq -\phi(V) + C \tag{3.2.5}$$

can be verified in a very straightforward and transparent way. However, here we meet the second difficulty mentioned above: assuming eq. (3.2.5) with *concave* ϕ, we are not able to derive directly for $v(x,t) = \mathsf{E}_x V(X_t)$ an inequality similar to eq. (3.2.2):

$$v(x,t_2) - v(x,t_1) \le \int_{t_1}^{t_2} \left(-\phi\big(v(x,s)\big) + C\right) \, ds, \quad t_1 < t_2.$$

Indeed, even if we knew that M^V is a (true) martingale, to get the required inequality we would need that

$$-\mathsf{E}_x \phi\big(V(X_s)\big) \ge -\phi\big(\mathsf{E}_x V(X_s)\big),$$

which is just the Jensen inequality, valid for *convex* ϕ. This explains that the linear function $\phi(v) = av$, which is both concave and convex, is a special case, and in a general setting the argument should be modified. Such a modification, which also takes into account localization and integrability issues, is given in the following theorem.

For a given $\phi : [1,+\infty) \to (0,\infty)$, denote

$$\Phi(v) = \int_1^v \frac{1}{\phi(w)} \, dw, \quad v \in [1,\infty)$$

and

$$H(t,v) = \Phi^{-1}\big(t + \Phi(v)\big), \quad t \ge 0, \quad v \ge 1;$$

see eqs. (2.8.13) and (2.8.14).

Theorem 3.2.3. *Assume that for a given Markov process X there exist a continuous function $V : \mathbb{X} \to [1,+\infty)$ from the domain of the extended generator, a nondecreasing function $\phi : [1,+\infty) \to (0,\infty)$ which admits a concave, nonnegative extension to $[0,\infty)$, and a constant $C \ge 0$ such that eq. (3.2.5) holds true.*

Then for any $h > 0$

$$\mathsf{E}_x V^h(X_1^h) - V^h(x) \le -\phi^h\big(V^h(x)\big) + C^h \tag{3.2.6}$$

with

$$V^h(x) = \Phi^{-1}\Big(h + \Phi\big(V(x)\big)\Big) - \Phi^{-1}(h) + 1, \quad x \in \mathbb{X}, \tag{3.2.7}$$

$$\phi^h(v) = v - 1 + \Phi^{-1}(h) - \Phi^{-1}\Big(-h + \Phi\big(v - 1 + \Phi^{-1}(h)\big)\Big), \quad v \ge 1, \tag{3.2.8}$$

$$C^h = C \int_0^h \frac{\phi(H(s,1))}{\phi(1)} \, ds. \tag{3.2.9}$$

The function $\phi^h : [1,\infty) \to (0,\infty)$ is increasing and admits a concave, nonnegative extension to $[0,\infty)$.

The proof is based on the following statement.

Lemma 3.2.4. *Let for a given Markov process X there exist a continuous function $V : \mathbb{X} \to [1, +\infty)$ from the domain of the extended generator, a nondecreasing function $\phi : [0, +\infty) \to [0, \infty)$ with $\phi(1) > 0$, and $C_V \geq 0$ such that*

$$\mathscr{A}V \leq -\phi(V) + C_V.$$

Then,

$$E_x H\big(t, V(X_t)\big) \leq V(x) + C_V \int_0^t \frac{\phi(H(s,1))}{\phi(1)}\, ds. \tag{3.2.10}$$

Proof. For the reader's convenience, let us briefly outline the idea the proof is based on, before proceeding with the details. The vague idea is to apply the Itô formula in order to clarify the semimartingale structure of the process $H\big(t, V(X_t)\big)$. The crucial point here is that the function $H(t, v)$ is concave in v, hence the additional "stochastic" term which appears in the Itô formula will be nonpositive. Since the structure of the local martingale M_t^V is not specified, it is not easy to justify the use of the Itô formula in such a general setting (of course, in numerous particular cases this difficulty does not occur). We will use time partitioning and change integrals to integral sums; actually, this is just a repetition of the proper part of the proof of the Itô formula. This trick is quite standard, for example, Ref. [38, Chapter 4, Lemma 3.4].

Fix x, and choose a localizing sequence $\tau_N, n \geq 1$ w.r.t. P_x for the local martingale

$$M_t^V = V(X_t) - \int_0^t \mathscr{A}V(X_s)\, ds;$$

that is, a sequence of stopping times such that $\tau_N \to \infty, N \to \infty$ P_x-a.s. and for each N

$$M_{t\wedge\tau_N}^V, \quad t \geq 0$$

is a martingale w.r.t. P_x. Recall that $\{X_t\}$ was assumed to have càdlàg trajectories and V is continuous, hence the trajectories of $\{V(X_t)\}$ are càdlàg, as well. Hence the sequence $\tau_N, N \geq 1$ can be taken such that

$$V(X_t) \leq N, \quad t < \tau_N. \tag{3.2.11}$$

Next, denote for $n, N \geq 1$

$$t_{k,n} = tk/n, \quad \tau_{k,n}^N = t_{k,n} \wedge \tau_N, \quad k = 0, \dots, n$$

By inequalities (2.8.17) and (2.8.18) we have for $k \geq 2$

$$\begin{aligned} H\big(\tau^N_{k-1,n}, V(X_{\tau^N_{k,n}})\big) - H\big(\tau^N_{k-2,n}, V(X_{\tau^N_{k-1,n}})\big) & \\ \leq (\tau^N_{k-1,n} - \tau^N_{k-2,n})\phi\Big(H\big(\tau^N_{k-1,n}, V(X_{\tau^N_{k,n}})\big)\Big) & \\ + \frac{\phi\Big(H\big(\tau^N_{k-1,n}, V(X_{\tau^N_{k-1,n}})\big)\Big)}{\phi\big(V(X_{\tau^N_{k-1,n}})\big)}\big(V(X_{\tau^N_{k,n}}) - V(X_{\tau^N_{k-1,n}})\big). & \end{aligned} \tag{3.2.12}$$

By the martingale property of M^V, stopped at τ_N, and Doob's optional sampling theorem,

$$\mathsf{E}_x\big(V(X_{\tau^N_{k,n}}) - V(X_{\tau^N_{k-1,n}}) \big| \mathscr{F}_{\tau^N_{k-1,n}}\big) \leq \mathsf{E}_x\left(\int\limits_{\tau^N_{k-1,n}}^{\tau^N_{k,n}} \Big(-\phi(V(X_s)) + C\Big)\, \mathrm{d}s \Big| \mathscr{F}_{\tau^N_{k-1,n}} \right).$$

Since the term

$$\frac{\phi\Big(H\big(\tau^N_{k-1,n}, V(X_{\tau^N_{k-1,n}})\big)\Big)}{\phi\big(V(X_{\tau^N_{k-1,n}})\big)}$$

is $\mathscr{F}_{\tau^N_{k-1,n}}$-measurable, this implies

$$\begin{aligned} &\mathsf{E}_x\Big(H\big(\tau^N_{k-1,n}, V(X_{\tau^N_{k,n}})\big) - H\big(\tau^N_{k-2,n}, V(X_{\tau^N_{k-1,n}})\big)\Big) \\ &\leq \mathsf{E}_x\left(\int\limits_{\tau^N_{k-2,n}}^{\tau^N_{k-1,n}} \phi\Big(H\big(\tau^N_{k-1,n}, V(X_{\tau^N_{k,n}})\big)\Big)\, \mathrm{d}s \right) \\ &+ \mathsf{E}_x\left(\int\limits_{\tau^N_{k-1,n}}^{\tau^N_{k,n}} \frac{\phi\Big(H\big(\tau^N_{k-1,n}, V(X_{\tau^N_{k-1,n}})\big)\Big)}{\phi\big(V(X_{\tau^N_{k-1,n}})\big)} \Big(-\phi(V(X_s)) + C\Big)\, \mathrm{d}s \right). \end{aligned}$$

Recall that $V(X_t)$ is bounded on $[0, \tau_N)$, hence there exists a constant C_N such that each of the integrals in the right-hand side of the previous inequality is bounded by C_N/n. Then, taking the sum over $k = 2, \dots, n$, we get after a simple rearrangement

$$\begin{aligned}
\mathsf{E}_x\Big(H\big(\tau^N_{n-1,n}, V(X_{\tau^N_{n,n}})\big) - V(X_{\tau^N_{1,n}})\Big) &\le \frac{2C_N}{n} \\
&+ \mathsf{E}_x \sum_{k=2}^{n-1} \int\limits_{\tau^N_{k-1,n}}^{\tau^N_{k,n}} \Bigg(\phi\Big(H\big(\tau^N_{k,n}, V(X_{\tau^N_{k+1,n}})\big)\Big) \\
&\qquad - \frac{\phi(V(X_s))\phi\Big(H\big(\tau^N_{k-1,n}, V(X_{\tau^N_{k-1,n}})\big)\Big)}{\phi\big(V(X_{\tau^N_{k-1,n}})\big)}\Bigg)\, ds \\
&+ C\mathsf{E}_x \sum_{k=2}^{n-1} \int\limits_{\tau^N_{k-1,n}}^{\tau^N_{k,n}} \frac{\phi\Big(H\big(\tau^N_{k-1,n}, V(X_{\tau^N_{k-1,n}})\big)\Big)}{\phi\big(V(X_{\tau^N_{k-1,n}})\big)}\, ds \\
&=: \frac{2C_N}{n} + J^1_{n,N} + J^2_{n,N}.
\end{aligned}$$

Note that on the set $\{\tau_N \le t_{1,n}\}$ we have $\tau^N_{n,n} = \tau^N_{1,n}$ and thus

$$H\big(\tau^N_{n-1,n}, V(X_{\tau^N_{n,n}})\big) \ge V(X_{\tau^N_{1,n}}).$$

This gives finally the bound

$$\mathsf{E}_x H\big(\tau^N_{n-1,n}, V(X_{\tau^N_{n,n}})\big)1_{\tau_N > t_{1,n}} \le \mathsf{E}_x V(X_{\tau^N_{1,n}})1_{\tau_N > t_{1,n}} + \frac{2C_N}{n} + J^1_{n,N} + J^2_{n,N}. \tag{3.2.13}$$

Now we pass to the limit in this bound as $n \to \infty$. The term

$$V(X_{\tau^N_{1,n}})1_{\tau_N > t_{1,n}}$$

converges in probability to $V(x)1_{\tau_N > 0}$ and is uniformly bounded, hence the limit of corresponding expectations is bounded by $V(x)$. Similarly, the term $J^1_{n,N}$ can be written in the form

$$J^1_{n,N} = \mathsf{E}_x \int\limits_0^t \Xi_{n,N}(s)\, ds,$$

where the family $\{\Xi_{n,N}(s)\}$ is uniformly bounded and for each s

$$\Xi_{n,N}(s) \to 0, \quad n \to \infty.$$

in probability. Recall that $H(s, v)$ is concave in v, and thus

$$H'_v(s, v) = \frac{\phi(H(s, v))}{\phi(v)} \le \frac{\phi(H(s, 1))}{\phi(1)}, \quad v \ge 1.$$

Hence the term $J^2_{n,N}$ satisfies the obvious bound

$$J^2_{n,N} \le C \int_0^t \frac{\phi(H(s+1/n,1))}{\phi(1)}\,ds \to C\int_0^t \frac{\phi(H(s,1))}{\phi(1)}\,ds, \quad n\to\infty.$$

On the other hand,

$$H\big(\tau^N_{n-1,n}, V(X_{\tau^N_{n,n}})\big)1_{\tau_N > t_{1,n}} \to \Big(H\big(t\wedge\tau_N, V(X_{t\wedge\tau_N})\big)1_{\tau_N>0}$$

in probability, hence by the Fatou lemma

$$\mathsf{E}_x H\big(t\wedge\tau_N, V(X_{t\wedge\tau_N})\big)1_{\tau_N>0} \le C\int_0^t \frac{\phi(H(s,1))}{\phi(1)}\,ds.$$

Taking the limit as $N \to \infty$ and applying the Fatou lemma once again, we obtain eq. (3.2.10). □

Proof of Theorem 3.2.3. Relation (3.2.10) with $t = h$ has the form

$$\mathsf{E}_x H\big(h, V(X_h)\big) \le V(x) + C^h, \tag{3.2.14}$$

which after a proper change of notation will give the required Lyapunov-type condition eq. (3.2.6) for the skeleton chain. Note that the domains where the functions Φ, Φ^{-1} and H are well-defined can be naturally extended. Namely, Φ is well-defined on $(0,\infty)$ and takes values in (κ,∞) with

$$\kappa = \int_1^0 \frac{1}{\phi(v)}\,dv \in [-\infty, 0).$$

Hence Φ^{-1} is well-defined on (κ,∞). Finally, H is well-defined on the set of the pairs (t,v) such that $t + \Phi(v) > \kappa$.

Denote $\tilde V^h(x) = H\big(h, V(x)\big)$. Then $V(x) = H\big(-h, \tilde V^h(x)\big)$ and eq. (3.2.14) can be written in the form

$$\mathsf{E}_x \tilde V^h(X_h) - \tilde V^h(x) \le -\tilde\phi^h\big(\tilde V(x)\big) + C^h$$

with

$$\tilde\phi^h(v) = v - H(-h, v).$$

The function $\tilde V^h(x)$ takes its values in $[H(h,1),\infty)$, and $H(h,1) = \Phi^{-1}\big(h+\Phi(1)\big) = \Phi^{-1}(h)$. Then the function

$$V^h(x) = \tilde V^h(x) - \Phi^{-1}(h) + 1$$

takes its values in $[1,\infty)$ and eq. (3.2.6) is just eq. (3.2.14) written in terms of V^h. Note that

$$\phi^h(v) = \tilde{\phi}^h\big(v - 1 + \Phi^{-1}(h)\big).$$

To finalize the proof, we need to prove the required properties for ϕ^h. Observe that the function $H(-h,\cdot)$ is *convex* on its natural domain $\{v : h + \Phi(v) > \kappa\}$; the proof here is the same used for the concavity of $H(t,\cdot)$ in the proof of Theorem 2.8.1. Hence $\tilde{\phi}^h$ is concave on its natural domain. Because

$$H_v'(-h,v) = \phi\big(H(-h,v)\big)/\phi(v) \le 1,$$

$\tilde{\phi}^h$ is nondecreasing. Since ϕ^h is just $\tilde{\phi}^h$ with a shift in argument, ϕ^h is also convex and nondecreasing on its natural domain.

Finally, recall that ϕ is well-defined and increasing on $[0,\infty)$. Hence, because $\Phi^{-1}(h) > 1$,

$$\begin{aligned}
-h + \Phi\big(-1 + \Phi^{-1}(h)\big) &= -\Phi\big(\Phi^{-1}(h)\big) + \Phi\big(-1 + \Phi^{-1}(h)\big) \\
&= -\int_1^{\Phi^{-1}(h)} \frac{dw}{\phi(w)} + \int_1^{-1+\Phi^{-1}(h)} \frac{dw}{\phi(w)} \\
&= -\int_{-1+\Phi^{-1}(h)}^{\Phi^{-1}(h)} \frac{dw}{\phi(w)} > -\int_0^1 \frac{dw}{\phi(w)} = \kappa.
\end{aligned}$$

Therefore the natural domain for ϕ^h contains $[0,\infty)$. Clearly, one has

$$\begin{aligned}
\phi^h(0) &= -1 + \Phi^{-1}(h) - \Phi^{-1}\Big(-h + \Phi\big(-1 + \Phi^{-1}(h)\big)\Big) \\
&> -1 + \Phi^{-1}(h) - \Phi^{-1}\Big(\Phi\big(-1 + \Phi^{-1}(h)\big)\Big) = 0,
\end{aligned}$$

and therefore ϕ^h is positive, convex and nondecreasing on $[0,\infty)$. □

Let us give three typical examples where Theorem 3.2.3 applies.

Example 3.2.5. Let eq. (3.2.5) hold true with $\phi(v) = av$. Then,

$$\Phi(v) = \log\frac{v}{a}, \quad \Phi^{-1}(t) = e^{at}, \quad H(t,v) = e^{at}v,$$

and

$$\bar{\phi}^h(v) = (1 - e^{-ah})v, \quad \phi^h(v) = (1 - e^{-ah})v + (e^{ah} - 1)(1 - e^{-ah}),$$

$$V^h(x) = e^{ah}V(x) - (e^{ah} - 1), \quad C^h = C\int_0^t e^{as}\, ds = C\frac{e^{ah} - 1}{a}.$$

Then inequality eq. (3.2.6) can be written as

$$\mathsf{E}_x V^h\big(X_1^h\big) - V^h(x) \le -(1 - e^{-ah})V^h(x) + (e^{ah} - 1)(1 - e^{-ah}) + C\frac{e^{ah} - 1}{a}, \tag{3.2.15}$$

that is the skeleton chain X^h satisfies a linear Lyapunov-type condition eq. (2.8.1) with the Lyapunov function and the constants properly modified. Note that now we can re-write eq. (3.2.15) in the terms of the initial function V:

$$e^{ah}\mathsf{E}_x V\big(X_1^h\big) - e^{ah}V(x) \le -(1 - e^{-ah})e^{ah}V(x) + C\frac{e^{ah} - 1}{a},$$

which just coincides with eq. (3.2.3).

The above example makes explicit the fact (which we had already discussed in the preamble to Theorem 3.2.3) that a “differential” Lyapunov-type condition (3.2.5) with linear $\phi(v)$ provides a linear Lyuapunov-type condition for the skeleton chain. In the next two examples we show that the same feature hold true for the Lyapunov-type conditions from Corollaries 2.8.10 and 2.8.12, which provide polynomial and sub-exponential stabilization rates, respectively.

Example 3.2.6. Let eq. (3.2.5) hold true with $\phi(v) = av^\sigma, v \ge 0$. Then,

$$\Phi(v) = \frac{1}{a(1-\sigma)}\left(v^{1-\sigma} - 1\right), \quad \Phi^{-1}(t) = \Big(1 + a(1-\sigma)t\Big)^{1/(1-\sigma)},$$

$$H(t, v) = \Big(a(1-\sigma)t + v^{1-\sigma}\Big)^{1/(1-\sigma)},$$

and

$$\bar{\phi}^h(v) = v - \Big(v^{1-\sigma} - a(1-\sigma)h\Big)^{1/(1-\sigma)}.$$

We have

$$\bar{\phi}^h(v) = v\left[1 - \Big(1 - a(1-\sigma)hv^{-1+\sigma}\Big)^{1/(1-\sigma)}\right] \sim ahv^\sigma, \quad v \to \infty.$$

Note also that the function

$$\phi^h(v) = \tilde{\phi}^h\big(v - 1 + \Phi^{-1}(h)\big)$$

is nondecreasing, continuous, and satisfies

$$\phi^h(1) = \tilde{\phi}^h\big(\Phi^{-1}(h)\big) = \Phi^{-1}(h) - \Phi^{-1}(-h + \Phi(\Phi^{-1}(h))) = \Phi^{-1}(h) - \Phi^{-1}(0) > 0.$$

Then there exists $a^h > 0$ such that

$$\phi^h(v) \ge a^h v^\sigma, \quad v \ge 1.$$

Finally,

$$V^h(x) = \Big(a(1-\sigma)h + V^{1-\sigma}(x)\Big)^{1/(1-\sigma)} - \Big(1 + a(1-\sigma)h\Big)^{1/(1-\sigma)} + 1,$$

and there exist $C_1, C_2 > 0$ such that

$$C_1 V(x) \le V^h(x) \le C_2 V(x).$$

Example 3.2.7. Let eq. (3.2.5) hold true with $\phi(v) = a(v+b)\log^{-\sigma}(v+b)$, where $a > 0$, $\sigma \in (0,1)$, and b is sufficiently large, such that ϕ has a nonnegative concave extension to $[0,\infty)$; see Corollary 2.8.12. We have

$$\Phi(v) = \frac{1}{a(1+\sigma)}\Big(\log^{1+\sigma}(v+b) - \log^{1+\sigma}(1+b)\Big),$$

$$\Phi^{-1}(t) = \exp\left\{\Big(\log^{1+\sigma}(1+b) + a(1+\sigma)t\Big)^{1/(1+\sigma)}\right\} - b,$$

$$H(t,v) = \exp\left\{\Big(\log^{1+\sigma}(v+b) + a(1+\sigma)t\Big)^{1/(1+\sigma)}\right\} - b,$$

and

$$\begin{aligned}\tilde{\phi}^h(v) &= v + b - \exp\left\{\Big(\log^{1+\sigma}(v+b) - a(1+\sigma)h\Big)^{1/(1+\sigma)}\right\}\\ &= (v+b)\left[1 - \exp\left\{\Big(\log^{1+\sigma}(v+b) - a(1+\sigma)h\Big)^{1/(1+\sigma)} - \log(v+b)\right\}\right].\end{aligned}$$

Since

$$\begin{aligned}z &- \Big(z^{1+\sigma} - a(1+\sigma)h\Big)^{1/(1+\sigma)}\\ &= z\left[1 - \Big(1 - a(1+\sigma)hz^{-1-\sigma}\Big)^{1/(1+\sigma)}\right] \sim ahz^{-\sigma}, \quad z \to \infty,\end{aligned}$$

we have

$$\tilde{\phi}^h(v) \sim ah(v+b)\log^{-\sigma}(v+b), \quad v \to \infty.$$

In addition, the function

$$\phi^h(v) = \tilde{\phi}^h\left(v - 1 + \Phi^{-1}(h)\right)$$

is nondecreasing, continuous, and satisfies

$$\phi^h(1) = \tilde{\phi}^h\left(\Phi^{-1}(h)\right) = \Phi^{-1}(h) - \Phi^{-1}(0) > 0.$$

Then there exists $a^h > 0$ such that

$$\phi^h(v) \geq a^h(v+b)\log^{-\sigma}(v+b), \quad v \geq 1.$$

Likewise to the previous example, we also have $C_1, C_2 > 0$ such that

$$C_1 V(x) \leq V^h(x) \leq C_2 V(x)$$

(we leave the detailed proof for a reader).

3.3 Diffusion Processes

In this and the subsequent sections, we consider two particular Markov models, where the general methods developed above are well applicable. These models have considerable interest themselves, and, on the other hand, illustrate clearly the general technique.

In this section, $X_t, t \geq 0$ is a *diffusion* in $\mathbb{R}^m$; that is, a Markov process solution to an SDE

$$dX_t = a(X_t)\,dt + b(X_t)\,dW_t \tag{3.3.1}$$

driven by a k-dimensional Wiener process $W_t, t \geq 0$. Let the coefficients $a : \mathbb{R}^m \to \mathbb{R}^m$ and $b : \mathbb{R}^m \to \mathbb{R}^{m\times k}$ satisfy usual conditions for a (weak) solution to exist uniquely (e.g., Ref. [61]); then this solution X is a time-homogeneous Markov process in $\mathbb{R}^m$.

3.3.1 Lyapunov-Type Conditions

Below, we give explicit conditions in terms of the coefficients of eq. (3.3.1), which are sufficient for eq. (3.2.5) to hold for a particular ϕ. Denote for $f \in C^2(\mathbb{R}^m)$

$$\mathscr{L}f(x) = \sum_{i=1}^m a_i(x)\,\partial_{x_i}f(x) + \frac{1}{2}\sum_{i,j=1}^m B_{ij}(x)\,\partial^2_{x_ix_j}f(x) \tag{3.3.2}$$

with $B(x) = b(x)b^*(x)$. By the Itô formula (e.g., Ref. [61, Chapter II.5]), the process

$$M_t^f = f(X_t) - \int_0^t \mathscr{L}f(X_s)\,ds$$

is a local martingale w.r.t. every P_x, $x \in \mathbb{R}^m$; that is, each $f \in C^2(\mathbb{R}^m)$ belongs to the domain of the extended generator $\mathscr{A}$ for X, and

$$\mathscr{A}f = \mathscr{L}f.$$

Hence a very natural way to get eq. (3.2.5) with some ϕ is just to construct Lyapunov function $V \in C^2(\mathbb{R}^m)$ such that

$$\mathscr{L}V \le -\phi(V) + C. \tag{3.3.3}$$

For particular process, both the choice of the Lyapunov function V and the function ϕ strongly depend on the properties of the coefficients $a(x)$ and $b(x)$. To simplify the exposition, we reduce the variety of possibilities and assume that

- coefficient $b(x)$ is bounded;
- coefficient $a(x)$ is locally bounded and for some $\kappa \in \mathbb{R}$

$$\limsup_{|x|\to\infty}\left(a(x), \frac{x}{|x|^{\kappa+1}}\right) = -A_\kappa \in [-\infty, 0). \tag{3.3.4}$$

The *drift condition* eq. (3.3.4) is quite transparent: it requires that the radial part of the drift is negative far from the origin; that is, the drift pushes the diffusive point toward the origin when this point stays far from the origin. The index κ controls the growth rate of the absolute value of the radial part at ∞ (actually its decay rate, if $\kappa < 0$), and hence indicates how powerful is the radial part of the drift.

We give Lyapunov-type conditions under three various assumptions on κ: $\kappa \ge 0$, $\kappa \in (-1, 0)$, and $\kappa = -1$. Denote

$$\|B(x)\| = \sup_{l\in\mathbb{R}^m\setminus\{0\}} |l|^{-1}|B(x)l| = \sup_{l\in\mathbb{R}^m\setminus\{0\}} |l|^{-2}|b(x)l|^2,$$

$$|||B||| = \sup_{x\in\mathbb{R}^m} \|B(x)\|.$$

Proposition 3.3.1. *Let either $\kappa > 0$ and $\alpha > 0$ is arbitrary, or $\kappa = 0$ and $\alpha > 0$ satisfies*

$$\alpha < \frac{2A_0}{|||B|||}.$$

Let $V \in C^2(\mathbb{R}^m)$ be such that

$$V(x) = e^{\alpha|x|}, \quad |x| \geq 1.$$

Then there exist $a, C > 0$ such that eq. (3.3.3) holds true with $\phi(v) = av$.

Proof. Because coefficients $a(x), b(x)$ are locally bounded and $V \in C^2(\mathbb{R}^m)$, the function $\mathscr{L}(V)$ is locally bounded, as well. Hence we only have to verify

$$\mathscr{L}V(x) \leq -aV(x) \tag{3.3.5}$$

outside a large ball in $\mathbb{R}^m$.

For $V(x) = e^{\alpha|x|}$ we have on the set $\{|x| \geq 1\}$

$$\begin{aligned} \partial_{x_i} V(x) &= \alpha e^{\alpha|x|} \frac{x_i}{|x|}, \\ \partial^2_{x_i x_j} V(x) &= \alpha^2 e^{\alpha|x|} \frac{x_i x_j}{|x|^2} + \alpha e^{\alpha|x|} \left(\frac{\delta_{ij}}{|x|} - \frac{x_i x_j}{|x|^3} \right), \end{aligned} \tag{3.3.6}$$

where $\delta_{ij} = 1_{i=j}$ is the Kroenecker symbol. Then

$$\mathscr{L}V(x) = e^{\alpha|x|} \left[\alpha \left(a(x), \frac{x}{|x|} \right) + \frac{\alpha}{2} \sum_{i,j=1}^{m} B_{ij}(x) \left(\alpha \frac{x_i x_j}{|x|^2} + \frac{\delta_{ij}}{|x|} - \frac{x_i x_j}{|x|^3} \right) \right].$$

We have

$$\sum_{i,j=1}^{m} B_{ij}(x) x_i x_j = |b(x)x|^2 \leq |||B||| |x|^2,$$

$$\sum_{i,j=1}^{m} B_{ij}(x) \delta_{ij} = \text{Trace}\, B(x) = \sum_{i=1}^{m} |b(x) l_i|^2 \leq m |||B|||,$$

where $l_i, i = 1, \ldots, m$ denote the basic coordinate vectors in $\mathbb{R}^m$. Then

$$\mathscr{L}V(x) \leq V(x) \left[\alpha \left(a(x), \frac{x}{|x|} \right) + \frac{\alpha}{2} \left(\alpha + \frac{m+1}{|x|} \right) |||B||| \right].$$

If $\kappa = 0$, we have

$$\limsup_{|x| \to \infty} \left[\alpha \left(a(x), \frac{x}{|x|} \right) + \frac{\alpha}{2} \left(\alpha + \frac{m+1}{|x|} \right) |||B||| \right] = -\alpha A_0 + \frac{\alpha^2}{2} |||B||| < 0.$$

If $\kappa > 0$,

$$\limsup_{|x|\to\infty}\left[\alpha\left(a(x), \frac{x}{|x|}\right) + \frac{\alpha}{2}\left(\alpha + \frac{m+1}{|x|}\right)|||B|||\right] = -\infty.$$

In both these cases, eq. (3.3.5) holds true outside some large ball. □

Proposition 3.3.2. *Let $\kappa \in (-1, 0)$ and $V \in C^2(\mathbb{R}^m)$ be such that $V \geq 1$ and*

$$V(x) = e^{\alpha|x|^{1+\kappa}}, \quad |x| \geq 1$$

with the constant

$$\alpha < \frac{2A_\kappa}{(1+\kappa)|||B|||}.$$

Then there exist $a, b, C > 0$ such that eq. (3.3.3) holds true with $\phi(v) = a\log^{-\sigma}(v+b)$,

$$\sigma = -\frac{2\kappa}{1+\kappa} > 0.$$

Proof. The argument here is completely the same as in the above proof, so we just give explicit calculation. For $V(x) = e^{\alpha|x|^{1+\kappa}}$ we have

$$\partial_{x_i}V(x) = \alpha(1+\kappa)|x|^\kappa V(x)\frac{x_i}{|x|},$$

$$\begin{aligned}\partial^2_{x_ix_j}V(x) &= \alpha^2(1+\kappa)^2|x|^{2\kappa}V(x)\frac{x_ix_j}{|x|^2}\\ &\quad + \alpha(1+\kappa)|x|^{\kappa-1}V(x)\left(\delta_{ij} + (-1+\kappa)\frac{x_ix_j}{|x|^2}\right).\end{aligned}$$

Then,

$$\begin{aligned}\mathscr{L}V(x) \leq V(x)\bigg[&\alpha(1+\kappa)|x|^{2\kappa}\left(a(x), \frac{x}{|x|^{\kappa+1}}\right)\\ &+ \frac{\alpha(1+\kappa)}{2}|x|^{2\kappa}\left(\alpha(1+\kappa) + \frac{m+|\kappa-1|}{|x|^{1+\kappa}}\right)|||B|||\bigg].\end{aligned}$$

We have

$$\begin{aligned}\limsup_{|x|\to\infty}&\left[\left(a(x), \frac{x}{|x|^{\kappa+1}}\right) + \frac{1}{2}\left(\alpha(1+\kappa) + \frac{m+|\kappa-1|}{|x|^{1+\kappa}}\right)|||B|||\right]\\ &= A_\kappa + \frac{1}{2}\alpha(1+\kappa)|||B||| < 0,\end{aligned}$$

hence there exists $c > 0$ such that

$$\mathscr{L}V(x) \le -cV(x)|x|^{2\kappa}$$

outside some large ball. Since

$$|x| = \left(\frac{1}{\alpha}\log V(x)\right)^{\frac{1}{1+\kappa}}, \quad |x| \ge 1,$$

this gives

$$\mathscr{L}V(x) \le -cV(x)\log^{-\sigma} V(x)$$

outside a large ball. □

Proposition 3.3.3. *Let* $\kappa = -1$ *and*

$$2A_{-1} > \sup_x \left(\operatorname{Trace} B(x)\right).$$

Let $V \in C^2(\mathbb{R}^m)$ *be such that*

$$V(x) = |x|^p, \quad |x| \ge 1,$$

where $p > 2$ *satisfies*

$$2A_{-1} > \sup_x \left(\operatorname{Trace} B(x) + (p-2)\|B(x)\|\right).$$

Then eq. (3.3.3) *holds true for* $\phi(v) = av^{1-2/p}$ *with some* $a, C > 0$.

Proof. Again, we only give a short calculation. For $V(x) = |x|^p$ we have

$$\begin{aligned}
\partial_{x_i} V(x) &= p|x|^{p-1}\frac{x_i}{|x|}, \\
\partial^2_{x_i x_j} V(x) &= p(p-1)|x|^{p-2}\frac{x_i x_j}{|x|^2} + p|x|^{p-1}\left(\frac{\delta_{ij}}{|x|} - \frac{x_i x_j}{|x|^3}\right) \\
&= p|x|^{p-2}\left(\delta_{ij} + (p-2)\frac{x_i x_j}{|x|^2}\right).
\end{aligned}$$

Then

$$\begin{aligned}
\mathscr{L}V(x) &= p|x|^{p-2}\left[(a(x), x) + \frac{1}{2}\left(\operatorname{Trace} B(x) + (p-2)\left(B(x)\frac{x}{|x|}, \frac{x}{|x|}\right)\right)\right] \\
&\le p|x|^{p-2}\left[(a(x), x) + \left(\operatorname{Trace} B(x) + (p-2)\|B(x)\|\right)\right],
\end{aligned}$$

and hence

$$\mathscr{L}V(x) \le -c|x|^{p-2} = -cV(x)^{1-2/p}$$

outside some large ball. □

3.3.2 Dobrushin Condition

Below, we briefly outline several natural and well-developed ways to verify the local Dobrushin condition for skeleton chains of diffusions. First, by the analytical approach which dates back to Kolmogorov, the transition probability density $p_t(x,y)$ of a diffusion process can be treated as the fundamental solution to the Cauchy problem for the parabolic second-order PDE

$$\partial_t - \mathscr{L} = 0.$$

Assuming, for instance, that a and B are Hölder continuous and B is *uniformly elliptic*:

$$\inf_{x\in\mathbb{R}^m,\, l\in\mathbb{R}^m\setminus\{0\}} |l|^{-2}\big(B(x)l, l\big) > 0, \tag{3.3.7}$$

we are able to apply the classical PDE theory (e.g., Ref. [43]), which in particular provides that $p_t(x,y)$ is a continuous function on $(0,\infty)\times\mathbb{R}^m\times\mathbb{R}^m$. Using this continuity, it is easy to arrange the decomposition of the transition probability required in the uniform version of Proposition 2.9.1: take $\nu(dy) = dy$ (the Lebesgue measure), and put

$$P_c(x,dy) = \psi(y)P(x,dy) = p_1(x,y)\psi(y)\,dy,$$

where $\psi > 0$ is a continuous function such that

$$\int_{\mathbb{R}^m} \psi(y)\,dy < \infty.$$

Then the mapping

$$x \mapsto \frac{P_c(x,dy)}{\nu(dy)} \in L_1$$

is continuous by the dominated convergence theorem, and

$$P_c(x,\mathbb{R}^m) > 0, \quad x\in\mathbb{R}^m$$

because ψ is strictly positive. Hence, applying the uniform version of Proposition 2.9.1, we get that for any skeleton chain for X the local Dobrushin condition holds true on any set $B = K\times K$ with a compact K.

The same conclusion can be made, under more demanding regularity assumptions on the coefficients, for diffusions which fail to be uniformly elliptic, but satisfy the Hörmander hypoellipticity condition instead. We refer to Ref. [61] for a detailed exposition, based on the use of the Malliavin calculus, which is a "probabilistic" alternative to the analytical methods mentioned above.

These general methods provide *continuity* of the transition probability density $p_t(x,y)$ and thus, exactly the same argument which we used to prove the local Dobrushin condition, actually leads to local minorization condition. These methods, however, are not applicable in various complicated situations when the process is degenerate and its coefficients lack the regularity. Some specific probabilistic tools are still available in these situations, and it is still realistic to prove the Dobrushin condition. We do not go into deeper details here, just referring to Ref. [1] for an example of degenerate diffusion with nonsmooth coefficients, for which an approach based on the Girsanov transformation leads to the Dobrushin condition (but not to the minorization one).

3.3.3 Summary

Summarizing the above analysis, we give sufficient conditions for exponential, sub-exponential, and polynomial ergodic rates for a diffusion process. We restrict ourselves to the case where X is a diffusion process with Hölder continuous coefficients $a(x)$ and $b(x)$, bounded and uniformly elliptic diffusion matrix $B(x) = b(x)b^*(x)$, and the drift coefficient $a(x)$ which satisfies the drift condition eq. (3.3.4) with the index κ.

Theorem 3.3.4. *Let either $\kappa > 0$ and $\alpha > 0$ is arbitrary, or $\kappa = 0$ and $\alpha > 0$ satisfies*

$$\alpha < \frac{2A_0}{|||B|||}.$$

Then there exist $c_1, c_2 > 0$ such that, for any $x, y \in \mathbb{R}^m$,

$$\left\|P^t(x,\cdot) - P^t(y,\cdot)\right\|_{TV} \le c_1 e^{-c_2 t}\left(e^{\alpha|x|} + e^{\alpha|y|}\right), \quad t \ge 0. \tag{3.3.8}$$

In addition, there exists a unique IPM μ for X, this measure satisfies

$$\int_{\mathbb{R}^m} e^{\alpha|x|}\,\mu(dx) < \infty,$$

and, for every $x \in \mathbb{R}^m$,

$$\left\|P^t(x,\cdot) - \mu\right\|_{TV} \le c_1 e^{-c_2 t}\left(e^{\alpha|x|} + \int_{\mathbb{R}^m} e^{\alpha|y|}\,\mu(dy)\right), \quad t \ge 0. \tag{3.3.9}$$

Proof. The Hölder continuity of the coefficients and the uniform ellipticity of $B(x)$ guarantee the local Dobrushin condition on each set $B = K \times K$ with a compact K; see Section 3.3.2. Hence the statement follows directly from Corollary 2.8.3, Theorem 3.2.3, and Proposition 3.3.1. □

Theorem 3.3.5. *Let $\kappa \in (-1, 0)$. Then for any positive*

$$\alpha < \frac{2A_\kappa}{(1+\kappa)|||B|||}.$$

there exist c_1, $c_2 > 0$ such that, for any x, $y \in \mathbb{R}^m$,

$$\|P^t(x,\cdot) - P^t(y,\cdot)\|_{TV} \le c_1 e^{-c_2 t^{(1+\kappa)/(1-\kappa)}} \left(e^{\alpha|x|^{1+\kappa}} + e^{\alpha|y|^{1+\kappa}}\right), \quad t \ge 0. \tag{3.3.10}$$

In addition, there exists a unique IPM μ for X, this measure satisfies

$$\int_{\mathbb{R}^m} e^{\alpha|x|^{1+\kappa}} \mu(dx) < \infty,$$

and, for every $x \in \mathbb{R}^m$,

$$\left\|P^t(x,\cdot) - \mu\right\|_{TV} \le c_1 e^{-c_2 t^{(1+\kappa)/(1-\kappa)}} \left(e^{\alpha|x|^{1+\kappa}} + \int_{\mathbb{R}^m} e^{\alpha|y|^{1+\kappa}} \mu(dy)\right), \quad t \ge 0. \tag{3.3.11}$$

Proof. Again, the local Dobrushin condition holds true on each set $B = K \times K$ with a compact K. By Proposition 3.3.2 and Theorem 3.2.3, the skeleton chain X^h satisfies eq. (3.2.6) with $\phi^h(v) = a(v+b)\log^{-\sigma}(v+b)$, where

$$\sigma = -\frac{2\kappa}{1+\kappa}$$

and a, b are some positive constants (see Example 3.2.7). In addition,

$$C_1 V(x) \le V^h(x) \le C_2 V(x), \quad V(x) = e^{\alpha|x|^{1+\kappa}}.$$

We have

$$\frac{1}{1-\sigma} = \frac{1+\kappa}{1-\kappa},$$

and we deduce the required statement from Corollary 2.8.12. □

Theorem 3.3.6. *Let $\kappa = -1$ and*

$$2A_{-1} > \sup_x \left(\text{Trace}\, B(x)\right).$$

Denote

$$\eta = \frac{1}{2|||B|||}\left(2A_{-1} - \sup_x \left(\operatorname{Trace} B(x)\right)\right) > 0.$$

Then for any

$$p \in (2, 2 + 2\eta)$$

there exist c_1, $c_2 > 0$ *such that, for any* x, $y \in \mathbb{R}^m$,

$$\left\|P^t(x,\cdot) - P^t(y,\cdot)\right\|_{TV} \le c_1(1 + c_2 t)^{-p/2}\left(1 + |x|^p + |y|^p\right), \quad t \ge 0. \tag{3.3.12}$$

In addition, there exists a unique IPM μ *for* X, *this measure satisfies*

$$\int_{\mathbb{R}^m} |x|^{p-2}\,\mu(dx) < \infty,$$

and, for every $x \in \mathbb{R}^m$,

$$\left\|P^t(x,\cdot) - \mu\right\|_{TV} \le c_1(1 + c_2 t)^{1-p/2}\left(1 + |x|^{p-2} + \int_{\mathbb{R}^m} |y|^{p-2}\,\mu(dy)\right), \quad t \ge 0. \tag{3.3.13}$$

Proof. Again, the local Dobrushin condition holds true on each set $B = K \times K$ with a compact K. By Proposition 3.3.3 and Theorem 3.2.3, the skeleton chain X^h satisfies eq. (3.2.6) with $\phi^h(v) = av^\sigma$, where $\sigma = 1 - 2/p$ and a is a positive constant (see Example 3.2.6). In addition,

$$C_1 V(x) \le V^h(x) \le C_2 V(x), \quad V(x) = |x|^p.$$

We have

$$\frac{1}{1-\sigma} = \frac{p}{2}, \quad \frac{\sigma}{1-\sigma} = \frac{p}{2} - 1$$

and we deduce the required statement from Corollary 2.8.10. □

3.4 Solutions to Lévy-Driven SDEs

In this section, X is a solution to the SDE

$$dX_t = a(X_t)\,dt + b(X_{t-})\,dZ_t \tag{3.4.1}$$

where Z is now a *Lévy process* in $\mathbb{R}^m$ which has the *Itô–Lévy decomposition*

$$Z_t = \int_0^t \int_{|u|\le 1} u\,\tilde{N}(ds, du) + \int_0^t \int_{|u|>1} u\,N(ds, du).$$

Here $N(ds, du)$ is a *Poisson point measure* with the *intensity measure* $ds\nu(du)$, $\widetilde{N}(ds, du) = N(ds, du) - ds\nu(du)$ is corresponding *compensated* measure. The *Lévy measure* ν is, by definition, a measure on $\mathbb{R}^m$ satisfying

$$\int_{\mathbb{R}^m} \left(|u|^2 \wedge 1\right)\nu(du) < \infty.$$

Equation (3.4.1) is a particular case of a Lévy-driven SDE where, in general, instead of the term $b(X_{t-})\, dZ_t$, an expression of the form

$$\sigma(X_t)\, dW_t + \int_{|u|\le 1} c(X_{t-}, u)\,\widetilde{N}(dt, du) + \int_{|u|>1} c(X_{t-}, u)\,N(dt, du)$$

should appear (e.g., Ref. [60, Chapter IV.9]). In order to make comparison with the diffusive case more transparent, we restrict our consideration. We assume that Z does not contain a diffusive term and the jump coefficient $c(x, u)$ is linear w.r.t. the variable u, which corresponds to a jump amplitude; that is, $c(x, u) = b(x)u$. The coefficients $a(x)$ and $b(x)$ are assumed to be locally Lipschitz and to satisfy the linear growth condition, hence eq. (3.4.1) has a unique (strong) solution X, which defines a Markov process in $\mathbb{R}^m$ with càdlàg trajectories.

3.4.1 Lyapunov-Type Condition: "Light Tails" Case

The methodology, which allows one to verify Lyapunov-type condition for X given by eq. (3.4.1), is in general similar to the one we used in the diffusion setting. It is based on the Itô formula, which now can be written as

$$f(X_t) = f(x) + \int_0^t \mathscr{L}^{\mathrm{Levy}} f(X_s)\, ds + M_t^f,$$

where M^f is a local martingale and

$$\mathscr{L}^{\mathrm{Levy}} f(x) = \left(f'(x), a(x)\right) + \int_{\mathbb{R}^m} \left[f\left(x + b(x)u\right) - f(x) - 1_{|u|\le 1}\left(f'(x), b(x)u\right)\right] \nu(du) \quad (3.4.2)$$

(e.g., Ref. [61, Chapter II.5]). Our aim will be to construct a Lyapunov function $V \in C^2(\mathbb{R}^m)$ such that eq. (3.3.3) holds true with $\mathscr{L} = \mathscr{L}^{\mathrm{Levy}}$ and proper ϕ.

In the Lévy-driven setting, realization of this general plan meets some new difficulties. One of them is caused by *nonlocality* of the operator $\mathscr{L}^{\mathrm{Levy}}$, which means that the value of $\mathscr{L}^{\mathrm{Levy}} f$ at some point x involves the values of f in all other points. This brings new requirements which should be taken into account in the choice of V. The most evident requirement is that the function V should satisfy certain integrability condition w.r.t. the "tails" of the Lévy measure ν: otherwise the integral in eq. (3.4.2) will just "blow up", and eq. (3.3.3) will fail. That is, in the Lévy-driven setting, the analogues of Proposition 3.3.1 to Proposition 3.3.3 should take into account the "tail structure" of the Lévy measure ν. In this section, we derive such analogues in the "light tails" case; namely, here we assume that there exists $\beta > 0$ such that

$$\int_{|u|>1} e^{\beta|u|}\,\nu(\mathrm{d}u) < \infty. \tag{3.4.3}$$

To keep visible the analogy with the diffusion setting, we assume coefficient $b(x)$ to be bounded and the coefficient $a(x)$ to be locally bounded. Because of condition eq. (3.4.3), we have

$$\int_{|u|>1} |u|\nu(\mathrm{d}u) < \infty,$$

and we can rewrite the process Z to the form

$$Z_t = \int_0^t \int_{\mathbb{R}^m} u\,\widetilde{N}(\mathrm{d}s, \mathrm{d}u) + ct, \quad c = \int_{|u|>1} u\nu(\mathrm{d}u).$$

Changing notation $a(x)$ to $a(x)+cb(x)$, we now can rewrite the SDE eq. (3.4.1) to the form

$$\mathrm{d}X_t = a(X_t)\,\mathrm{d}t + b(X_{t-}) \int_{\mathbb{R}^m} u\,\widetilde{N}(\mathrm{d}s, \mathrm{d}u). \tag{3.4.4}$$

Correspondingly, operator eq. (3.4.2) will be rewritten as

$$\mathscr{L}^{\mathrm{Levy}} f(x) = \big(f'(x), a(x)\big) + \int_{\mathbb{R}^m} \big[f\big(x + b(x)u\big) - f(x) - \big(f'(x), b(x)u\big)\big]\,\nu(\mathrm{d}u). \tag{3.4.5}$$

We assume the (new) drift coefficient $a(x)$ to satisfy the drift condition eq. (3.3.4) with the index κ, and denote

$$|||b||| = \sup_x \|b(x)\|.$$

Proposition 3.4.1. *Let $\kappa \geq 0$ and $V \in C^2(\mathbb{R}^m)$ be such that*

$$V(x) = e^{\alpha|x|}, \quad |x| \geq 1$$

and

$$V(x) \leq e^{\alpha|x|}, \quad |x| < 1. \tag{3.4.6}$$

Here $\alpha \in (0, \beta/|||b|||)$ is arbitrary if $\kappa > 0$, and satisfies an additional assumption

$$-A_0 + \int_{\mathbb{R}} |||b||||u|\left(e^{\alpha|||b||||u|} - 1\right)\nu(du) < 0$$

if $\kappa = 0$.

Then there exist $a, C > 0$ such that eq. (3.3.3) holds true with $\phi(v) = av$.

Proof. We temporarily fix $\varepsilon \in (0,1)$, and for $|x| > \varepsilon^{-1}$ decompose

$$\begin{aligned}
\mathcal{L}^{\text{Levy}} V(x) &= \mathcal{L}^{\text{drift}} V(x) + \mathcal{L}^{\text{small}} V(x) + \mathcal{L}^{\text{large}} V(x) + \mathcal{L}^{\text{huge}} V(x),\\
\mathcal{L}^{\text{drift}} V(x) &= \left(V'(x), a(x)\right),\\
\mathcal{L}^{\text{small}} V(x) &= \int_{|u|\le 1} \left[V\left(x + b(x)u\right) - V(x) - \left(V'(x), b(x)u\right)\right] \nu(\mathrm{d}u)\\
\mathcal{L}^{\text{large}} V(x) &= \int_{1<|u|\le\varepsilon|x|} \left[V\left(x + b(x)u\right) - V(x) - \left(V'(x), b(x)u\right)\right] \nu(\mathrm{d}u)\\
\mathcal{L}^{\text{huge}} V(x) &= \int_{|u|>\varepsilon|x|} \left[V\left(x + b(x)u\right) - V(x) - \left(V'(x), b(x)u\right)\right] \nu(\mathrm{d}u).
\end{aligned}$$

We analyze these four terms separately. The "drift" term can be given explicitly:

$$\mathcal{L}^{\text{drift}} V(x) = \alpha e^{\alpha|x|}\left(a(x), \frac{x}{|x|}\right), \quad |x| \ge 1.$$

For the "small jumps" term, the bounds, similar to the "diffusion" term in the proof of Proposition 3.3.1 can be given. Namely, we have

$$\begin{aligned}
V\left(x + b(x)u\right) &- V(x) - \left(V'(x), b(x)u\right)\\
&= \int_0^1 (1-s)\left(V''\left(x + sb(x)u\right)b(x)u, b(x)u\right) \mathrm{d}s.
\end{aligned} \tag{3.4.7}$$

We have

$$|x + sb(x)u| > |x| - |||b|||, \quad s \in [0,1], \quad |u| \le 1.$$

That is, if $|x| > 1 + |||b|||$, all these points are located outside of the unit ball in $\mathbb{R}^m$, centered at origin. The matrix V'' outside the unit ball is given by eq. (3.3.6), and a straightforward calculation shows that its matrix norm possesses the bound

$$\|V''(x)\| \le \alpha e^{\alpha|x|}\left(\alpha + \frac{1}{|x|}\right), \quad |x| \ge 1.$$

Therefore

$$\mathscr{L}^{\text{small}} V(x) \le \left(\alpha + \frac{1}{|x| - |||b|||}\right) \int_0^1 \int_{|u|\le 1} (1-s)\alpha e^{\alpha|x+sb(x)u|} |||b|||^2 |u|^2 \, \nu(\mathrm{d}u)\, \mathrm{d}s.$$

We have

$$\begin{aligned}\int_0^1 (1-s)\alpha e^{\alpha|x+sb(x)u|} |||b|||^2|u|^2 \,\mathrm{d}s &\le e^{\alpha|x|}\int_0^1 \alpha e^{\alpha s|||b||||u|} |||b|||^2|u|^2\,\mathrm{d}s\\ &= |||b||||u|\left(e^{\alpha|||b||||u|} - 1\right).\end{aligned}$$

This gives

$$\mathscr{L}^{\text{small}} V(x) \le C^{\text{small}}(x) V(x)$$

with

$$\limsup_{|x|\to\infty} C^{\text{small}}(x) \le \alpha \int_{|u|\le 1} |||b||||u|\left(e^{\alpha|||b||||u|} - 1\right)\nu(\mathrm{d}u).$$

The estimate for the "large jumps" term is similar, with the second-order Taylor expansion eq. (3.4.7) replaced by the first-order one

$$V\big(x + b(x)u\big) - V(x) - \big(V'(x), b(x)u\big) = \int_0^1 \Big(V'\big(x + sb(x)u\big) - V'(x), b(x)u\Big)\,\mathrm{d}s.$$

We have

$$|x + sb(x)u| \ge |x|(1 - \varepsilon|||b|||), \quad s \in [0,1], \quad |u| \le \varepsilon|x|,$$

hence if $\varepsilon|||b||| < 1$ and $|x| \ge (1 - \varepsilon|||b|||)^{-1}$, all these points are located outside the unit ball. Thus

$$V'\big(x + sb(x)u\big) - V'(x) = \alpha e^{\alpha|x+sb(x)u|}\frac{x + sb(x)u}{|x + sb(x)u|} - \alpha e^{\alpha|x|}\frac{x}{|x|}.$$

We have

$$e^{\alpha|x+sb(x)u|} \le e^{\alpha|x|}e^{\alpha|||b||||u|}$$

and

$$\left|\frac{x + sb(x)u}{|x + sb(x)u|} - \frac{x}{|x|}\right| \le \frac{2\varepsilon|||b|||}{1 - \varepsilon|||b|||}, \tag{3.4.8}$$

which gives the bound

$$\mathscr{L}^{\text{large}} V(x) \le C^{\text{large}}(x) V(x)$$

with

$$\limsup_{|x|\to\infty} C^{\text{large}}(x) \le \alpha \int_{|u|>1} |||b||| |u| \left(e^{\alpha |||b||| |u|} - 1 \right) \nu(du) + \alpha \frac{2\varepsilon |||b|||^2}{1 - \varepsilon |||b|||} \int_{|u|>1} |u| \, \nu(du).$$

For the "huge jumps" term, unlike for those considered previously, we cannot specify the location of $x + b(x)u$ in terms of the position of x; this is exactly the point where the nonlocality of $\mathscr{L}^{\text{Levy}}$ is most evident, and which actually motivates convention eq. (3.4.6). By this convention, we have for $|x| > 1$

$$\begin{aligned} \mathscr{L}^{\text{huge}} V(x) &\le \int_{|u|>\varepsilon|x|} \left(e^{\alpha|x+b(x)u|} - e^{\alpha|x|} - \alpha e^{\alpha|x|} \frac{(x, b(x)u)}{|x|} \right) \nu(du) \\ &\le C^{\text{huge}}(x) V(x), \end{aligned}$$

where

$$C^{\text{huge}}(x) = \int_{|u|>\varepsilon|x|} \left(e^{\alpha |||b||| |u|} + \alpha |||b||| |u| \right) \nu(du) \to 0, \quad |x| \to \infty.$$

Summarizing the above calculations, we get inequality

$$\mathscr{L}^{\text{Levy}} V(x) \le C(x) V(x)$$

with

$$\limsup_{|x|\to\infty} C(x) = -\infty$$

if $\kappa > 0$ and

$$\begin{aligned} \limsup_{|x|\to\infty} C(x) = &- \alpha A_0 + \alpha \int_{\mathbb{R}} |||b||| |u| \left(e^{\alpha |||b||| |u|} - 1 \right) \nu(du) \\ &+ \alpha \frac{2\varepsilon |||b|||^2}{1 - \varepsilon |||b|||} \int_{|u|>1} |u| \, \nu(du). \end{aligned}$$

if $\kappa = 0$. Since $\varepsilon > 0$ in the above estimates can be taken arbitrarily small, this shows that, for $|x|$ large enough, inequality

$$\mathscr{L}^{\text{Levy}} V(x) \le -a V(x)$$

holds true for some $a > 0$. Similar calculations easily show that $\mathscr{L}^{\text{Levy}} V$ is locally bounded, which completes the proof. □

Proposition 3.4.2. *Let $\kappa \in (-1, 0)$ and $V \in C^2(\mathbb{R}^m)$ satisfy*

$$V(x) = e^{\alpha|x|^{1+\kappa}}, \quad |x| > 1, \quad V(x) \le e^{\alpha|x|^{1+\kappa}}, \quad |x| \le 1$$

with the constant $\alpha > 0$ such that

$$\frac{\alpha(1+\kappa)}{2} \int_{|u|\le 1} |||b|||^2|u|^2\, \nu(du) + \int_{|u|>1} |||b|||\,|u| \left(e^{\alpha|||b|||^{1+\kappa}|u|^{1+\kappa}} - 1\right) \nu(du) < A_\kappa.$$

Then there exist $a, b, C > 0$ such that eq. (3.3.3) holds true with $\phi(v) = a\log^{-\sigma}(v + b)$,

$$\sigma = -\frac{2\kappa}{1+\kappa} > 0.$$

Proof. The argument is essentially the same as in the proof of Proposition 3.4.1, hence we just outline the calculations. We have

$$\mathscr{L}^{\text{drift}} V(x) = C^{\text{drift}}(x)|x|^{2\kappa} V(x), \quad |x| \ge 1$$

with

$$\limsup_{|x|\to\infty} C^{\text{drift}}(x) \le -\alpha(\kappa+1)A_\kappa.$$

Next,

$$\mathscr{L}^{\text{small}} V(x) \le \frac{1}{2} \sup_{|y|\le 1} \|V''(x+y)\| \int_{|u|\le 1} |||b|||^2|u|^2\, \nu(\mathrm{d}u).$$

Together with the explicit formula for V'' (see Proposition 3.3.2), this gives

$$\mathscr{L}^{\text{small}} V(x) \le C^{\text{small}}(x)|x|^{2\kappa} V(x)$$

with

$$\limsup_{|x|\to\infty} C^{\text{small}}(x) \le \frac{\alpha^2(1+\kappa)^2}{2} \int_{|u|\le 1} |||b|||^2|u|^2\, \nu(\mathrm{d}u).$$

For $|x|$ large enough, we have for all $s \in [0, 1]$, $|u| \le \varepsilon|x|$

$$V'\left(x + sb(x)u\right) = \alpha(1+\kappa)e^{\alpha|x+sb(x)u|^{1+\kappa}} \frac{x + sb(x)u}{|x + sb(x)u|^{1+\kappa}},$$

and the similar estimate as we had for the "large jumps" term of the previous proof leads to inequality

$$\mathscr{L}^{\text{large}}V(x) \le C^{\text{large}}(x)|x|^{2\kappa}V(x)$$

with

$$\limsup_{|x|\to\infty} C^{\text{large}}(x) \le \alpha(1+\kappa)\int_{|u|>1} |||b||||u|\left(e^{\alpha|||b|||^{1+\kappa}|u|^{1+\kappa}} - 1\right)\nu(du) + \delta(\varepsilon),$$

where

$$\delta(\varepsilon) \to 0, \quad \varepsilon \to 0.$$

Since

$$\nu\{u : |u| > |x|\} \le Ce^{-\beta|x|}, \qquad \int_{|u|>x} e^{\alpha|||b|||^{1+\kappa}|u|^{1+\kappa}}\nu(du) \le Ce^{-\beta|x|/2},$$

we have

$$\mathscr{L}^{\text{huge}}V(x) \le C^{\text{huge}}(x)|x|^{2\kappa}V(x),$$

where

$$C^{\text{huge}}(x) = |x|^{-2\kappa}\int_{|u|>\varepsilon|x|}\left(e^{\alpha|||b|||^{1+\kappa}|u|^{1+\kappa}} - \alpha(1+\kappa)\frac{(x, b(x)u)}{|x|^{1-\kappa}}\right)\nu(du)$$

and

$$C^{\text{huge}}(x) \to 0, \quad |x| \to \infty.$$

Summarizing the above calculations, we get inequality

$$\mathscr{L}^{\text{Levy}}V(x) \le C(x)V(x)$$

with

$$\begin{aligned}\limsup_{|x|\to\infty} C(x) = &-\alpha(1+\kappa)A_\kappa + \frac{\alpha^2(1+\kappa)^2}{2}\int_{|u|\le 1} |||b|||^2|u|^2\nu(du)\\ &+ \alpha(1+\kappa)\int_{|u|>1} |||b||||u|\left(e^{\alpha|||b|||^{1+\kappa}|u|^{1+\kappa}} - 1\right)\nu(du) + \delta(\varepsilon).\end{aligned}$$

Since $\varepsilon > 0$ in the above estimates can be taken arbitrarily small, this shows that, for $|x|$ large enough, inequality

$$\mathscr{L}^{\mathrm{Levy}} V(x) \le -a|x|^{2\kappa} V(x)$$

holds true for some $a > 0$. Similar calculations easily show that $\mathscr{L}^{\mathrm{Levy}} V$ is locally bounded, which completes the proof. □

Proposition 3.4.3. *Let* $\kappa = -1$ *and*

$$2A_{-1} > |||b|||^2 \int_{\mathbb{R}} |u|^2 \nu(du).$$

Let $V \in C^2(\mathbb{R}^m)$ *satisfy*

$$V(x) = |x|^p, \quad |x| \ge 1, \quad V(x) \le |x|^p, \quad |x| < 1,$$

where $p > 2$ *is such that*

$$2A_{-1} > (p-1)|||b|||^2 \int_{\mathbb{R}} |u|^2 \nu(du)$$

Then eq. (3.3.3) *holds true for* $\phi(v) = av^{1-2/p}$ *with some* $a, C > 0$.

Proof. We change slightly the decomposition of $\mathscr{L}^{\mathrm{Levy}} V(x)$: we introduce a new parameter $R > 1$ and define

$$\mathscr{L}^{\mathrm{small}} V(x) = \int_{|u|\le R} \left[V\big(x + b(x)u\big) - V(x) - \big(V'(x), b(x)u\big)\right] \nu(\mathrm{d}u),$$

$$\mathscr{L}^{\mathrm{large}} V(x) = \int_{R<|u|\le\varepsilon|x|} \left[V\big(x + b(x)u\big) - V(x) - \big(V'(x), b(x)u\big)\right] \nu(\mathrm{d}u),$$

the terms $\mathscr{L}^{\mathrm{drift}} V$ and $\mathscr{L}^{\mathrm{huge}} V$ remain the same as above. We have

$$\mathscr{L}^{\mathrm{drift}} V(x) = C^{\mathrm{drift}}(x)|x|^{p-2}, \quad |x| \ge 1$$

with

$$\limsup_{|x|\to\infty} C^{\mathrm{drift}}(x) \le -pA_{-1}.$$

Next,

$$\mathscr{L}^{\mathrm{small}}V(x) \le \frac{1}{2}\sup_{|v|\le R}\|V''(x+v)\| \int_{|u|\le R} |||b|||^2|u|^2\,\nu(\mathrm{d}u).$$

Together with the explicit formula for V'' (see Proposition 3.3.3), this gives

$$\mathscr{L}^{\mathrm{small}}V(x) \le C^{\mathrm{small}}(x)|x|^{p-2}, \quad |x|\ge 1$$

with

$$\limsup_{|x|\to\infty} C^{\mathrm{small}}(x) \le \frac{p(p-1)}{2}\int_{|u|\le R} |||b|||^2|u|^2\,\nu(\mathrm{d}u).$$

For $|x|$ large enough, we have for all $s \in [0,1]$, $|u|\le \varepsilon|x|$

$$V'\big(x+sb(x)u\big) = p|x+sb(x)u|^{p-1}\frac{x+sb(x)u}{|x+sb(x)u|^{1+\kappa}}.$$

By the Taylor formula,

$$|x+sb(x)u|^{p-1} \le |x|^{p-1} + (p-1)|x|^{p-2}\big(1+\varepsilon|||b|||\,|u|\big)^{p-2}s|||b|||\,|u|,$$

and the following more precise version of eq. (3.4.8) holds true:

$$\left|\frac{x+sb(x)u}{|x+sb(x)u|} - \frac{x}{|x|}\right| \le \frac{2s|||b|||\,|u|}{|x|(1-\varepsilon|||b|||)}.$$

Then for $|x|$ large enough

$$\begin{aligned}\mathscr{L}^{\mathrm{large}}V(x) &= \int_{R<|u|\le\varepsilon|x|}\int_0^1 \Big(V'\big(x+sb(x)u\big) - V'(x), b(x)u\Big)\,\mathrm{d}s\nu(\mathrm{d}u)\\ &\le C^{\mathrm{large}}(x)|x|^{p-2},\end{aligned}$$

where

$$\begin{aligned}C^{\mathrm{large}}(x) &= p\left((p-1)\big(1+\varepsilon|||b|||\big) + \frac{2}{(1-\varepsilon|||b|||)}\right)\int_{R<|u|\le\varepsilon|x|}\int_0^1 s|||b|||^2|u|^2\,\mathrm{d}s\nu(\mathrm{d}u)\\ &\le \frac{p(p+1)}{2}\big(1+\delta(\varepsilon)\big)\int_{1<|u|\le\varepsilon|x|} |||b|||^2|u|^2\,\nu(\mathrm{d}u),\end{aligned}$$

and $\delta(\varepsilon)\to 0, \varepsilon\to 0$.

Finally, likewise to the above proof we have

$$\mathscr{L}^{\mathrm{huge}}V(x) \le C^{\mathrm{huge}}(x)|x|^{p-2}, \quad |x| \ge 1,$$

where

$$C^{\mathrm{huge}}(x) \to 0, \quad |x| \to \infty.$$

Summarizing the above calculations, we get inequality

$$\mathscr{L}^{\mathrm{Levy}}V(x) \le C(x)|x|^{p-2}, \quad |x| \ge 1$$

with

$$\limsup_{|x|\to\infty} C(x) = -pA_{-1} + \frac{p(p-1)}{2} \int_{|u|\le R} |||b|||^2 |u|^2\, \nu(du) + \frac{p(p+1)}{2}\Big(1+\delta(\varepsilon)\Big) \int_{|u|>R} |||b|||^2 |u|^2\, \nu(du).$$

Since in the above estimates $R > 1$ can be taken arbitrarily large and $\varepsilon > 0$ is arbitrary, this shows that, for $|x|$ large enough, inequality

$$\mathscr{L}^{\mathrm{Levy}}V(x) \le -aV^{(p-2)/p}(x)$$

holds true for some $a > 0$. Similar calculations easily show that $\mathscr{L}^{\mathrm{Levy}}V$ is locally bounded, which completes the proof. □

3.4.2 Lyapunov-Type Condition: "Heavy Tails" Case

In this section, we derive the Lyapunov-type conditions in the case where the Lévy measure of the noise is "heavy tailed". Namely, here we assume that there exist $\alpha, C > 0$ such that

$$\nu(|u| > r) \le Cr^{-\alpha}, \quad r \ge 1. \tag{3.4.9}$$

This framework includes particularly important class of α-stable processes, see Ref. [120, 121]. Indeed, for arbitrary α-stable process

$$\nu(|u| > r) = cr^{-\alpha}, \quad r > 0$$

with some $c > 0$. Note that for a stable process $\alpha \in (0, 2)$; we do not impose any assumptions of that type, and $\alpha > 0$ in eq. (3.4.9) is arbitrary.

Unless $\alpha > 1$, we cannot re-arrange the original SDE to the form eq. (3.4.4). Thus, in order not to restrict generality, we formulate the results below for SDE eq. (3.4.1) directly. We assume that coefficient $b(x)$ is bounded, coefficient $a(x)$ is locally bounded, and $a(x)$ satisfies the drift condition eq. (3.3.4).

Proposition 3.4.4. *Let $\kappa \geq 1$.*

Then for any $p \in (0, \alpha)$ and a function $V \in C^2(\mathbb{R}^m)$ such that

$$V(x) = |x|^p, \quad |x| \geq 1, \quad V(x) \leq |x|^p, \quad |x| < 1,$$

there exist $a, C > 0$ such that eq. (3.3.3) *holds true with $\phi(v) = av$.*

Proof. The general structure of the proof is the same as above, but the decomposition of $\mathscr{L}^{\text{Levy}} V$ is different. We fix $\varepsilon \in (0, 1)$, and for $|x| > \varepsilon^{-1}$ decompose

$$\mathscr{L}^{\text{Levy}} V(x) = \mathscr{L}^{\text{drift}} V(x) + \mathscr{L}^{\text{small}} V(x) + \mathscr{L}^{\text{large}} V(x),$$

$$\mathscr{L}^{\text{drift}} V(x) = \big(V'(x), a(x) + q(x)\big),$$

$$\mathscr{L}^{\text{small}} V(x) = \int\limits_{|u| \leq \varepsilon |x|} \big[V\big(x + b(x)u\big) - V(x) - \big(V'(x), b(x)u\big)\big] \, \nu(du)$$

$$\mathscr{L}^{\text{large}} V(x) = \int\limits_{|u| > \varepsilon |x|} \big[V\big(x + b(x)u\big) - V(x)\big)\big] \, \nu(du),$$

where we denote

$$q(x) = \int\limits_{1 < |u| \leq \varepsilon |x|} b(x)u \, \nu(du).$$

We have

$$|q(x)| \leq |||b||| \int\limits_{1 < |u| \leq \varepsilon |x|} |u| \, \nu(du),$$

and

$$\int\limits_{1 < |u| \leq \varepsilon |x|} |u| \, \nu(du) = \int\limits_1^{\varepsilon |x|} r \, \nu\big(\{u : |u| \in dr\}\big) = -\int\limits_1^{\varepsilon |x|} r \, d\nu\big(\{u : |u| \geq r\}\big)$$

$$= -\varepsilon |x| \nu(|u| \geq \varepsilon |x|) + \nu(|u| \geq 1) + \int\limits_1^{\varepsilon |x|} \nu\big(\{u : |u| \geq r\}\big) \, dr.$$

Hence, under condition (3.4.9), there exists C such that, for $|x| \geq 1$,

$$|q(x)| \leq \begin{cases} C, & \alpha > 1; \\ C\big(1 + \log|x|\big), & \alpha = 1; \\ C|x|^{1-\alpha}, & \alpha \in (0, 1). \end{cases} \tag{3.4.10}$$

In each of these cases,

$$|x|^{-\kappa}|q(x)| \to 0, \quad |x| \to \infty,$$

hence

$$\mathscr{L}^{\rm drift}V(x) = p|x|^{p-1}\frac{(a(x), x) + (q(x), x)}{|x|} = C^{\rm drift}(x)|x|^{p-1+\kappa}, \quad |x| \geq 1$$

with

$$\limsup_{|x|\to\infty} C^{\rm drift}(x) = -pA_\kappa.$$

Next, by eq. (3.4.7) we have for $|x| > (1-\varepsilon)^{-1}$

$$\mathscr{L}^{\rm small}V(x) \leq \frac{1}{2}\sup_{|v|\leq\varepsilon|x|}\|V''(x+v)\| \int\limits_{|u|\leq\varepsilon|x|} \|\|b\|\|^2|u|^2\,\nu(du).$$

Clearly,

$$\sup_{|v|\leq\varepsilon|x|}\|V''(x+v)\| \leq C|x|^{p-2}, \quad |x| > (1-\varepsilon)^{-1}.$$

On the other hand, similar to eq. (3.4.10), we have for $|x| \geq 1$

$$\int\limits_{|u|\leq\varepsilon|x|} |u|^2\,\nu(du) \leq \begin{cases} C, & \alpha > 2; \\ C\big(1 + \log|x|\big), & \alpha = 2; \\ C|x|^{2-\alpha}, & \alpha \in (0, 2). \end{cases} \tag{3.4.11}$$

That is,

$$\mathscr{L}^{\rm small}V(x) \leq C|x|^{p-\alpha}, \quad |x| > (1-\varepsilon)^{-1}.$$

Finally, we have simply

$$\mathscr{L}^{\text{large}} V(x) \le \int\limits_{|u|>\varepsilon|x|} |x + b(x)u|^p \, \nu(du)$$

$$\le (2^{p-1} \wedge 1) \left(|x|^p \nu(|u| > \varepsilon|x|) + \int\limits_{|u|>\varepsilon|x|} \|\|b\|\| |u|^p \, \nu(du) \right).$$

Similar to eq. (3.4.10), we have

$$\int\limits_{|u|>\varepsilon|x|} |u|^p \, \nu(du) \le c|x|^{p-\alpha},$$

hence

$$\mathscr{L}^{\text{large}} V(x) \le C|x|^{p-\alpha}, \quad |x| \ge 1.$$

Recall that $\kappa \ge 1$, hence

$$p - 1 + \kappa \ge p > p - \alpha.$$

That is, the essential behavior of $\mathscr{L}^{\text{Levy}} V(x)$ as $|x| \to \infty$ is determined by the term $\mathscr{L}^{\text{drift}} V(x)$:

$$\mathscr{L}^{\text{Levy}} V(x) = C(x)|x|^{p-1+\kappa}, \quad |x| \ge 1 \tag{3.4.12}$$

with

$$\limsup_{|x|\to\infty} C(x) = -pA_\kappa < 0. \tag{3.4.13}$$

Since $p - 1 + \kappa \ge p$, this shows that, for $|x|$ large enough, inequality

$$\mathscr{L}^{\text{Levy}} V(x) \le -aV(x)$$

holds true for some $a > 0$. Similar calculations show that $\mathscr{L}^{\text{Levy}} V$ is locally bounded, which completes the proof. □

Essentially the same calculation can be applied in the case $\kappa < 1$, as well. One can easily see that the proof of eqs. (3.4.12) and (3.4.13) relies only on eq. (3.4.9) and the inequality

$$\alpha + \kappa > 1. \tag{3.4.14}$$

If $\kappa < 1$, we consider eq. (3.4.14) as an additional assumption, and call it the *balance condition*.

Proposition 3.4.5. *Let $\kappa < 1$ and the balance condition* (3.4.14) *hold.*
Then for any $p \in (1-\kappa, \alpha)$ and a function $V \in C^2(\mathbb{R}^m)$ such that

$$V(x) = |x|^p, \quad |x| \geq 1, \quad V(x) \leq |x|^p, \quad |x| < 1,$$

there exist $a, C > 0$ such that eq. (3.3.3) *holds true with $\phi(v) = av^{1-(1-\kappa)/p}$.*

Proof. Repeating literally the previous proof, we get eqs. (3.4.12) and (3.4.13). This means that, for $|x|$ large enough, inequality

$$\mathscr{L}^{\mathrm{Levy}} V(x) \leq -aV^{1-(1-\kappa)/p}(x)$$

holds true for some $a > 0$. Since $\mathscr{L}^{\mathrm{Levy}} V$ is locally bounded, this completes the proof. □

3.4.3 Dobrushin Condition

Likewise to the diffusive case, Dobrushin condition for X is naturally related to local properties of the transition probabilities, that is, existence and regularity of the transition probability density $p_t(x, y)$. For a solution to a Lévy-driven SDE, these properties are in general more delicate than in the diffusive case; below we briefly outline several methods applicable in that concern.

Within an analytic approach, $p_t(x, y)$ is treated as a fundamental solution to the Cauchy problem for *pseudo-differential operator*

$$\partial_t - \mathscr{L}^{\mathrm{Levy}}.$$

For $p_t(x, y)$ to be specified in this setting, an analogue of the classical *parametrix method* [43] has to be developed in a Lévy noise setting. Such an analogue typically requires nontrivial "structural" assumptions on the noise; a well studied model here concerns the case of a Lévy process being a mixture of α-stable processes. For an overview of the topic, details and further bibliography, we refer a reader to the monograph Ref. [36].

As an alternative to the analytical approach, a variety of "stochastic calculus of variations" methods are available, cf. Ref. [9, 116] or a more recent exposition in Ref. [5, 63]; these are just a few references from an extensively developing field, which we cannot discuss in detail here. These methods are based either on the integration by parts formula (in the Malliavin calculus case) or the duality formula (in the Picard approach) and typically provide existence and *continuity* (or, moreover, smoothness) of the transition probability density $p_t(x, y)$. The cost is that relatively strong assumptions on the Lévy measure of the noise should be required; a typical requirement here is that for some $\alpha \in (0, 2)$ and $c_1, c_2 > 0$,

$$c_1\varepsilon^{2-\alpha}|l|^2 \le \int_{|u|\le\varepsilon} (u,l)^2 \mu(du) \le c_2\varepsilon^{2-\alpha}|l|^2, \quad l \in \mathbb{R}^m. \tag{3.4.15}$$

Condition eq. (3.4.15) is a kind of a frequency regularity assumption on the Lévy measure and heuristically means that the intensity of small jumps is comparable with that for an α-stable noise. When this assumption fails, genuinely new effects may appear, which is illustrated by the following simple example, cf. Ref. [11, Example 2].

Example 3.4.6. Consider an eq. (3.4.1) with $d = m = 1$, $a(x) = cx$ with $c \neq 0$, $b(x) \equiv 1$ and the process Z of the form $Z_t = \sum_{k=1}^{\infty}(1/k!)N_t^k$, where $\{N^k\}$ are independent copies of a Poisson process. Then the solution to eq. (3.4.1) possesses the transition probability density $p_t(x,y)$, but for every t and x the function

$$p_t(x,\cdot) \in L_1(\mathbb{R})$$

does not belong to any $L_{p,\mathrm{loc}}(\mathbb{R})$ and therefore is not continuous.

This example shows that, when the intensity of small jumps is low, it may happen that the transition probability density $p_t(x,y)$ exists, but is highly irregular. In this framework another type of stochastic calculus of variations is highly appropriate, based on Davydov's *stratification method*, cf. Ref. [84] for a version of this method specially designed for Lévy-driven SDEs with minimal requirements for the Lévy measure of the noise. The crucial point is that this method is well designed to give continuity of the function

$$x \mapsto p_t(x,\cdot)$$

in the integral form; that is, as a mapping $\mathbb{R}^m \to L_1(\mathbb{R}^m)$. This continuity is exactly the key ingredient for the proof of the local Dobrushin condition for the process X, hence, following this general line, it is possible to obtain the following sufficient condition (we necessarily omit the numerous technical details, referring the reader to Ref. [82, Theorem 1.3 and Section 4]).

Proposition 3.4.7. *Let coefficients a and b in eq.* (3.4.1) *belong to $C^1(\mathbb{R}^m)$ and $C^1(\mathbb{R}^{d\times d})$, respectively, the Lévy measure μ satisfy $\int_{|u|\le 1} |u|\,\mu(du) < \infty$, and at some point x_* the matrix*

$$\nabla a(x_*)b(x_*) - \nabla b(x_*)a(x_*)$$

is nondegenerate. Assume also that for the measure μ the following cone condition holds: for every $l \in \mathbb{R}^m \setminus \{0\}$ and $\varepsilon > 0$, there exists a cone

$$V_{l,\rho} = \{u : (u,l) \ge \rho|u||l|\}, \quad \rho \in (0,1),$$

such that

$$\mu\left(V_{l,\rho} \cap \{|u| \le \varepsilon\}\right) > 0.$$

Then the Markov process solution to eq. (3.4.1) *X satisfies the local Dobrushin condition on every compact set in* $\mathbb{R}^m$.

Observe that if one intends to verify the irreducibility in the form of the minorization condition, this would require assumptions on the Lévy measure of the noise similar to eq. (3.4.15), which is much more restrictive than the cone condition used in Proposition 3.4.7. This well illustrates the above-mentioned point that in Markov models with comparatively complicated local structure, the Dobrushin condition is more practical than the minorization one.

3.4.4 Summary

Here we summarize the above calculations and provide sufficient conditions for exponential, sub-exponential, and polynomial ergodic rates for solutions to Lévy-driven SDEs. We restrict ourselves to the case where the coefficients $a(x)$ and $b(x)$ of SDE eq. (3.4.1) are Lipschitz continuous. The coefficient $b(x)$ is assumed to bounded. The process X is assumed to satisfy the local Dobrushin condition; one possible sufficient condition for that is provided by Proposition 3.4.7.

We first consider the "light tails" case, where eq. (3.4.3) holds true for some $\beta > 0$. We rewrite the original SDE to form eq. (3.4.4) and assume that the drift coefficient $a(x)$ satisfies the drift condition eq. (3.3.4) with the index κ and the constant A_κ.

Theorem 3.4.8. *Let either* $\kappa > 0$ *and*

$$\alpha \in \left(0, \frac{\beta}{|||b|||}\right),$$

or $\kappa = 0$ *and* $\alpha > 0$ *satisfy or*

$$-A_0 + \int_{\mathbb{R}} |||b||| |u| \left(e^{\alpha |||b||| |u|} - 1\right) \mu(du) < 0.$$

Then there exist $c_1, c_2 > 0$ *such that, for any* $x, y \in \mathbb{R}^m$,

$$\left\| P^t(x,\cdot) - P^t(y,\cdot) \right\|_{TV} \le c_1 e^{-c_2 t}\left(e^{\alpha|x|} + e^{\alpha|y|}\right), \quad t \ge 0. \tag{3.4.16}$$

In addition, there exists unique IPM μ *for X, this measure satisfies*

$$\int_{\mathbb{R}^m} e^{\alpha|x|}\, \mu(dx) < \infty,$$

and, for every $x \in \mathbb{R}^m$,

$$\left\|P^t(x,\cdot) - \mu\right\|_{TV} \le c_1 e^{-c_2 t}\left(e^{\alpha|x|} + \int_{\mathbb{R}^m} e^{\alpha|y|}\,\mu(dy)\right), \quad t \ge 0. \tag{3.4.17}$$

The proof repeats, with obvious changes, the proof of Theorem 3.3.4, and is omitted.

Theorem 3.4.9. *Let* $\kappa \in (-1, 0)$. *Then for any* $\alpha > 0$ *such that*

$$\frac{\alpha(1+\kappa)}{2}\int_{|u|\le 1} |||b|||^2|u|^2\,\nu(du) + \int_{|u|>1} |||b||||u|\left(e^{\alpha|||b|||^{1+\kappa}|u|^{1+\kappa}} - 1\right)\nu(du) < A_\kappa$$

there exist $c_1, c_2 > 0$ *such that, for any* $x, y \in \mathbb{R}^m$,

$$\left\|P^t(x,\cdot) - P^t(y,\cdot)\right\|_{TV} \le c_1 e^{-c_2 t^{(1+\kappa)/(1-\kappa)}}\left(e^{\alpha|x|^{1+\kappa}} + e^{\alpha|y|^{1+\kappa}}\right), \quad t \ge 0. \tag{3.4.18}$$

In addition, there exists unique IPM μ *for* X, *this measure satisfies*

$$\int_{\mathbb{R}^m} e^{\alpha|x|^{1+\kappa}}\,\mu(dx) < \infty,$$

and, for every $x \in \mathbb{R}^m$,

$$\left\|P^t(x,\cdot) - \mu\right\|_{TV} \le c_1 e^{-c_2 t^{(1+\kappa)/(1-\kappa)}}\left(e^{\alpha|x|^{1+\kappa}} + \int_{\mathbb{R}^m} e^{\alpha|y|^{1+\kappa}}\,\mu(dy)\right), \quad t \ge 0. \tag{3.4.19}$$

The proof repeats, with obvious changes, the proof of Theorem 3.3.5, and is omitted.

Theorem 3.4.10. *Let* $\kappa = -1$ *and*

$$2A_{-1} > |||b|||^2\int_{\mathbb{R}} |u|^2\,\nu(du).$$

Then for any $p > 2$ *such that*

$$2A_{-1} > (p-1)|||b|||^2\int_{\mathbb{R}} |u|^2\,\nu(du),$$

there exist $c_1, c_2 > 0$ *such that, for any* $x, y \in \mathbb{R}^m$,

$$\left\|P^t(x,\cdot) - P^t(y,\cdot)\right\|_{TV} \le c_1(1+c_2 t)^{-p/2}\left(1 + |x|^p + |y|^p\right), \quad t \ge 0. \tag{3.4.20}$$

In addition, there exists unique IPM μ for X, this measure satisfies

$$\int_{\mathbb{R}^m} |x|^{p-2}\,\mu(dx) < \infty,$$

and, for every $x \in \mathbb{R}^m$,

$$\left\|P^t(x,\cdot) - \mu\right\|_{TV} \le c_1(1+c_2t)^{1-p/2}\left(1+|x|^{p-2}+\int_{\mathbb{R}^m}|y|^{p-2}\,\mu(dy)\right),\quad t \ge 0. \tag{3.4.21}$$

The proof repeats, with obvious changes, the proof of Theorem 3.3.6, and is omitted.

The last two theorems concern the "heavy tails" case. In these theorems, eq. (3.4.9) is assumed to hold true with some $\alpha > 0$ and the drift coefficient a of the original SDE eq. (3.4.1) is assumed to satisfy the drift condition eq. (3.3.4) with the index κ.

Theorem 3.4.11. *Let $\kappa \ge 1$.*

Then for any $p \in (0, \alpha)$ there exist $c_1, c_2 > 0$ such that, for any $x, y \in \mathbb{R}^m$,

$$\left\|P^t(x,\cdot) - P^t(y,\cdot)\right\|_{TV} \le c_1e^{-c_2t}\left(|x|^p + |y|^p\right),\quad t \ge 0. \tag{3.4.22}$$

In addition, there exists unique IPM μ for X, this measure satisfies

$$\int_{\mathbb{R}^m} |x|^{p}\,\mu(dx) < \infty,$$

and, for every $x \in \mathbb{R}^m$,

$$\left\|P^t(x,\cdot) - \mu\right\|_{TV} \le c_1e^{-c_2t}\left(|x|^{p}+\int_{\mathbb{R}^m}|y|^{p}\,\mu(dy)\right),\quad t \ge 0. \tag{3.4.23}$$

The proof repeats, with obvious changes, the proof of Theorem 3.3.4, and is omitted.

Theorem 3.4.12. *Let $\kappa < 1$ be such that eq. (3.4.14) hold.*

Then for any $p \in (1-\kappa, \alpha)$ there exist c_1, $c_2 > 0$ such that, for any x, $y \in \mathbb{R}^m$,

$$\left\|P^t(x,\cdot) - P^t(y,\cdot)\right\|_{TV} \le c_1(1+c_2t)^{-p/(1-\kappa)}\left(1+|x|^p + |y|^p\right),\quad t \ge 0. \tag{3.4.24}$$

In addition, there exists unique IPM μ for X, this measure satisfies

$$\int_{\mathbb{R}^m} |x|^{p+\kappa-1}\,\mu(dx) < \infty,$$

and, for every $x \in \mathbb{R}^m$,

$$\|P^t(x,\cdot) - \mu\|_{TV} \le c_1(1+c_2t)^{-(p+\kappa-1)/(1-\kappa)} \times \left(1 + |x|^{p+\kappa-1} + \int_{\mathbb{R}^m} |y|^{p+\kappa-1}\,\mu(dy)\right), \quad t \ge 0. \tag{3.4.25}$$

Proof. The proof repeats, with minimal changes, the proof of Theorem 3.3.6. Namely, by Proposition 3.4.5 and Theorem 3.2.3, the skeleton chain X^h satisfies eq. (3.2.6) with $\phi^h(v) = av^\sigma$, where $\sigma = 1 - (1-\kappa)/p$ and a is a positive constant. In addition,

$$C_1V(x) \le V^h(x) \le C_2V(x), \quad V(x) = |x|^p.$$

We have

$$\frac{1}{1-\sigma} = \frac{p}{1-\kappa}, \quad \frac{\sigma}{1-\sigma} = \frac{p}{1-\kappa} - 1 = \frac{p+\kappa-1}{1-\kappa},$$

and we deduce the required statement from Corollary 2.8.10. □

3.5 Comments

1. The skeleton chain approach, developed in this section, is not the only possible one in the continuous-time setting. Another natural possibility is just to repeat, with proper modifications, the coupling construction used in the proof of Theorem 2.8.6 directly for the continuous-time process, and then perform the similar calculation based on the decomposition of the trajectory of the coupled process into coupling segments and excursions away from the irreducibility set. Such an approach is more direct and typically leads to shorter proofs. It is also more flexible, because it allows one to stop the excursion immediately once the irreducibility set is reached, on the contrary to the skeleton chain approach, which actually reduces the range of the corresponding hitting times to the grid $\{kh, k \ge 0\}$. In some sophisticated models such a possibility is quite important, see Ref. [18], where exponential stabilization was established for a diffusive system of hard balls. A hidden difficulty in such an approach is that the construction of the coupling now should be more sophisticated; typically, the "excursion away from the irreducibility set" stage is adopted to be an independent coupling. It is far from being clear whether this construction can be further extended to cover the *weak* ergodicity, as well: the problem here is that, in this more general setting, one has to take care that the coordinates of the coupled process do not diverge too far during the "excursion" stage. This is the reason for us to focus our exposition at the skeleton chain approach, which has a direct extension to the weak ergodicity setting, see Chapter 4.

2. Lemma 3.2.4 is our main tool to link the discrete- and the continuous-time settings. In its current form, it is an improved version of the corresponding part of the proof of Ref. [85, Theorem 1.23], which in turn was motivated by Ref. [30]. The main improvement of Lemma 3.2.4 compared to Ref. [85, Theorem 1.23] is actually contained in Definition 3.2.1, where process eq. (3.2.4) is allowed to be a *local* martingale, while Ref. [85, Definition 1.21] requires the true martingale property. Such a modification appears quite substantial and useful; for instance, it allows us to apply the Itô formula in Sections 3.3 and 3.4 directly, without checking auxiliary uniform integrability conditions. For similar statements, see Refs. [53, Section 4.1.2] and [37, Lemma 2].
3. For diffusion processes, stability of transition probabilities was first established by R. Has'minskii, see Ref. [73]. Has'minskii's stability result assumes that, for the hitting time τ_B of a bounded ball by the diffusion process, the function

$$\mathsf{E}_x\tau_B, \quad x \in \mathbb{R}^m$$

is locally bounded; note that this condition has a clear relation with the positive recurrence condition from Theorem 1.2.5. Has'minskii's result does not specify any rate of stability, and this has a further relation with the type of limit theorems obtained as an application, see a more detailed discussion in Section 5.4. Various types of ergodic rates for diffusions were systematically studied by A. Veretennikov, see the series of papers Ref. [76, 131, 132] and the survey in Ref. [133]. The aim of Section 3.3 is to make a systematic presentation of these results in three most important cases, that is, for exponential, sub-exponential, and polynomial ergodic rates. The skeleton chain approach, based on Theorem 3.2.3, appears to be strong enough to cover, in a unified way, all these cases, and furthermore to provide a similar set of results for solutions to Lévy-driven SDEs. We note that essentially the same argument remains applicable in for stochastic delay equations, see Section 4.6.
4. Both Has'minskii's stability result and Veretennikov's ergodic rates for diffusions are formulated under the moment assumptions for the hitting times, similar to the recurrence condition **R** in Theorem 2.7.2. In "regular" models similar to those studied in Section 3.3, there is no actual difference between such "integral-type" conditions and the "differential-type" Lyapunov condition eq. (3.2.5). The latter condition is now widely adopted in the literature, though, one has to keep in mind that in more sophisticated settings the difference between these conditions can become substantial. For instance, for the system of diffusive hard balls in Ref. [18], Lyapunov condition eq. (3.2.5) appears to be hardly verifiable because of presence of the local time terms, which correspond to possible collisions of the balls, while an "integral-type" recurrence condition can be verified efficiently.

4 Weak Ergodic Rates

In many cases of interest, the theory developed in the previous chapters is not applicable because the target process X does not satisfy the Dobrushin condition (or other irreducibility-type assumptions discussed in Section 2.9), which is a principal ingredient in the entire approach. However, it still may happen that the process is ergodic, but in a weaker sense; that is, it possesses a unique IPM, and its transition probabilities weakly converge as $t \to \infty$ to this IPM. This chapter is devoted to the study of this "weak ergodicity" property.

4.1 Markov Models with Intrinsic Memory

In this section, we consider two examples which motivate the subsequent studies.

Example 4.1.1. Let $X = \left(X^1, X^2\right)$ be a process in $\mathbb{R}^2$ which solves the system of SDEs

$$\begin{cases} dX_t^1 = -aX_t^1\, dt + dW_t \\ dX_t^2 = -aX_t^2\, dt + dW_t \end{cases},$$

where $a > 0$, the components X^1, X^2 have different initial values x^1, x^2 and the Wiener process W is the same for both components. The process X can be given explicitly:

$$\begin{pmatrix} X_t^1 \\ X_t^2 \end{pmatrix} = e^{-at}\begin{pmatrix} x^1 \\ x^2 \end{pmatrix} + \left(\int_0^t e^{-at+as}\, dW_s\right)\begin{pmatrix} 1 \\ 1 \end{pmatrix}.$$

From this formula, one can easily derive two principal properties of the system. First, if we take two initial points $x = (x^1, x^2)$, $y = (y^1, y^2)$ with

$$x^1 - x^2 \neq y^1 - y^2,$$

then, for every $t > 0$, the respective transition probabilities $P^t(x, \cdot)$, $P^t(y, \cdot)$ are supported by the following disjoint sets of $\mathbb{R}^2$

$$\left\{z = \left(z^1, z^2\right) : z^1 - z^2 = e^{-at}\left(x^1 - x^2\right)\right\}, \qquad \left\{z = \left(z^1, z^2\right) : z^1 - z^2 = e^{-at}\left(y^1 - y^2\right)\right\}.$$

Hence, the total variation distance between them remains equal to 2. Next, because $e^{-at} \to 0$, $t \to \infty$, for *any* $x = (x^1, x^2)$, we have

$$\begin{pmatrix} X_t^1 \\ X_t^2 \end{pmatrix} \Rightarrow \left(\int_0^\infty e^{-as}\, dW_s\right)\begin{pmatrix} 1 \\ 1 \end{pmatrix}.$$

DOI 10.1515/9783110458930-005

This means that, for every $x \in \mathbb{R}^2$, transition probabilities $P^t(x, \cdot)$ converge weakly as $t \to \infty$ to the (unique) IPM, which is concentrated on the diagonal in $\mathbb{R}^2$.

Note that if $x^1 \neq x^2$, the above argument also shows that $P^t(x, \cdot)$ and the IPM are mutually singular; hence, the convergence does not hold true in the sense of the total variation distance.

One can consider the SDE for $X = (X^1, X^2)$ as a particular case of the SDE (3.3.1). Then $m = 2$, $k = 1$,

$$a(x) = -a\begin{pmatrix} x^1 \\ x^2 \end{pmatrix} \qquad b(x) = \begin{pmatrix} 1 \\ 1 \end{pmatrix},$$

and a possible "technical" explanation of the lack of ergodicity in the total variation norm is that the system is not irreducible because of the degeneracy of the diffusion matrix

$$B(x) = \begin{pmatrix} 1 & 1 \\ 1 & 1 \end{pmatrix}.$$

Heuristically, the system contains some partial "intrinsic memory": no matter how much time t has passed, the value X_t keeps the partial information about the initial point x; namely, the difference $x^1 - x^2$ can be recovered completely given the value X_t:

$$x^1 - x^2 = e^{at}(X_t^1 - X_t^2).$$

This simple example gives a natural insight into understanding the ergodic properties for more general and sophisticated Markov models. The process X, in fact, represents the two-point motion for the *stochastic flow* which naturally corresponds to the Ornstein–Uhlenbeck process

$$\mathrm{d}U_t = -aU_t\,\mathrm{d}t + \mathrm{d}W_t;$$

for example, Ref. [91]. Hence one can naturally expect that the ergodic properties observed in the above example should also be valid for stochastic flows generated by SDEs; in particular, because of the degeneracy of SDEs for N-point motions, the whole system typically would contain some partial intrinsic memory.

Example 4.1.2. This beautiful example comes from Ref. [123], see also the introduction to Ref. [55]. Consider the following real-valued *stochastic differential delay equation* (SDDE):

$$\mathrm{d}X_t = -a(X_t)\,\mathrm{d}t + b(X_{t-r})\,\mathrm{d}W_t \tag{4.1.1}$$

with a fixed $r > 0$. We will not go into the basics of the theory of SDDEs, referring the reader to Ref. [107]. We just mention that, in order to determine the values X_t, $t \geq 0$, one should initially specify the values X_s, $s \in [-r, 0]$, since otherwise the term $b(X_{t-r})$ is not well-defined for small positive t. Assuming $a(x)$, $b(x)$ to be Lipschitz continuous, one has that for every function $f \in C([-r, 0], \mathbb{R}^m)$, there exists a unique (strong) solution to eq. (4.1.1) with $X_s = f_s$, $s \in [-r, 0]$, which has continuous trajectories. In general, the process X_t, $t \geq 0$, unlike the diffusion process solution to eq. (3.3.1), is not Markov; instead, the *segment process* $\mathbf{X} = \{\mathbf{X}(t),\ t \geq 0\}$

$$\mathbf{X}(t) = \left\{X_{t+s}, s \in [-r, 0]\right\} \in C([-r, 0], \mathbb{R}^m), \qquad t \geq 0$$

possesses the Markov property.

Denote by $\mathcal{F}_{=t}$ the completion of $\sigma\big(\mathbf{X}(t)\big)$. Any random variable measurable w.r.t. $\mathcal{F}_{=t}$ a.s. equals a Borel-measurable functional on $C([-r, 0], \mathbb{R}^m)$ applied to $X(t)$. We fix $t > 0$ and apply to the segment X_{t+s}, $s \in [-r, 0]$ the well-known statistical procedure, which makes it possible, given a trajectory of an Itô-type process, to estimate consistently the variance of the martingale part of the process. Namely, we put

$$V_{t,n}(s) = \sum_{k=1}^{[2^n(-s/r)]} \left(X_{t-k2^{-n}r} - X_{t-(k-1)2^{-n}r}\right)^2, \qquad s \in [-r, 0],$$

and obtain that, with probability 1,

$$V_{t,n}(s) \to V_t(s) = \int_{t+s}^{t} b^2(X_{v-r})\, dv, \qquad s \in [-r, 0].$$

Consequently, for every $s \in [-r, 0]$,

$$b^2(X_{t+s-r}) = \lim_{\varepsilon \to 0} \frac{V_{t,n}(s+\varepsilon) - V_{t,n}(s)}{\varepsilon}$$

belongs to $\mathcal{F}_{=t}$. If we assume that b is positive and strictly monotone, then the above argument shows that every value X_{t+s-r}, $s \in [-r, 0]$ of the segment $\mathbf{X}(t-r)$ is $\mathcal{F}_{=t}$-measurable. This means that $\mathbf{X}(t-r)$ can be recovered uniquely given the value $\mathbf{X}(t)$. Repeating this argument, we obtain that the initial value $\mathbf{X}(0)$ of the segment process can be recovered uniquely given the value $\mathbf{X}(t)$; hence, any two transition probabilities for the segment process $\mathbf{X}$ with different initial values are mutually singular. This means that the Markov system described by $\mathbf{X}$ contains the full "intrinsic memory," which clearly prohibits this system from converging to an invariant distribution in the total variation norm.

The above argument does not prohibit $\mathbf{X}$ to be ergodic in a weaker sense, likewise to the process from Example 4.1.1. We postpone to the subsequent sections a detailed discussion of the available methods, which allow one to prove such a weak ergodicity. Here we just mention that the presence of an "intrinsic memory" is a typical feature for

Markov systems with "complicated" state spaces, like $\mathbb{X} = C([-r, 0], \mathbb{R}^m)$ in the above example. The same effect can be observed for stochastic partial differential equations (SPDEs), SDEs driven by fractional noises, and so on. Example 4.1.1 indicates one possible heuristic reason for such an effect: for systems with "complicated" state spaces, there are many possibilities for the noise to degenerate, in a sense, and therefore for the whole system to not be irreducible.

4.2 Dissipative Stochastic Systems

In this section, we discuss one important particular situation, where weak ergodicity can be derived directly, with the argument based on the stochastic calculus tools, only.

4.2.1 "Warm-up" Calculation: Diffusions

Consider SDE (3.3.1), and assume its coefficients $a(x), b(x)$ to be Lipschitz continuous. Then this equation well defines a diffusion process X. Since the *strong* solution to SDE (3.3.1) is uniquely defined, for any $x, y \in \mathbb{R}^m$ we can consider two solutions X, Y to eq. (3.3.1) with the same Wiener process W and $X_0 = x, Y_0 = y$.

Proposition 4.2.1. *For any $p \geq 1$, the pair X, Y specified above satisfies*

$$\mathsf{E}|X_t - Y_t|^p \leq |x - y|^p e^{\theta_p t},$$

where

$$\theta_p = p \sup_{x \neq y} \frac{1}{|x - y|^2} \left[\big(a(x) - a(y), x - y\big) + \frac{m(p-1)}{2} \|b(x) - b(y)\|^2 \right].$$

Proof. Denote $\Delta_t = X_t - Y_t$, then

$$\mathrm{d}\Delta_t = \big(a(X_t) - a(Y_t)\big)\, \mathrm{d}t + \big(b(X_t) - b(Y_t)\big)\, \mathrm{d}W_t.$$

If $p \geq 2$, then the Itô formula applied to the C^2-function $F(x) = |x|^p$ gives

$$\begin{aligned} \mathrm{d}|\Delta_t|^p &= p|\Delta_t|^{p-2}\big(a(X_t) - a(Y_t), X_t - Y_t\big)\, \mathrm{d}t \\ &\quad + \frac{p}{2}|\Delta_t|^{p-2} \sum_{i,j=1}^m \sum_{l=1}^k \big(b_{il}(X_t) - b_{il}(Y_t)\big)\big(b_{jl}(X_t) - b_{jl}(Y_t)\big) \\ &\quad \times \left(\delta_{ij} + (p-2)\frac{\Delta_t^i \Delta_t^j}{|\Delta_t|^2} \right) \mathrm{d}t + \mathrm{d}M_t, \end{aligned}$$

where M is a local martingale. For any $m\times k$-matrices B,C and $m\times m$ matrix D, we have

$$\left|\sum_{i,j=1}^{m}\sum_{l=1}^{k}B_{il}C_{jl}D_{ij}\right|\leq m\|B\|\|C\|\|D\|. \tag{4.2.1}$$

Since

$$\left\|\left(\delta_{ij}+(p-2)\frac{\Delta_s^i\Delta_s^j}{|\Delta_s|^2}\right)_{i,j=1}^{m}\right\|\leq p-1, \tag{4.2.2}$$

we conclude that

$$\mathrm{d}|\Delta_t|^p=\eta_t|\Delta_t|^p\,\mathrm{d}t+\mathrm{d}M_t$$

with a local martingale M and

$$\eta_t\leq\theta_p,\quad t\geq 0.$$

Applying the Itô formula once again, we get that

$$e^{-\theta_p t}|\Delta_t|^p$$

is a local super-martingale. This means that, for the corresponding localizing sequence $\{\tau_n\}$, one has

$$\mathsf{E}\left(e^{-\theta_p(t\wedge\tau_n)}|\Delta_{t\wedge\tau_n}|^p\right)\leq|x-y|^p,\quad t\geq 0.$$

Taking $n\to\infty$ and applying the Fatou lemma, we prove the required inequality.

For $p\in[1,2)$, essentially the same argument applies, with the minor difference caused by the fact that the function $F(x)=|x|^p$ does not belong to $C^2(\mathbb{R}^m)$. However, its only irregularity point is 0, and once Δ_t hits 0, it stays equal 0. The latter follows by the strong Markov property and uniqueness of the solution to eq. (3.3.1). The required statement then follows by the same argument, where the Itô formula should be applied locally, that is, up to the first time for Δ to hit 0; we omit further details here. □

Proposition 4.2.1 shows that, for any diffusion process with Lipschitz coefficients, the L_p-distance between the realizations of the process, with different initial conditions and the same noise, do not expand faster than at an exponential rate. In some cases we can guarantee that

$$\theta_p<0, \tag{4.2.3}$$

and then Proposition 4.2.1 actually gives a *contraction* of the L_p-distance at exponential rate. Namely, assume the coefficient a to satisfy the following "drift dissipativity condition":

$$\sup_{x \neq y} \frac{(a(x) - a(y), x - y)}{|x - y|^2} < 0. \tag{4.2.4}$$

Then for any Lipschitz continuous b we have eq. (4.2.3) for $p > 1$ sufficiently close to 1.

This effect has the following natural interpretation. Consider eq. (3.3.1) as a deterministic dynamical system, defined by the ODE

$$\mathrm{d}X_t = a(X_t)\,\mathrm{d}t, \tag{4.2.5}$$

perturbed by a stochastic term $b(X_t)\,\mathrm{d}W_t$. Condition (4.2.4) is a simplest one which provides an exponential contraction between the trajectories of system (4.2.5), and adding a Lipschitz continuous stochastic perturbation to the system do not spoil such a contraction (which now, however, should be understood in the sense of proper L_p-distance). This effect is not model-specific: once one has a dissipative (in a proper sense) deterministic model, it is typical that a (Lipschitz continuous) stochastic perturbation do not spoil the dissipativity property. For infinite-dimensional SDEs such results can be found in Refs. [20, Chapter 11.5] and [115, Chapter 16.2]. In the next section, we prove such a result for an SDDE, thus giving one possible way to justify a weak ergodicity claimed in Example 4.1.2.

4.2.2 Dissipativity for a Solution to SDDE

Consider SDDE (4.1.1) with Lipschitz continuous coefficients

$$a : \mathbb{R}^m \to \mathbb{R}^m, \quad b : \mathbb{R}^m \to \mathbb{R}^{m \times k}.$$

To make the notation consistent, in what follows we denote the points of the state space $C([-r, 0], \mathbb{R}^m)$ of the segment process $\mathbf{X}$ by $\mathbf{x}, \mathbf{y}, \ldots$ rather than $f, g, \ldots$.

Proposition 4.2.2. *For every $p \geq 4$, there exists a positive C such that, for any $\mathbf{x}, \mathbf{y} \in C([-r, 0], \mathbb{R}^m)$, respective solutions X, Y to eq. (4.1.1) satisfy*

$$\mathsf{E}\big\|\mathbf{X}(t) - \mathbf{Y}(t)\big\|_C^p \leq C e^{\theta_p t} \|\mathbf{x} - \mathbf{y}\|_C^p, \qquad t \geq 0, \tag{4.2.6}$$

where

$$\|\mathbf{x}\|_C = \sup_{s \in [-r, 0]} |\mathbf{x}_s|$$

and

$$\theta_p = p \sup_{x \neq y} \frac{1}{|x-y|^2} \Big[\big(a(x) - a(y), x - y \big) + \frac{m(p-1)}{2} \big\| b(x) - b(y) \big\|^2 \Big].$$

Proof. The first part of the proof essentially repeats the proof of Proposition 4.2.1. Namely, we apply the Itô formula to the process $\Delta_t = X_t - Y_t$, $t \geq 0$ with the function $F(t, x) = |x|^p$:

$$\begin{aligned}
|\Delta_t|^p = |\Delta_0|^p &+ p \int_0^t |\Delta_s|^{p-2} \big(a(X_s) - a(Y_s), \Delta_s \big) \,\mathrm{d}s \\
&+ \frac{p}{2} \int_0^t |\Delta_s|^{p-2} \sum_{i,j=1}^m \sum_{l=1}^k \big(b_{il}(X_{s-r}) - b_{il}(Y_{s-r}) \big) \\
&\qquad \times \big(b_{jl}(X_{s-r}) - b_{jl}(Y_{s-r}) \big) \left(\delta_{ij} + (p-2) \frac{\Delta_s^i \Delta_s^j}{|\Delta_s|^2} \right) \mathrm{d}s + \mathrm{d}M_t,
\end{aligned}$$

where M is a local martingale. We have by eqs. (4.2.1) and (4.2.2), and the Young inequality

$$\begin{aligned}
|\Delta_s|^{p-2} &\sum_{i,j=1}^m \sum_{l=1}^k \big(b_{il}(X_{s-r}) - b_{il}(Y_{s-r}) \big) \\
&\qquad \times \big(b_{jl}(X_{s-r}) - b_{jl}(Y_{s-r}) \big) \left(\delta_{ij} + (p-2) \frac{\Delta_s^i \Delta_s^j}{|\Delta_s|^2} \right) \qquad (4.2.7) \\
&\leq m(p-1) |\Delta_s|^{p-2} \| b(X_{s-r}) - b(Y_{s-r}) \|^2 \\
&\leq m(p-1) \|b\|_{\mathrm{Lip}}^2 \left(\frac{(p-2)|\Delta_s|^p}{p} + \frac{2|\Delta_{s-r}|^p}{p} \right).
\end{aligned}$$

In addition,

$$\int_0^t |\Delta_{s-r}|^p \,\mathrm{d}s = \int_{-r}^{t-r} |\Delta_s|^p \,\mathrm{d}s \leq \int_0^t |\Delta_s|^p \,\mathrm{d}s + \int_{-r}^0 |\mathbf{x}_s - \mathbf{y}_s|^p \,\mathrm{d}s.$$

Summarizing the above inequalities, we get

$$|\Delta_t|^p = |\mathbf{x}_0 - \mathbf{y}_0|^p + m(p-1)\|b\|_{\mathrm{Lip}}^2 \int_{-r}^0 |\mathbf{x}_s - \mathbf{y}_s|^p \,\mathrm{d}s + \int_0^t \eta_s |\Delta_s|^p \,\mathrm{d}s + M_t \qquad (4.2.8)$$

with

$$\eta_s \leq \theta_p.$$

Then the same argument which we used in the proof of Proposition 4.2.1, based on the localization procedure and the Fatou lemma, gives

$$\begin{aligned} \mathsf{E}|\Delta_t|^p &\le e^{\theta_p t}\left(|\mathbf{x}_0 - \mathbf{y}_0|^p + m(p-1)\|b\|^2_{\mathrm{Lip}} \int_{-r}^{0} |\mathbf{x}_s - \mathbf{y}_s|^p \, \mathrm{d}s \right) \\ &\le \left(1 + rm(p-1)\|b\|^2_{\mathrm{Lip}}\right) e^{\theta_p t} \|\mathbf{x} - \mathbf{y}\|_{C(-r,0)}. \end{aligned} \tag{4.2.9}$$

In the second part of the proof, we use the "point-wise" inequality (4.2.9) in order to give the bound for the $\|\cdot\|_{C(-r,0)}$-norm for the difference $\mathbf{X}(t) - \mathbf{Y}(t)$. We have

$$\sup_{s\in[-r,0]} |\Delta_{t+s}| \le \Delta_{t-r} + \sup_{s\in[-r,0]} |\Delta_{t+s} - \Delta_{t-r}|.$$

By the Itô formula, we have for $t \ge r$, $s \in [-r, 0]$

$$|\Delta_{t+s} - \Delta_{t-r}|^2 = \int_{t-r}^{t+s} \alpha_v \, \mathrm{d}v + \int_{t-r}^{t+s} \beta_v \, \mathrm{d}W_v,$$

where

$$|\alpha_v| \le 2\|a\|_{\mathrm{Lip}}|\Delta_v|^2 + m\|b\|_{\mathrm{Lip}}|\Delta_{v-r}|^2, \quad |\beta_v| \le 2\|b\|_{\mathrm{Lip}}|\Delta_v||\Delta_{v-r}|. \tag{4.2.10}$$

By the Hölder inequality,

$$\mathsf{E} \sup_{s\in[-r,0]} \left| \int_{t-r}^{t+s} \alpha_v \, \mathrm{d}v \right|^{p/2} \le r^{p/2-1} \int_{t-r}^{t} \mathsf{E}|\alpha_v|^{p/2} \, \mathrm{d}v,$$

and by the Burkholder–Davis–Gundy inequality (e.g., Ref. [69, Theorem 23.12])

$$\mathsf{E} \sup_{s\in[-r,0]} \left| \int_{t-r}^{t+s} \beta_v \, \mathrm{d}W_v \right|^{p/2} \le \mathsf{E} \left| \int_{t-r}^{t} \beta_v^2 \, \mathrm{d}v \right|^{p/4}.$$

Applying the Hölder inequality once again (this is the point, where we need that $p \ge 4$), we get

$$\mathsf{E} \sup_{s\in[-r,0]} \left| \int_{t-r}^{t+s} \beta_v \, \mathrm{d}W_v \right|^{p/2} \le r^{p/4-1} \int_{t-r}^{t} \mathsf{E}|\beta_v|^{p/2} \, \mathrm{d}v.$$

Combined with eqs. (4.2.9) and (4.2.10), this completes the proof of eq. (4.2.6) for $t \ge r$. For $t \in [0, r)$ the proof is similar and is simpler, and thus we omit the details. □

The discussion made at the end of the previous section well suits to the current SDDE setting, as well. Note that, since the second part of the proof requires $p \ge 4$, the drift dissipativity condition (4.2.4) itself is not sufficient to provide contraction of the L_p-distance between the solutions to the SDDE (4.1.1) with the same W. Such a contraction holds true, if

$$\theta_4 = 4 \sup_{x \neq y} \frac{1}{|x-y|^2} \left[\left(a(x) - a(y), x - y \right) + \frac{3m}{2} \left\| b(x) - b(y) \right\|^2 \right] < 0. \tag{4.2.11}$$

In that case we have an effect, which looks very similar to the "exponential loss of memory" one, which we had in Chapter 2 (see eq. (2.3.2)). It is natural to expect that such "weak loss of memory" bound should lead to "weak ergodic rates", likewise to the exponential ergodic rate (2.3.3), which was derived from eq. (2.3.2); see Theorem 2.3.1. In the subsequent sections, we develop the proper tools for the study of such effects.

4.3 Coupling Distances for Probability Measures

To develop analogues of "loss of memory" bound (2.3.2) and ergodic rate (2.3.3), valid in a weak sense, the first natural step would be to quantify weak convergence; that is, to specify a distance on the set $\mathscr{P}(\mathbb{X})$ of probability measures on $\mathbb{X}$, adapted to weak convergence in $\mathscr{P}(\mathbb{X})$. Respective theory of *probability metrics* is well-developed, and we do not pretend to expose its constructions and ideas here in details, referring a reader to Ref. [31, Chapter 11], or Ref. [136, Chapter 1]. However, the part of this theory related to *coupling* (or *minimal*) probability metrics will be crucial for our subsequent constructions, that is why we outline this topic in this section.

With a slight abuse of terminology and notation, we will call a pair of random variables ξ, η, defined on a common probability space and taking values in $\mathbb{X}$, a *coupling* for $\mu, \nu \in \mathscr{P}(\mathbb{X})$, and denote $(\xi, \eta) \in C(\mu, \nu)$, if

$$\mathrm{Law}(\xi) = \mu, \quad \mathrm{Law}(\eta) = \nu$$

That is, a pair (ξ, η) belongs to $C(\mu, \nu)$ if, and only if, its joint distribution κ is a coupling for μ, ν. Using the same name and notation both for variables and their laws will not cause misunderstanding, and in certain situations will be very convenient.

We call a *distance-like function* any measurable function $d : \mathbb{X} \times \mathbb{X} \to [0, \infty)$, which is symmetric and satisfies

$$d(x, y) = 0 \quad \Longleftrightarrow \quad x = y.$$

Denote by the same letter d the respective *coupling distance* on the class $\mathscr{P}(\mathbb{X})$, defined by

$$d(\mu, \nu) = \inf_{\kappa \in C(\mu, \nu)} \int_{\mathbb{X}} d(x, y) \kappa(dx, dy), \qquad \mu, \nu \in \mathscr{P}(\mathbb{X}). \tag{4.3.1}$$

We use the term "coupling distance" instead of "coupling metric" because, in general, $d : \mathscr{P}(\mathbb{X}) \times \mathscr{P}(\mathbb{X}) \to [0, \infty]$ may fail to satisfy the triangle inequality. However, the following statement shows that, once the initial distance-like function satisfies the triangle inequality or its weaker version, the coupling distance inherits the same property.

We say that a distance-like function is a *c-quasi-metric* for some $c \geq 1$ such that

$$d(x,z) \leq c\big(d(x,y) + d(y,z)\big), \qquad x,\ y,\ z \in \mathbb{X}.$$

Proposition 4.3.1. *If the distance-like function d is a c-quasi-metric, then the respective coupling distance d is a c-quasi-metric, as well.*

Proof. For any $\varepsilon > 0$, there exist $(\xi_\varepsilon, \eta_\varepsilon) \in C(\mu, \nu)$, $(\xi'_\varepsilon, \eta'_\varepsilon) \in C(\nu, \lambda)$ such that

$$\mathsf{E}d(\xi_\varepsilon, \eta_\varepsilon) \leq d(\mu, \nu) + \varepsilon, \qquad \mathsf{E}d(\xi'_\varepsilon, \eta'_\varepsilon) \leq d(\nu, \lambda) + \varepsilon.$$

The following useful fact is well known (e.g., Ref. [31, Problem 11.8.8]); for the reader's convenience, we sketch its proof after completing the proof of the proposition.

Lemma 4.3.2. *Let (ξ, η) and (ξ', η') be two pairs of random elements valued in a Borel-measurable space $(\mathbb{X}, \mathscr{X})$ such that η and ξ' have the same distribution. Then on a proper probability space, there exist three random elements $\zeta_1, \zeta_2, \zeta_3$ such that the law of (ζ_1, ζ_2) in $(\mathbb{X}\times\mathbb{X}, \mathscr{X}\otimes\mathscr{X})$ coincides with the law of (ξ, η) and the law of (ζ_2, ζ_3) coincides with the law of (ξ', η').*

Applying this fact, we get a triple $\zeta_1, \zeta_2, \zeta_3$, defined on a common probability space, such that

$$\mathsf{E}d(\zeta_1, \zeta_2) \leq d(\mu, \nu) + \varepsilon, \qquad \mathsf{E}d(\zeta_2, \zeta_3) \leq d(\nu, \lambda) + \varepsilon.$$

In addition, the laws of ζ_1, ζ_3 are equal to μ, λ, respectively. Hence

$$d(\mu, \lambda) \leq \mathsf{E}d(\zeta_1, \zeta_3) \leq c\Big(d(\mu, \nu) + d(\nu, \lambda) + 2\varepsilon\Big).$$

Because $\varepsilon > 0$ is arbitrary, this gives the required inequality

$$d(\mu, \lambda) \leq c\big(d(\mu, \nu) + d(\nu, \lambda)\big), \qquad \mu,\ \nu,\ \lambda \in \mathscr{P}(\mathbb{X}). \qquad \square$$

Proof of Lemma 4.3.2. We construct the joint law of the triple $(\zeta_1, \zeta_2, \zeta_3)$ using the representation of the laws of the pairs based on the *disintegration formula*. Since $\mathbb{X}$ is assumed have a measurable bijection to $[0, 1]$ with a measurable inverse, we can consider the case $\mathbb{X} = \mathbb{R}$, only. For any pair of random variables ξ, η, there exists a *regular version of the conditional probability* $P_{\eta|\xi}(x, dy)$ (e.g., Ref. [31, Chapter 10.2]), which is measurable w.r.t. x, is a probability measure w.r.t. dy, and satisfies

$$P(\xi \in A, \eta \in B) = \int_A P_{\eta|\xi}(x, B)\,\mu(dx), \quad A, B \in \mathscr{X},$$

where μ denotes the law of ξ. Let us define the joint law κ for the triple $(\zeta_1, \zeta_2, \zeta_3)$ by

$$\kappa(A_1 \times A_2 \times A_3) = \int_{A_2} P_{\xi|\eta}(x, A_1) P_{\eta'|\xi'}(x, A_3)\,\mu(dx), \qquad A_1, A_2, A_3 \in \mathscr{X},$$

where μ now denotes the same distribution for η and ξ'. This corresponds to the choice of the conditional probability $P_{(\zeta_1,\zeta_3)|\zeta_2}$ equal to the product measure

$$P_{\xi|\eta} \otimes P_{\eta'|\xi'}.$$

It is straightforward to verify that such a triple $(\zeta_1, \zeta_2, \zeta_3)$ satisfies the required properties. □

The definition of the coupling distance is strongly related to the classical Monge-Kantorovich *mass transportation problem* (e.g., Ref. [117]). Namely, given two mass distributions μ, ν and the transportation cost $d : \mathbb{X} \times \mathbb{X} :\to \mathbb{R}_+$, the coupling distance $d(\mu, \nu)$ introduced above represents exactly the minimal cost required to transport μ to ν. An important fact is that the "optimal transportation plan" in this problem exists under some natural topological assumptions on the model. In terms of couplings (which is just another name for transportation plans) and coupling distances, this fact can be formulated as follows.

Proposition 4.3.3. *Let* $\mathbb{X}$ *be a Polish space and the distance-like function* d *be lower semi-continuous; that is, for any sequences* $x_n \to x$, $y_n \to y$

$$d(x, y) \le \liminf_n d(x_n, y_n).$$

Then for any $\mu, \nu \in \mathscr{P}(\mathbb{X})$, *there exists a coupling* $(\xi_*, \eta_*) \in C(\mu, \nu)$ *such that*

$$d(\mu, \nu) = \mathsf{E} d(\xi_*, \eta_*). \tag{4.3.2}$$

In other words, "inf" in eq. (4.3.1) in fact can be replaced by "min." We call any pair $(\xi_*, \eta_*) \in C(\mu, \nu)$ which satisfies eq. (4.3.2) a *d-optimal coupling* and denote the class of the laws of d-optimal couplings by $C_{d,\mathrm{opt}}(\mu, \nu)$.

Proof of Proposition 4.3.3. Observe that the family of measures $C(\mu, \nu)$ is *tight* (e.g., Ref. [10]). Indeed, to construct a compact set $K \subset \mathbb{X} \times \mathbb{X}$ such that, for a given ε,

$$\kappa(K) \ge 1 - \varepsilon, \quad \kappa \in C(\mu, \nu)$$

one can simply choose two compact sets $K_1, K_2 \subset \mathbb{X}$ such that

$$\mu(K_1) \geq 1 - \frac{\varepsilon}{2}, \qquad \nu(K_2) \geq 1 - \frac{\varepsilon}{2},$$

and then take $K = K_1 \times K_2$.

Consider a sequence of pairs $\{(\xi_n, \eta_n)\} \subset C(\mu, \nu)$ such that

$$\mathsf{E}d(\xi_n, \eta_n) \leq d(\mu, \nu) + \frac{1}{n}.$$

Then by the Prokhorov theorem, there exists a subsequence $\{(\xi_{n_k}, \eta_{n_k})\}$ which converges in law to some pair (ξ_*, η_*). Then both sequences of components $\{\xi_{n_k}\}$, $\{\eta_{n_k}\}$ also converge in law to ξ_*, η_* respectively, and hence $(\xi_*, \eta_*) \in C(\mu, \nu)$. Next, by Skorokhod's "common probability space" principle (e.g., Ref. [31, Theorem 11.7.2]), there exists a sequence $\{(\tilde{\xi}_k, \tilde{\eta}_k)\}$ and a pair $(\tilde{\xi}_*, \tilde{\eta}_*)$, defined on the same probability space, such that the laws of respective pairs (ξ_{n_k}, η_{n_k}) and $(\tilde{\xi}_k, \tilde{\eta}_k)$ coincide and

$$\left(\tilde{\xi}_k, \tilde{\eta}_k\right) \to \left(\tilde{\xi}_*, \tilde{\eta}_*\right)$$

with probability 1. Then by the lower semi-continuity of d, one has

$$d\left(\tilde{\xi}_*, \tilde{\eta}_*\right) \leq \liminf_k d\left(\tilde{\xi}_k, \tilde{\eta}_k\right),$$

and hence the Fatou lemma gives

$$\mathsf{E}d\left(\tilde{\xi}_*, \tilde{\eta}_*\right) \leq \liminf_k \mathsf{E}d\left(\tilde{\xi}_k, \tilde{\eta}_k\right) = \liminf_k \mathsf{E}d(\xi_k, \eta_k) = d(\mu, \nu).$$

Because $\left(\tilde{\xi}_*, \tilde{\eta}_*\right)$ has the same law as $(\xi_*, \eta_*) \in C(\mu, \nu)$, this completes the proof. □

Let us give several typical examples. In what follows, $\mathbb{X}$ is a Polish space with the metric ρ.

Example 4.3.4. Let $d(x, y) = \rho(x, y)$; we denote the respective coupling distance by $W_{\rho,1}$ and discuss its properties. First, since ρ satisfies the triangle inequality, so does $W_{\rho,1}$. Next, $W_{\rho,1}$ is symmetric and nonnegative. Finally, it possesses the identification property:

$$W_{\rho,1}(\mu, \nu) = 0 \quad \Longleftrightarrow \quad \mu = \nu.$$

The part "$\Leftarrow$" of this statement is trivial; to prove the "$\Rightarrow$" part, we just notice that there exists an optimal coupling (ξ_*, η_*) for μ, ν: because d is continuous, we can apply Proposition 4.3.3. For this coupling, we have

$$\mathsf{E}d(\xi_*, \eta_*) = W_{\rho,1}(\mu, \nu) = 0,$$

and because d has the identification property, this means that, in fact, $\xi_* = \eta_*$ a.s. Hence their laws coincide.

We have just seen that for the coupling distance $W_{\rho,1}$ all the axioms of a metric hold true; the one detail which may indeed cause $W_{\rho,1}$ to not be a metric is that $W_{\rho,1}$, in general, may take value ∞. If ρ is *bounded*, this does not happen, and $W_{\rho,1}$ is a metric on $\mathscr{P}(\mathbb{X})$.

Example 4.3.5. Let $p > 1$ and $d(x,y) = \rho^p(x,y)$; we denote the respective coupling distance $W^p_{\rho,p}$ (this notation and related terminology will be discussed at the end of this section). Similarly to the case $d = 1$, this coupling distance is symmetric, nonnegative and possesses the identification property. The triangle inequality for ρ and the elementary inequality

$$(a+b)^p \le 2^{p-1}(a^p + b^p), \quad a, b \ge 0$$

yield that $d(x,y) = \rho^p(x,y)$ is a 2^{p-1}-quasi-metric. By Proposition 4.3.1, $W^p_{\rho,p}$ is a 2^{p-1}-quasi-metric, as well.

The coupling distance $W^p_{\rho,p}$ is very convenient in various situations similar to those considered in Section 4.2. Indeed, let X, Y be the solutions to SDE (3.3.1) with the same Wiener process and $X_0 = x$, $Y_0 = y$. Clearly, for any t the pair (X_t, Y_t) is a coupling for

$$\mu = P^t(x,\cdot), \quad \nu = P^t(y,\cdot).$$

Thus Proposition 4.2.1 immediately leads to the following bound for the transition probabilities of the process X:

$$W^p_{\rho,p}\Big(P^t(x,\cdot), P^t(y,\cdot)\Big) \le e^{\theta_p t}\rho^p(x,y), \quad \rho(x,y) = |x-y|.$$

Similarly, Proposition 4.2.2 gives the following bound for the transition probabilities of the segment process $\mathbf{X}$:

$$W^p_{\rho,p}\Big(P^t(\mathbf{x},\cdot), P^t(\mathbf{y},\cdot)\Big) \le Ce^{\theta_p t}\rho^p(\mathbf{x},\mathbf{y}), \quad \rho(\mathbf{x},\mathbf{y}) = \|\mathbf{x}-\mathbf{y}\|_C.$$

Example 4.3.6. Let $d(x,y) = 1_{x\neq y}$ be the *discrete metric* on $\mathbb{X}$. Then, by the Coupling lemma,

$$d(\mu,\nu) = \frac{1}{2}\|\mu-\nu\|_{TV}.$$

Hence the total variation distance is a particular representative of the class of coupling distances.

Note that the function $d(x,y) = 1_{x \neq y}$ is lower semi-continuous; hence, Proposition 4.3.3 covers this particular case and actually should be understood as an extended version of the Coupling lemma. In what follows, we will apply essentially the same coupling approach, which was developed in previous chapters, in order to obtain the weak ergodic rates. This would naturally require a construction of a "d-greedy" Markov coupling for a chain with given transition probability $P(x, dy)$; that is, a two-component Markov chain such that its transition probability satisfies

$$Q\big((x,y), \cdot\big) \in C_{d,\mathrm{opt}}\big(P(x, \cdot), P(y, \cdot)\big).$$

In the case $d(x,y) = 1_{x \neq y}$ such a construction, given in Theorem 2.2.4, was comparatively simple, since in the proof of the Coupling lemma (Theorem 2.2.2) the construction of the optimal coupling was given explicitly. The current setting is more sophisticated, since the proof of Proposition 4.3.3 is not a constructive one. We postpone the detailed discussion of this important topic to the next section.

At the end of this section, we briefly discuss some important and closely related topics, which however will not be used in the subsequent exposition. Our construction of a coupling (minimal) distance is in some aspects more restrictive than the one available in the literature. Generally, such a distance is defined as

$$\inf_{(\xi,\eta) \in C(\mu,\nu)} H(\xi, \eta),$$

where H is an analogue of a distance-like function on a class of random variables, defined on a common probability space. The particular choice

$$H(\xi, \eta) = \mathsf{E} d(\xi, \eta)$$

leads to the definition used above. Another natural choice of H is the L_p-distance

$$H(\xi, \eta) = \Big(\mathsf{E} d^p(\xi, \eta)\Big)^{1/p}, \quad p > 1.$$

Observe that such a distance H possesses the triangle inequality if d does as well, and hence, the respective coupling distance inherits this property; the proof here is the same as in Proposition 4.3.1. The probability metric

$$W_{p,p}(\mu, \nu) = \inf_{\kappa \in C(\mu,\nu)} \left(\int d^p(x,y)\, \kappa(dx, dy)\right)^{1/p}$$

is called the L_p-Kantorovich(-Wasserstein) distance on $\mathscr{P}(\mathbb{X})$.

One good reason to use coupling distances is their convenience for estimation purposes. We have already seen this in the previous chapter: to bound the total variation distance between the laws of X_n with various starting points, we have constructed the pair of processes with the prescribed law of the components, and thus transformed the initial problem to estimating the probability $\mathsf{P}(X_n^1 \neq X_n^2)$. A similar argument appears to be practical in other frameworks and for other coupling distances. This is the reason why it is very useful that some natural probability metrics possess a coupling representation. We have already seen one such example: the coupling representation given by the Coupling lemma for the total variation distance. Another example is the *Lipschitz metric*

$$d_{\mathrm{Lip}}(\mu,\nu) = \sup_{f:\|f\|_{\mathrm{Lip}}=1} \left| \int f\, d\mu - \int f\, d\nu \right|;$$

the coupling representation here is given by the *Kantorovich–Rubinshtein (duality) theorem*, which states that

$$d_{\mathrm{Lip}} = W_{\rho,1}.$$

Finally, for the classical *Lévy–Prokhorov metric*

$$d_{\mathrm{LP}}(\mu,\nu) = \inf\left\{ \varepsilon > 0 : \mu(A) \le \nu\big(\{y : \rho(y,A) \le \varepsilon\}\big) + \varepsilon,\ A \in \mathscr{X} \right\},$$

the *Strassen theorem* gives the coupling representation

$$d_{\mathrm{LP}}(\mu,\nu) = \inf_{(\xi,\eta)\in C(\mu,\nu)} H_{\mathrm{KF}}(\xi,\eta),$$

where H_{KF} stands for the *Ky Fan metric*

$$H_{\mathrm{KF}}(\xi,\eta) = \inf\left\{ \varepsilon > 0 : \mathsf{P}\big(\rho(\xi,\eta) > \varepsilon\big) < \varepsilon \right\}.$$

For a detailed exposition of this topic, we refer to Ref. [31, Chapter 11].

4.4 Measurable Selections and General Coupling Lemma for Probability Kernels

In what follows, we always assume d to be lower semi-continuous, which makes it sure that for any pair $\mu,\nu \in \mathscr{P}(\mathbb{X})$ there exists a d-optimal coupling; see Proposition 4.3.1. Our aim in this section is, given a probability kernel P on $\mathbb{X}$, to construct a probability kernel Q on $\mathbb{X}\times\mathbb{X}$ such that, for each x, y, the measure $Q((x,y),\cdot)$ is a d-optimal coupling for $P(x,\cdot), P(y,\cdot)$. Since the set of such d-optimal couplings for given x, y may contain more than one point, the choice of $Q((x,y),\cdot)$ is not unique, and the main difficulty is that the required function should be *measurable* in (x,y).

This problem can be specified as follows. Consider the spaces $\mathbb{S} = \mathscr{P}(\mathbb{X}) \times \mathscr{P}(\mathbb{X})$, $\mathbb{S}' = \mathscr{P}(\mathbb{X} \times \mathbb{X})$ and the set-valued mapping $\psi : \mathbb{S} \ni s \mapsto \psi(s) \subset \mathbb{S}'$,

$$\psi\big((\mu, \nu)\big) = C_{d,\mathrm{opt}}(\mu, \nu). \tag{4.4.1}$$

The aim is to choose a *selector* ϕ for this mapping, that is, a function $\phi : \mathbb{S} \to \mathbb{S}'$ such that

$$\phi(s) \in \psi(s), \qquad s \in \mathbb{S},$$

which in addition should be measurable.

In general, the *measurable selection* problem may be is quite complicated, and in some cases it even may fail to have a solution. We refer to Ref. [34, Appendix 3] for a short and transparent, but very informative exposition of the measurable selection topic, and to §3 therein for a counterexample where the measurable selector does not exist. We also refer a reader, who is deeply interested in the general measurable selection topic, to an excellent survey paper [134].

Fortunately, the particular spaces $\mathbb{S}$, $\mathbb{S}'$ and the set-valued mapping ψ, which we have introduced, possess fine topological properties, which make it possible to solve the measurable selection problem. Below, we briefly recall such properties, referring for more details to Ref. [31, Chapter 11].

Let $(\mathbb{X}, \rho)$ be a *Polish* space; that is, a separable and complete metric space. Replacing, if needed, ρ by $\rho \wedge 1$, we can assume ρ to be bounded. We denote by the same symbol ρ the respective coupling distance $W_{\rho,1}$ on $\mathscr{P}(\mathbb{X})$; see Example 4.3.4.

Proposition 4.4.1.
1. *$(\mathscr{P}(\mathbb{X}), \rho)$ is a Polish space.*
2. *Convergence w.r.t. ρ in $\mathscr{P}(\mathbb{X})$ is equivalent to the weak convergence.*

Proof. By Proposition 4.3.1, $(\mathscr{P}(\mathbb{X}), \rho)$ is a metric space. If $\mu_n \Rightarrow \mu$, then the "common probability space" principle and the Lebesgue theorem on dominated convergence provide $\rho(\mu_n, \mu) \to 0$. On the other hand, if $\rho(\mu_n, \mu) \to 0$, then for any Lipschitz continuous function $f : \mathbb{X} \to \mathbb{R}$ and any $(\xi, \eta) \in C(\mu, \nu)$

$$\left|\int f\, d\mu_n - \int f\, d\mu\right| = |\mathsf{E}f(\xi) - \mathsf{E}f(\eta)| \le \mathsf{E}|f(\xi) - f(\eta)| \le \|f\|_{\mathrm{Lip}} \mathsf{E}\rho(\xi, \eta),$$
$$\|f\|_{\mathrm{Lip}} = \sup_{x \ne y} \frac{|f(x) - f(y)|}{\rho(x, y)}.$$

This yields

$$\left|\int f\, d\mu_n - \int f\, d\mu\right| \le \|f\|_{\mathrm{Lip}} \rho(\mu_n, \mu) \to 0$$

for any Lipschitz continuous function f and proves $\mu_n \Rightarrow \mu$; cf. Ref. [10, Theorem 2.1]. This completes the proof of the second statement.

Next, if $\{x_i\}$ is a separability set in $\mathbb{X}$, then a countable set of all finite sums of the form

$$\sum_k c_k \delta_{x_{i_k}}, \qquad \{c_k\} \subset \mathbb{Q} \cap [0,\infty), \qquad \sum_k c_k = 1$$

is dense in $\big(\mathscr{P}(\mathbb{X}), \rho\big)$. This proves separability of $(\mathscr{P}(\mathbb{X}), \rho)$. To prove its completeness, we first observe that any sequence $\{\mu_n\}$, which is Cauchy w.r.t. ρ is tight. Then by the Prokhorov theorem, this sequence has a weakly convergent subsequence. By the second statement, this subsequence converges in $\big(\mathscr{P}(\mathbb{X}), \rho\big)$, as well, and then the entire Cauchy sequence converges to the same limit in $\big(\mathscr{P}(\mathbb{X}), \rho\big)$. □

For the further reference convenience, we extend slightly the setting of the previous section, and assume $d : \mathbb{X} \to [0,\infty)$ to be lower semi-continuous, only. Proposition 4.3.3 still applies in this case, and we denote by the same symbol $C_{d,\mathrm{opt}}(\mu,\nu)$ the (nonempty) set of $\kappa \in C(\mu,\nu)$ such that

$$\int_{\mathbb{X}\times\mathbb{X}} d(x,y)\kappa(dx,dy) = \inf_{\chi \in C(\mu,\nu)} \int_{\mathbb{X}\times\mathbb{X}} d(x,y)\chi(dx,dy).$$

Proposition 4.4.2. *$C(\mu,\nu)$, $C_{d,opt}(\mu,\nu)$ are compact subsets of $\mathscr{P}(\mathbb{X}\times\mathbb{X})$.*

Proof. For any sequence $\{\kappa_n\} \subset C(\mu,\nu)$, there exists a subsequence which weakly converges to a measure $\kappa \in C(\mu,\nu)$; see the proof of Proposition 4.3.3. Hence by statement 2 of Proposition 4.4.1, $C(\mu,\nu)$ is a compact set in $\mathscr{P}(\mathbb{X}\times\mathbb{X})$.

Next, if $\{\kappa_n\} \subset C_{d,\mathrm{opt}}(\mu,\nu)$ weakly converges to some κ, then by the "common probability space" principle and the lower semi-continuity of d one has

$$\int_{\mathbb{X}\times\mathbb{X}} d(x,y)\kappa(dx,dy) \le \liminf_n \int_{\mathbb{X}\times\mathbb{X}} d(x,y)\kappa_n(dx,dy) = \inf_{\chi\in C(\mu,\nu)} \int_{\mathbb{X}\times\mathbb{X}} d(x,y)\chi(dx,dy).$$

That is, the set $C_{d,\mathrm{opt}}(\mu,\nu)$ is closed in $\mathscr{P}(\mathbb{X}\times\mathbb{X})$. Since it is a subset of the compact set $C(\mu,\nu)$, it is compact, as well. □

Now we are ready to formulate the main result of this section. Recall that a probability kernel P is called *Feller* if the mapping

$$\mathbb{X} \ni x \mapsto P(x,dy) \in \mathscr{P}(\mathbb{X})$$

is continuous w.r.t. weak convergence in $\mathscr{P}(\mathbb{X})$.

Theorem 4.4.3 (The general Coupling lemma for probability kernels). *Let $\mathbb{X}$ be a Polish space and $\mathscr{X}$ be the Borel σ-algebra. Then for any Feller probability kernels P_1, P_2 on*

$(\mathbb{X}, \mathscr{X})$ *and lower semi-continuous function* $d : \mathbb{X} \to [0, \infty)$, *there exists a coupling kernel on* $(\mathbb{X} \times \mathbb{X}, \mathscr{X} \otimes \mathscr{X})$ *such that*

$$Q\Big((x, y), \cdot\Big) \in C_{d,opt}\Big(P_1(x, \cdot), P_2(y, \cdot)\Big), \quad (x, y) \in \mathbb{X} \times \mathbb{X}.$$

Remark 4.4.4. Theorem 4.4.3 considerably generalizes the Coupling lemma for probability kernels (Theorem 2.2.4): while the function $d(x, y) = 1_{x \neq y}$ in Theorem 2.2.4 is specific, here $d : \mathbb{X} \to [0, \infty)$ is an arbitrary lower semi-continuous function. This explains the name we use for this theorem. To get such an extension, we had to impose the Feller condition; however, this additional assumption is not restrictive and is satisfied in most cases of interest.

Our proof of Theorem 4.4.3 is based on measurability and measurable selection results collected in Ref. [126, Chapter 12.1]. For a Polish space $(\mathbb{S}, \rho)$, denote by comp $(\mathbb{S})$ the space of all nonempty compact subsets of $\mathbb{S}$, endowed with the *Hausdorff metric*:

$$\rho_H(K_1, K_2) = \max\Big(\max_{x \in K_1} \min_{y \in K_2} \rho(x, y), \max_{x \in K_2} \min_{y \in K_1} \rho(x, y) \Big).$$

Theorem 4.4.5 [126, Theorem 12.1.10]. *Let* $(E, \mathscr{E})$ *be a measurable space and*

$$\Phi : E \to \operatorname{comp}(\mathbb{S})$$

be a measurable map.

Then there exists a measurable map $\phi : E \to \mathbb{S}$ *such that*

$$\phi(q) \in \Phi(q), \quad q \in E.$$

This theorem is a weaker version of the Kuratovskii and Ryll-Nardzevski's theorem on measurable selection for a set-valued mapping which takes values in the space of closed subsets of S; for example, Ref. [134].

We will use Theorem 4.4.5 in the following setting: $E = \mathbb{X} \times \mathbb{X}$, $\mathbb{S} = \mathscr{P}(\mathbb{X} \times \mathbb{X})$, and

$$\Phi\big((x, y)\big) = C_{\mathrm{d,opt}}(P_1(x, \cdot), P_2(y, \cdot)).$$

To verify the measurability property for a set-valued mapping Φ, two following sufficient conditions are very convenient.

Lemma 4.4.6 [126, Lemma 12.1.7]. *Let* $f(x)$ *be a real-valued upper semi-continuous function on* $\mathbb{S}$. *For* $K \in \operatorname{comp}(\mathbb{S})$, *set*

$$f_K = \sup_{x \in K} f(x)$$

and define $f : \mathrm{comp}\,(\mathbb{S}) \to \mathrm{comp}\,(\mathbb{S})$ *by*

$$f(K) = \{x : f(x) = f_K\}.$$

Then the maps $K \to f_K$ *and* $K \to f(K)$ *are Borel maps* $\mathrm{comp}\,(\mathbb{S}) \to \mathbb{R}$ *and* $\mathrm{comp}\,(\mathbb{S}) \to \mathrm{comp}\,(\mathbb{S})$, *respectively.*

Lemma 4.4.7 [126, Lemma 12.1.8]. Let $\mathbb{Y}$ be a metric space with the Borel σ-algebra $\mathscr{Y}$. Let $F : \mathbb{Y} \to \mathrm{comp}\,(\mathbb{S})$ be such that, for any sequences $y_n \to y$ and $x_n \in F(y_n)$, it is true that $\{x_n\}$ has a limit point in $F(y)$.

Then the map $F : \mathbb{Y} \to \mathrm{comp}\,(\mathbb{S})$ is measurable.

Proof of Theorem 4.4.3. In the setting described above, define the mapping $\Psi : E \to \mathrm{comp}\,(\mathbb{S})$ by

$$\Psi\big((x,y)\big) = C(P_1(x,\cdot), P_2(y,\cdot)), \quad (x,y) \in E.$$

This definition is formally correct, because each set $C(P_1(x,\cdot), P_2(y,\cdot))$ is nonempty and compact in $\mathscr{P}(\mathbb{X}\times\mathbb{X})$; see Proposition 4.4.2. If $(x_n, y_n) \to (x,y)$, then for any sequence $\varkappa_n \in C(P_1(x_n,\cdot), P_2(y_n,\cdot))$, $n \ge 1$ the marginal distributions of $\varkappa_n$ equal $P_1(x_n,\cdot), P_2(y_n,\cdot)$, which by the Feller property weakly converge to $P_1(x,\cdot), P_2(y,\cdot)$, respectively. Then sequences

$$\{P_1(x_n,\cdot)\} \quad \{P_2(y_n,\cdot)\}$$

are tight, and thus the sequence $\{\varkappa_n\}$ is tight, as well (the proof here is the same as in Proposition 4.3.3). Hence it has a limit point $\varkappa$, and because the projection maps on the first and second coordinates are continuous, the marginal distributions of $\varkappa$ equal

$$\lim_{n\to\infty} P_1(x_n,\cdot) = P_1(x,\cdot), \quad \lim_{n\to\infty} P_2(y_n,\cdot) = P_2(y,\cdot);$$

that is, $\varkappa \in C(P_1(x,\cdot), P_2(y,\cdot))$. Thus we can apply Lemma 4.4.7 and conclude that the map Ψ is measurable.

Next, define the function $f : \mathscr{P}(\mathbb{X}\times\mathbb{X}) \to [-\infty, 0]$ by

$$f(\varkappa) = -\int\limits_{\mathbb{X}\times\mathbb{X}} d(x,y)\,\varkappa(dx,dy).$$

Since $d(\cdot,\cdot)$ is lower semi-continuous, the function f is upper semi-continuous: this can be proved using the "common probability space" principle similarly to the proof of Proposition 4.3.3; we omit the details. Then the function

$$\Upsilon : \mathrm{comp}\Big(\mathscr{P}(\mathbb{X}\times\mathbb{X})\Big) \to \mathrm{comp}\Big(\mathscr{P}(\mathbb{X}\times\mathbb{X})\Big),$$

$$\Upsilon(K) = \left\{\kappa \in K : f(\kappa) = \sup_{\lambda\in K} f(\lambda)\right\}$$

is measurable by Lemma 4.4.6. We have

$$\Phi\big((x,y)\big) = \Upsilon\Big(\Psi\big((x,y)\big)\Big),$$

and thus Φ is measurable. Thus we can apply Theorem 4.4.5, which gives the required statement. □

4.5 Weak Ergodic Rates

In this section, we complete the main goal of the is chapter, and extend the ergodicity results, obtained in Chapter 2 for the total variation convergence, to the general setting with the convergence understood in the sense of a coupling probability distance. Within this section, d is a fixed lower semi-continuous distance-like function on $\mathbb{X}$, and the same symbol denotes respective coupling distance on $\mathscr{P}(\mathbb{X})$. The space $\mathbb{X}$ is assumed to be a Polish space with the metric ρ. We assume the Markov chain X to be Feller, in order to be able to apply the constructions from the previous section.

Our first result extends the Dobrushin theorem (Theorem 2.3.1): in the particular case

$$d(x,y) = 21_{x\neq y},$$

the statement of this theorem just repeats the Dobrushin theorem.

Theorem 4.5.1 (General Dobrushin theorem).

1. *Assume that, for some* $\upsilon \in [0,1)$,

$$d\Big(P(x,\cdot), P(y,\cdot)\Big) \le \upsilon d(x,y), \quad x,y \in \mathbb{X}. \tag{4.5.1}$$

   Then

$$d\Big(P^n(x,\cdot), P^n(y,\cdot)\Big) \le \upsilon^n d(x,y), \quad x,y \in \mathbb{X}. \tag{4.5.2}$$

2. *In addition, if*

$$\sup_{N\ge1} \frac{1}{N}\sum_{n=1}^N \int_{\mathbb{X}} d(x,y)\, P^n(x,dy) < \infty, \quad x \in \mathbb{X} \tag{4.5.3}$$

and there exist $p \geq 1, C > 0$ such that $Cd^{1/p}$ dominates the original metric ρ on $\mathbb{X}$, then there exists a unique IPM μ for $\mathbb{X}$,

$$U(x) = \int_{\mathbb{X}} d(x,y)\,\mu(dy) < \infty, \quad x \in \mathbb{X},$$

and for every $x \in \mathbb{X}$

$$d\big(P^n(x,\cdot),\mu\big) \leq \upsilon^n U(x), \qquad n \geq 1. \tag{4.5.4}$$

Proof. Consider the kernel Q on $\mathbb{X}\times\mathbb{X}$ constructed in Theorem 4.4.3. Let $Z = (X, Y)$ be a Markov process with the transition probability Q. We have

$$\begin{aligned}\mathsf{E}\Big(d\big(X_n, Y_n\big)\Big|\mathscr{F}^Z_{n-1}\Big) &= \int_{\mathbb{X}\times\mathbb{X}} d(y,y)Q\big((x',y'),dx\,dy\big)\Big|_{x'=X_{n-1},y'=Y_{n-1}} \\ &\times d\big(P(x',\cdot),P(y,\cdot)\big)\Big|_{x'=X_{n-1},y'=Y_{n-1}} \leq \upsilon d\big(X_{n-1},Y_{n-1}\big).\end{aligned}$$

Iterating this inequality, we get the bound

$$\mathsf{E}^Z_{(x,y)} d\big(X_n, Y_n\big) \leq \upsilon^n d(x,y), \qquad n \geq 1.$$

This proves the first statement, since the laws of X_n, Y_n w.r.t. $\mathsf{P}^Z_{(x,y)}$ equal $P^n(x,\cdot)$ and $P^n(y,\cdot)$, respectively.

The proof of the second statement is similar to the proof of Theorem 2.6.3, with minor changes caused by the different type of convergence in $\mathscr{P}(\mathbb{X})$. For the reader's convenience, we provide the entire proof.

For arbitrary x and $m \leq n$, consider the triple (ξ,η,ζ) defined by the following conventions:

(i) ζ has the law $P^{n-m}(x,\cdot)$;

(ii) conditioned by $\sigma(\zeta)$, the pair (ξ,η) has the law $Q\big((x,\zeta),\cdot\big)$.

Then $(\xi,\eta) \in C\big(P^n(x,\cdot),P^m(x,\cdot)\big)$, and thus by eq. (4.5.2)

$$\begin{aligned}d\big(P^n(x,\cdot),P^m(x,\cdot)\big) &\leq \int_{\mathbb{X}} \big(P^m(x,\cdot),P^m(x,\cdot)\big)\,P^{n-m}(x,dy) \\ &\leq \upsilon^m \int_{\mathbb{X}} d(x,y)P^{n-m}(x,dy).\end{aligned}$$

Repeating the same construction for $m \geq n$, we get a family

$$\kappa_{n,m} \in C\big(P^n(x,\cdot),P^m(x,\cdot)\big), \quad m,n \geq 1$$

such that

$$\int_{\mathbb{X}} d(x,y)\kappa_{n,m}(dx,dy) \le \upsilon^{n\wedge m}\int_{\mathbb{X}} d(x,y)P^{|n-m|}(x,dy).$$

Next, we observe that

$$\mu_x^{(N)} = \frac{1}{N}\sum_{n=1}^{N} P^n(x,\cdot)$$

can be represented as the law of the variable

$$\xi = \xi_n \text{ on the set } \{\varepsilon_N = n\}, \quad n = 1,\dots,N,$$

where ε_N takes values $1,\dots,N$ with probabilities $1/N$, the laws of $\xi_n, n \ge 1$ equal $P^n(x,\cdot), n \ge 1$, and $\{\xi_n\}, \varepsilon_N$ are independent. For given $N, M \ge 1$, define the pair

$$(\xi,\eta) = (\xi_{n,m},\eta_{n,m}) \text{ on the set } \{\varepsilon_N = n, \tilde{\varepsilon}_M = m\}, \quad n = 1,\dots,N, \quad m = 1,\dots,M,$$

where $\tilde{\varepsilon}_M$ has the same law with ε_M, each pair $(\xi_{n,m},\eta_{n,m})$ has the law $\kappa_{n,m}$ constructed above, and $\varepsilon_N, \tilde{\varepsilon}_M, \{(\xi_{n,m},\eta_{n,m})\}$ are independent. By the construction,

$$(\xi,\eta) \in C\left(\mu_x^{(N)},\mu_x^{(M)}\right),$$

and therefore

$$\begin{aligned} d\left(\mu_x^{(N)},\mu_x^{(M)}\right) &\le Ed(\xi,\eta) = \frac{1}{MN}\sum_{n=1}^{N}\sum_{m=1}^{M} Ed(\xi_{n,m},\eta_{n,m}) \\ &\le \frac{1}{MN}\sum_{n=1}^{N}\sum_{m=1}^{M}\upsilon^{n\wedge m}\int_{\mathbb{X}} d(x,y)P^{|n-m|}(x,dy). \end{aligned}$$

If $M < N$ and $m \in \{1,\dots,M\}, n \in \{1,\dots,N\}$, then $k = m\wedge n \le M$ and $l = |n-m| \le N-1$. Hence

$$d\left(\mu_x^{(N)},\mu_x^{(M)}\right) \le \frac{2}{MN}\sum_{k=1}^{M}\sum_{l=0}^{N-1}\upsilon^k\int_{\mathbb{X}} d(x,y)\,P^l(x,dy).$$

By eq. (4.5.3), the sequence

$$\frac{1}{N}\sum_{l=0}^{N-1}\int_{\mathbb{X}} d(x,y)\,P^l(x,dy)$$

is bounded, hence

$$d\left(\mu_x^{(N)}, \mu_x^{(M)}\right) \le \frac{C}{M} \sum_{k=1}^{M} \upsilon^k \to 0, \quad M \to \infty.$$

Recall that it is assumed that $Cd^{1/p}$ dominates ρ; without loss of generality we also can assume that $\rho \le 1$. Then, denoting by ρ the corresponding coupling distance in $\mathscr{P}(\mathbb{X})$, we get by the Jensen inequality

$$\rho\left(\mu_x^{(N)}, \mu_x^{(M)}\right) \le Cd\left(\mu_x^{(N)}, \mu_x^{(M)}\right)^{1/p} \to 0, \quad M, N \to \infty.$$

That is, the sequence $\mu_x^{(N)}, N \ge 1$ is fundamental w.r.t. ρ, and therefore weakly converges in $\mathscr{P}(\mathbb{X})$; see Proposition 4.4.1.

Applying the same argument and eq. (4.5.2) once more, we get

$$d\left(\mu_x^{(N)}, \mu_x^{(M)}\right) \le \frac{1}{N} \sum_{n=1}^{N} d\left(P^n(x, \cdot), P^n(y, \cdot)\right) \le \frac{C}{N} d(x, y) \sum_{n=1}^{N} \upsilon^n \to 0,$$

hence the limit μ of $\{\mu_x^{(N)}, N \ge 1\}$ does not depend on the initial point x. It is easy to show that μ is an invariant measure for X. Since $\mu_x^{(N)} \Rightarrow \mu$, we get using condition (2.6.9), the common probability space principle, lower semi-continuity of $d(x, y)$, and the Fatou lemma

$$\begin{aligned} U(x) &= \int_{\mathbb{X}} d(x, y)(x, y)\mu(dy) \\ &\le \liminf_N \frac{1}{N} \sum_{n=1}^{N} \int_{\mathbb{X}} d(x, y)(x, y)P^n(x, dy) < \infty, \quad x \in \mathbb{X}. \end{aligned}$$

Finally,

$$\begin{aligned} d\left(P^n(x, \cdot), \mu\right) &\le \int_{\mathbb{X}} d\left(P^n(x, \cdot), P^n(y, \cdot)\right) \mu(dy) \\ &\le \upsilon^n \int_{\mathbb{X}} d(x, y)\mu(dy) = \upsilon^n U(x). \end{aligned}$$ □

A new effect, which appears in Theorem 4.5.1 when compared to Theorem 2.3.1, is that we have to separate the "loss of memory"-type bound eq. (4.5.2) and the stabilization rate eq. (4.5.4). This effect is essentially the same we had discussed in Section 2.6: to get eq. (4.5.4) one has to guarantee that an IPM exists, and that the right-hand side in eq. (4.5.2) is well integrable w.r.t. the IPM. While considering the total variation distance, we have $d(x, y) = 21_{x \neq y}$ bounded, and unbounded terms appear only because of the recurrence properties of the chain. Hence such an effect appears in the Harris-type theorems, but is not visible in Theorem 2.3.1. In the current setting, the term $d(x, y)$ is possibly unbounded, and eq. (4.5.2) itself is not sufficient to guarantee existence of an IPM. A simple negative example here is provided by Example 2.6.4: taking two sequences X, Y with initial conditions x, y and the same $\varepsilon_n, \eta_n, n \geq 1$, we have

$$|X_n - Y_n| \leq |a|^n |x - y|, \quad n \geq 1.$$

That is, eq. (4.5.2) holds true for $d(x, y) = |x - y|$, but an IPM for X does not exist.

We note, however, that eq. (4.5.2) provides that the chain X has *at most one* IPM. Indeed, consider two IPMs μ, ν, and let the chain $Z = (X, Y)$ have the transition probability Q and the initial distribution $\mu \times \nu$. Fix $\varepsilon > 0$, $\delta > 0$ and take C and n large enough such that

$$\mathsf{P}\Big(d\big(X_0, Y_0\big) > C\Big) < \varepsilon, \qquad \upsilon^n < \varepsilon\delta.$$

Then

$$\begin{aligned}\mathsf{P}\Big(d\big(X_n, Y_n\big) > \delta\Big) &\leq \mathsf{P}\Big(d\big(X_0, Y_0\big) > C\Big) + \mathsf{P}\Big(d\big(X_n, Y_n\big) > \delta,\ d\big(X_0, Y_0\big) \leq C\Big) \\ &\leq \varepsilon + \frac{\upsilon^n}{\delta} \mathsf{E} d\big(X_0, Y_0\big) 1_{d(X_0, Y_0) \leq C} < 2\varepsilon.\end{aligned}$$

Because X_n, Y_n have the laws μ, ν, respectively, and ε and δ are arbitrary, this means that $d(\mu, \nu) = 0$ and therefore $\mu = \nu$.

The proof of the first part of Theorem 4.5.1 is one-to-one analogous to the proof of Theorem 2.3.1; this means that condition (4.5.1) should be understood as a condition for the chain to be "uniformly ergodic" w.r.t. the distance-like function d. We note that this condition can be easily verified for dissipative systems considered in Section 4.2; see Example 4.3.5. However, such a "global contraction" is a strong structural assumption. In the rest of this section, we localize this assumption in a way very similar to that explained in Sections 2.6 and 2.7; that is, we prove a general version of the Harris-type theorem with the convergence w.r.t. a general coupling distance.

We say that a distance-like function d is *contracting* (for a given Markov chain X) on a set $B \subset \mathbb{X} \times \mathbb{X}$ if there exists $\upsilon_B \in [0, 1)$ such that

$$d\big(P(x, \cdot), P(y, \cdot)\big) \leq \upsilon_B d(x, y), \quad (x, y) \in B.$$

We also say that a distance-like function d is *non-expanding* (for a given Markov chain X) if

$$d\big(P(x, \cdot), P(y, \cdot)\big) \leq d(x, y), \quad x, y \in \mathbb{X}$$

For a given Markov chain X and a distance-like function d, we denote by Z the "d-greedy coupling" for X; that is, Markov process $Z = (X, Y)$ with the transition probability Q, constructed in Theorem 4.4.3. We consider the corresponding hitting times

τ_B, θ_B for this process (see Section 2.6), and recall notation Λ for the family of monotonous submultiplicative functions (see Section 2.7). We also consider a family of new distance-like functions

$$d_p(x,y) = \big(d(x,y)\big)^{1/p}, \quad p > 1$$

and denote by $d_p, p > 1$ respective coupling distances on $\mathscr{P}(\mathbb{X})$.

Theorem 4.5.2 (General Harris-type theorem). *Let for some distance-like function d, set B, and function $W : \mathbb{X}\times\mathbb{X} \to [1,\infty)$, the following conditions hold true.*

I. *d is non-expanding, and d is contracting on B.*

R. *There exists function $\lambda \in \Lambda$ such that*

(i)

$$\mathsf{E}^Z_{(x,y)}\lambda(\tau_B) \le W(x,y), \quad (x,y) \in \mathbb{X}\times\mathbb{X};$$

(ii)

$$S_{\lambda(\cdot),B} = \sup_{(x,y)\in B} \mathsf{E}^Z_{(x,y)}\lambda(\theta_B) < \infty.$$

Then the following statements hold true.

1. *For any $p, q > 1$ with $1/p + 1/q = 1$ and*

$$\frac{1}{q} < \frac{\log \upsilon_B^{-1}}{\log \upsilon_B^{-1} + \log S_{\lambda(\cdot),B}}, \tag{4.5.5}$$

there exists $C < \infty$ such that

$$d_p\Big(P^n(x,\cdot), P^n(y,\cdot)\Big) \le C d_p(x,y)\Big(\lambda(n)\Big)^{-1/q} W^{1/q}(x,y). \tag{4.5.6}$$

2. *For arbitrary $p, q > 1$ with $1/p + 1/q = 1$, there exist $\gamma > 0$, $C < \infty$ such that*

$$d_p\Big(P^n(x,\cdot), P^n(y,\cdot)\Big) \le C d_p(x,y)\Big(\lambda(\gamma n)\Big)^{-1/q} W^{1/q}(x,y). \tag{4.5.7}$$

3. *Let eq. (4.5.7) hold true and*

$$\sup_{N\ge 1} \frac{1}{N}\sum_{n=1}^{N} \mathsf{E}_x d_p(x,X_n) W^{1/q}(x,X_n) < \infty, \quad x \in \mathbb{X}. \tag{4.5.8}$$

Assume also that and there exist $r \ge 1$, $C > 0$ such that $Cd^{1/r}$ dominates the original metric ρ on $\mathbb{X}$. Then there exists unique IPM μ for X, the function

$$U(x) = \int_{\mathbb{X}} d_p(x,y) W^{1/q}(x,y)\mu(dy), \quad x \in \mathbb{X}$$

takes finite values, and

$$d_p\Big(P^n(x,\cdot),\mu\Big) \le C\Big(\lambda(\gamma n)\Big)^{-1/q} U(x). \tag{4.5.9}$$

Proof. First, we modify properly the auxiliary construction developed in Section 2.4. Namely, we put

$$e(x,y) = \mathsf{E}^Z_{(x,y)}(X_1, Y_1), \quad \varrho(x,y) = \frac{d(x,y)}{e(x,y)} 1_{e(x,y)>0},$$

and consider the sequences $\{\mathscr{E}_n\}$, $\{M_n\}$ defined by

$$\mathscr{E}_n = \prod_{k=0}^{n-1} \varrho(X_k, Y_k), \quad n \ge 1, \quad \mathscr{E}_0 = 1, \quad M_n = d(X_n, Y_n)\mathscr{E}_n, \quad n \ge 0.$$

By the construction, $\{M_n\}$ is a super-martingale: for $n \ge 1$,

$$\begin{aligned}\mathsf{E}[M_n|\mathscr{F}_{n-1}] &= \mathsf{E}[d(X_n, Y_n)|\mathscr{F}_{n-1}]\mathscr{E}_n \\ &= e(X_{n-1}, Y_{n-1})\frac{d(X_{n-1}, Y_{n-1})}{e(X_{n-1}, Y_{n-1})} 1_{e(X_{n-1},Y_{n-1})>0}\mathscr{E}_{n-1} \\ &\le d(X_{n-1}, Y_{n-1})\mathscr{E}_{n-1} = M_{n-1}.\end{aligned}$$

Hence for given x, y

$$\mathsf{E}^Z_{(x,y)} M_n = \mathsf{E}^Z_{(x,y)} M_0 = d(x,y), \quad n \ge 1.$$

Under $\mathsf{P}^Z_{(x,y)}$, the pair X_n, Y_n gives a coupling for $P^n(x,\cdot), P^n(y,\cdot)$, hence

$$\begin{aligned}d_p\Big(P^n(x,\cdot), P^n(y,\cdot)\Big) &\le \mathsf{E}^Z_{(x,y)}\Big(d(X_n, Y_n)\Big)^{1/p} \\ &= \mathsf{E}^Z_{(x,y)} M_n^{1/p}\mathscr{E}_n^{-1/p} \le \Big(\mathsf{E}^Z_{(x,y)} M_n\Big)^{1/p}\Big(\mathsf{E}^Z_{(x,y)}\mathscr{E}_n^{-q/p}\Big)^{1/q} \\ &\le d_p(x,y)\Big(\mathsf{E}^Z_{(x,y)}\mathscr{E}_n^{-q/p}\Big)^{1/q}.\end{aligned} \tag{4.5.10}$$

By the assumptions on $d(x,y)$, we have

$$\varrho^{-1}(x,y) = \frac{e(x,y)}{d(x,y)} \le \begin{cases} \upsilon_B, & (x,y) \in B; \\ 1, & \text{otherwise.}\end{cases}$$

Therefore,

$$\mathsf{E}^Z_{(x,y)}\mathscr{E}_n^{-q/p} \le \mathsf{E}^Z_{(x,y)}\upsilon_B^{qN_n/p}, \tag{4.5.11}$$

where N_n denotes the number of visits of Z to B before time moment n. Now the rest of the proof of statement 1 repeats the proof of Theorem 2.7.2. Namely, by eq. (4.5.5) we have

$$\frac{p}{q} < \frac{\log \upsilon_B^{-1}}{\log S_{\lambda(\cdot),B}},$$

and thus

$$S_{\lambda(\cdot),B}\upsilon_B^{p/q} < 1.$$

Recall that by the condition **R** and the strong Markov property

$$\mathsf{E}^Z_{(x,y)}\lambda(\tau_B^k) \le S^{k-1}_{\lambda(\cdot),B}W(x,y).$$

Then

$$\mathsf{E}^Z_{(x,y)}\upsilon_B^{rN_n/q} \le \left(\upsilon_B^{-q/p}-1\right)\frac{1}{\lambda(n)}\sum_{k=0}^{\infty}\upsilon_B^{qk/p}S^{k-1}_{\lambda(\cdot),B}W(x,y) = \frac{C}{\lambda(n)}W(x,y).$$

Combining this estimate with eqs. (4.5.10) and (4.5.11), we get eq. (4.5.6).

The proof of statement 2 is literally the same with the proof of Corollary 2.7.4. The proof of statement 3 repeats the proof of statement 2 in Theorem 4.5.1. □

4.6 Application to Stochastic Delay Equations

In this section, we illustrate Theorem 4.5.2, applying it to the study of the ergodic properties of the $C([-r,0],\mathbb{R}^m)$-valued Markov process **X**, which is defined as the segment process corresponding to the solution of the SDDE (4.1.1). Such a study will be made in a way, which is remarkably similar to the one used in the analysis of the ergodic properties of diffusion processes; see Section 3.3. We will verify separately conditions **I** and **R** of Theorem 4.5.2. The latter one will be verified by means of a proper set of Lyapunov-type conditions; the argument here will not have substantial differences with the diffusion case. Condition **I** is more intrinsic, and in Section 4.6.2 we explain an algorithm for construction of the distance-like function $d(x,y)$, which possesses required properties, based on the notion of *generalized coupling*. Within the entire section, we assume that a, b are Lipschitz continuous, hence the solution to eq. (4.1.1) is well defined. We also assume that b is bounded and $B = bb^*$ satisfies the uniform ellipticity condition: for some $c > 0$,

$$\left(B(x)l,l\right) = |b(x)^*l|^2 \ge c|l|^2, \quad l \in \mathbb{R}^m.$$

4.6.1 Lyapunov-Type Conditions

Recall that we denote by

$$\mathbf{X}(t) = \{X_{t+s}, s \in [-r, 0]\} \in C([-r, 0], \mathbb{R}^m), \quad t \geq 0$$

the segment process $\mathbf{X}$, which corresponds to the solution of the SDDE (4.1.1). We will use literally the same argument which was developed in Section 3.3.1. Clearly, now the state space $C([-r, 0], \mathbb{R}^m)$ for the Markov process $\mathbf{X}$ is more complicated than $\mathbb{R}^m$ for a diffusion process, but for a certain class of functions on $\mathbb{X}$ a similar calculation is available, which is based on the Itô formula and leads to a Lyapunov-type condition. Namely, for $V \in C^2(\mathbb{R}^m)$, which will be specified later, we denote

$$\mathbf{V}(\mathbf{x}) = V(\mathbf{x}_0), \tag{4.6.1}$$

$$\mathscr{L}\mathbf{V}(\mathbf{x}) = \sum_{i=1}^{m} a_i(\mathbf{x}_0)\,\partial_{x_i} V(\mathbf{x}_0) + \frac{1}{2}\sum_{i,j=1}^{m} B_{ij}(\mathbf{x}_{-r})\,\partial^2_{x_i x_j} V(\mathbf{x}_0), \quad \mathbf{x} \in C([-r, 0], \mathbb{R}^m). \tag{4.6.2}$$

Then by the Itô formula Ref. [61, Chapter II.5], the process

$$M_t = V(X_t) - \int_0^t \mathscr{L}\mathbf{V}(X(s))\, ds = \mathbf{V}(X(t)) - \int_0^t \mathscr{L}\mathbf{V}(X(s))\, ds$$

is a local martingale w.r.t. every $\mathsf{P}_{\mathbf{x}}$, $\mathbf{x} \in C([-r, 0], \mathbb{R}^m)$; that is, $\mathbf{V}$ belongs to the domain of the extended generator $\mathscr{A}$ for $\mathbf{X}$, and

$$\mathscr{A}\,\mathbf{V} = \mathscr{L}\,\mathbf{V}.$$

The following proposition indicates that, for properly chosen V, ϕ, C,

$$\mathscr{L}\,\mathbf{V} \leq -\phi(\mathbf{V}) + C. \tag{4.6.3}$$

Proposition 4.6.1. *Let the drift coefficient $a(x)$ satisfy the drift condition* (3.3.4) *with some $\kappa \in \mathbb{R}$, $A_\kappa \in (0, \infty]$.*

(i) *Let either $\kappa > 0$ and $\alpha > 0$ be arbitrary, or $\kappa = 0$ and $\alpha > 0$ satisfy*

$$\alpha < \frac{2A_0}{|||B|||}.$$

Let $V \in C^2(\mathbb{R}^m)$ be such that

$$V(x) = e^{\alpha|x|}, \quad |x| \geq 1.$$

Then there exist $a, C > 0$ such that eq. (4.6.3) holds true with $\mathbf{V}$ *given by eq. (4.6.1) and* $\phi(v) = av$.

(ii) *Let* $\kappa \in (-1, 0)$ *and* $V \in C^2(\mathbb{R}^m)$ *be such that* $V \geq 1$ *and*

$$V(x) = e^{\alpha|x|^{1+\kappa}}, \quad |x| \geq 1$$

with the constant

$$\alpha < \frac{2A_\kappa}{(1+\kappa)|||B|||}.$$

Then there exist $a, b, C > 0$ such that eq. (4.6.3) holds true with $\mathbf{V}$ *given by eq. (4.6.1) and* $\phi(v) = a \log^{-\sigma}(v + b)$,

$$\sigma = -\frac{2\kappa}{1+\kappa} > 0.$$

(iii) *Let* $\kappa = -1$ *and*

$$2A_{-1} > \sup_x \big(\operatorname{Trace} B(x)\big).$$

Let $V \in C^2(\mathbb{R}^m)$ *be such that*

$$V(x) = |x|^p, \quad |x| \geq 1,$$

where $p > 2$ *satisfies*

$$2A_{-1} > \sup_x \big(\operatorname{Trace} B(x) + (p-2)\|B(x)\|\big).$$

Then there exist $a, C > 0$ such that eq. (4.6.3) holds true with $\mathbf{V}$ *given by eq. (4.6.1) and* $\phi(v) = av^{1-2/p}$.

The proof is completely analogous to the proofs of Propositions 3.3.1–3.3.3, and is based on the fact that, since the diffusion coefficient $b(x)$ is bounded, the principal part of the function $\mathscr{L}V$ is represented by the "drift part"

$$\sum_{i=1}^m a_i(x)\, \partial_{x_i} V(x);$$

we leave the details for the reader.

Using Theorem 3.2.3 and calculations from Examples 3.2.5–3.2.7, we get that for each $h > 0$ the h-skeleton chain $\mathbf{X}^h$ for $\mathbf{X}$ satisfies

$$\mathsf{E}_{\mathbf{x}} \mathbf{V}^h(\mathbf{X}^h) \leq \mathbf{V}^h(\mathbf{x}) - \phi^h(\mathbf{V}^h(\mathbf{x})) + C^h \tag{4.6.4}$$

with certain functions $\mathbf{V}^h, \phi^h$, well comparable with $\mathbf{V}, \phi$. Applying Theorem 2.8.6, we derive finally that condition **R** of Theorem 4.5.2 holds true for any

$$B = K_c \times K_c, \quad K_c = \{\mathbf{x} : \mathbf{V}(\mathbf{x}) \le c\}, \quad c > 0. \tag{4.6.5}$$

Note that now, unlike to the diffusion case, the level set

$$K_c = \{\mathbf{x} : V(\mathbf{x}_0) \le c\} \subset C([-r, 0]; \mathbb{R}^m)$$

is unbounded. This brings new challenges to the construction of the distance-like function $d(x, y)$, which in condition **I** is required to be contracting on B. Such a construction is developed in the next section.

4.6.2 Generalized Couplings and Construction of $d(x, y)$

By definition, *generalized coupling* for a pair of measures $\mu, \nu \in \mathscr{P}(\mathbb{X})$ is a pair of $\mathbb{X}$-valued random variables ξ, η such that

$$\mathrm{Law}\,(\xi) \ll \mu, \quad \mathrm{Law}\,(\eta) \ll \nu.$$

This definition extends the notion of the coupling, where the laws of ξ, η should be *equal* to μ, ν precisely. The class $\widehat{C}(\mu, \nu)$ of generalized couplings is much wider than the class $C(\mu, \nu)$ of (true) couplings, hence in general it is much easier to construct a pair (ξ, η) with the required properties. We illustrate this effect by the following typical calculation.

Example 4.6.2. Let for $\lambda > 0$, whose particular value will be specified later, a pair of processes X, Y be defined by SDDE

$$\begin{aligned} \mathrm{d}X_t &= -a(X_t)\,\mathrm{d}t + b(X_{t-r})\,\mathrm{d}W_t, \\ \mathrm{d}Y_t &= -a(Y_t)\,\mathrm{d}t + b(Y_{t-r})\,\mathrm{d}W_t - \lambda(Y_t - X_t)\,\mathrm{d}t \end{aligned} \tag{4.6.6}$$

with a given pair of initial conditions

$$\mathbf{X}(0) = \mathbf{x}, \quad \mathbf{Y}(0) = \mathbf{y}, \quad \mathbf{x}, \mathbf{y} \in C([-r, 0]; \mathbb{R}^m).$$

Here $\{X_t\}$ is just the (true) solution to the SDDE (4.1.1), hence $\mathrm{Law}(\mathbf{X}) = \mathsf{P}_{\mathbf{x}}$. The same fact is not true for the process $\{Y_t\}$, which contains an additional term

$$-\lambda(Y_t - X_t)\,\mathrm{d}t \tag{4.6.7}$$

in its definition. This term allows the following interpretation. Since $b(x)b^*(x)$ is nondegenerate for each x, there exists an $\mathbb{R}^{k\times m}$-valued function $b^{[-1]}(x)$ (which just gives for each x the pseudo-inverse matrix for $b(x)$), such that for any $l \in \mathbb{R}^m$

$$b(x)b^{[-1]}(x)l = l.$$

Then, we can write

$$Y_t - X_t = b(Y_{t-r})\beta_t, \quad \beta_t = b^{[-1]}(Y_{t-r})(Y_t - X_t).$$

That is, the SDDE for Y can be written in the following way:

$$dY_t = -a(Y_t)\,dt + b(Y_{t-r})\,d\widetilde{W}_t, \quad d\widetilde{W}_t = dW_t - \lambda\beta_t\,dt. \tag{4.6.8}$$

Now we can use the Girsanov theorem; for example, Ref. [98, Chapter 7]. Namely, if it is true that

$$\mathsf{P}\left(\int_0^\infty \beta_t^2\,dt < \infty\right) = 1, \tag{4.6.9}$$

then the law of $\widetilde{W}$ is absolutely continuous w.r.t. the law of W. The process Y can be understood as the strong solution to SDDE (4.1.1) with the noise $\widetilde{W}$, and thus

$$\mathrm{Law}(\mathbf{Y}) \ll \mathsf{P}_{\mathbf{y}},$$

provided that eq. (4.6.9) is satisfied.

On the other hand, the difference $\Delta = X - Y$ satisfies

$$d\Delta_t = -\lambda\Delta_t\,dt + \big(a(X_t) - a(Y_t)\big)\,dt + \big(b(X_{t-r}) - b(Y_{t-r})\big)\,dW_t.$$

Repeating literally the proof of Proposition 4.2.2, we get that, for any $p \geq 4$, there exists C such that

$$\mathsf{E}\|\mathbf{X}(t) - \mathbf{Y}(t)\|_C^p \leq Ce^{\theta_p(\lambda)t}\|f - g\|_C^p, \qquad t \geq 0,$$

where

$$\theta_p(\lambda) = -p\lambda + p\sup_{x\neq y}\frac{1}{|x-y|^2}\left[\big(a(x) - a(y), x - y\big) + \frac{m(p-1)}{2}\big\|b(x) - b(y)\big\|^2\right].$$

Since a, b are Lipschitz continuous, λ can be taken taken large enough to provide

$$\theta_p(\lambda) < 0:$$

for that, it is sufficient to take

$$\lambda > \|a\|_{\mathrm{Lip}} + \frac{m(p-1)}{2}\|b\|^2_{\mathrm{Lip}}. \tag{4.6.10}$$

Then,

$$\left\|\mathbf{X}(t) - \mathbf{Y}(t)\right\|_C \to 0, \quad t \to \infty$$

exponentially fast in L_p sense. Using the Borel–Cantelli lemma, one can show that, for some $c > 0$,

$$\sup_{t\geq 0} e^{ct}|X_t - Y_t| < \infty$$

with probability 1. Recall that $B(x) = b(x)b^*(x)$ is assumed to be uniformly elliptic, hence $b^{[-1]}(x)$ is uniformly bounded, and thus the previous inequality yields eq. (4.6.9). That is, the pair of segment processes $\mathbf{X}, \mathbf{Y}$, which correspond to the solution to eq. (4.6.6), gives a generalized coupling for $\mathsf{P}_{\mathbf{x}}, \mathsf{P}_{\mathbf{y}}$. Note that the distance between the values of these processes converge to 0 exponentially fast as $t \to \infty$, which looks similar to the dissipativity property established in Section 4.2.2. However, in Section 4.2.2, we required the drift dissipativity condition (4.2.4) (more precisely, eq. (4.2.11)) to hold true. In the current setting, we avoid this strong structural assumption, and just assume the coefficients to be Lipschitz continuous. This is the essence of the construction: by adding an extra term (4.6.7), we transform a general system into a dissipative one. The additional term can be naturally understood as a "stochastic control", which produces a desired dissipativity, and on the other hand keeps the law of the second component of the solution absolutely continuous w.r.t. $\mathsf{P}_{\mathbf{x}}$.

This example shows that, typically, a generalized coupling with desired properties can be constructed much easier than a (true) coupling. Clearly, generalized coupling keeps less information about the pair of the laws; however, several important conclusions can be made when a proper generalized coupling is well defined.

First, it is an easy argument based on the Birkhoff theorem that a Markov chain $\{X_n\}$ has at most one IPM if for any $x, y \in \mathbb{X}$ there exists a pair of sequences $X_n, Y_n, \geq 0$ such that

$$\mathrm{Law}\,(\{X_n\}) \ll \mathsf{P}_x, \quad \mathrm{Law}\,(\{Y_n\}) \ll \mathsf{P}_y,$$

and

$$d(X_n, Y_n) \to 0, \quad n \to \infty$$

with probability 1; see Ref. [55]. Using a more sophisticated argument, one can also show that if in addition it is assumed

$$\mathrm{Law}\,(\{X_n\}) \sim \mathsf{P}_x$$

and the chain is Feller, then the transition probabilities weakly converge to the IPM, assuming that the IPM exists; see Ref. [87]. We do not discuss these possibilities in details here, and focus on the other one, which gives a convenient way to verify the assumption **I** of Theorem 4.5.2.

The main idea of the construction can be simply explained as follows: likewise to Example 4.6.2, we will use a "stochastic control" argument in order to construct a generalized coupling for the pair of segment processes with required properties. Then we will use an explicit form of the Girsanov theorem and quantify, in a sense, the Girsanov weight used in the construction. This will make possible to make further rearrangement of the construction, which would lead to a true coupling with similar properties.

We develop this general plan in two different settings. First, we will consider the case where the distance $\|\mathbf{x}-\mathbf{y}\|_C$ is sufficiently small. For $\rho > 0$, whose particular value is yet to be specified, denote

$$d(\mathbf{x},\mathbf{y}) = \Big(\rho\|\mathbf{x}-\mathbf{y}\|_C \wedge 1\Big), \quad \mathbf{x},\mathbf{y} \in C([-r,0],\mathbb{R}^m). \tag{4.6.11}$$

Clearly, $d(\cdot,\cdot)$ is a metric on $C([-r,0],\mathbb{R}^m)$, equivalent to $\|\cdot - \cdot\|_C$.

Proposition 4.6.3. *There exist $h > r$ and $\rho > 0$ such that the metric $d(\cdot,\cdot)$ is contracting on the set*

$$\widehat{B} = \{(\mathbf{x},\mathbf{y}) : d(\mathbf{x},\mathbf{y}) < 1\} = \Big\{(\mathbf{x},\mathbf{y}) : \|\mathbf{x}-\mathbf{y}\|_C < \rho^{-1}\Big\}$$

for the h-skeleton chain for the segment process $\mathbf{X}$, which corresponds to the solution to eq. (4.1.1).

Proof. Take $p = 4$ and choose λ satisfying eq. (4.6.10). Define the pair of processes X, Y by eq. (4.6.6) with initial values $\mathbf{x},\mathbf{y}$, then for some $C_1, C_2 > 0$ corresponding segment processes satisfy

$$\mathsf{E}\Big\|\mathbf{X}(t)-\mathbf{Y}(t)\Big\|_C^4 \le C_1 e^{-C_2 t}\|\mathbf{x}-\mathbf{y}\|_C^4, \qquad t \ge 0; \tag{4.6.12}$$

see Example 4.6.2. Take

$$h > \frac{4}{C_2}\log(2C_1^{1/4}),$$

then by Jensen's inequality,

$$\mathsf{E}\big\|\mathbf{X}(h) - \mathbf{Y}(h)\big\|_C \le C_1^{1/4} e^{-C_2 h/4}\|\mathbf{x} - \mathbf{y}\|_C < \frac{1}{2}\|\mathbf{x} - \mathbf{y}\|_C. \tag{4.6.13}$$

We have by the construction,

$$\mathrm{Law}\,(\mathbf{X}(h)) = P^h(\mathbf{x}, \cdot).$$

However, similar identity for Law $(\mathbf{Y}(h))$ is not true, hence we can not use eq. (4.6.13) directly in order to estimate the coupling distance between $P^h(\mathbf{x}, \cdot), P^h(\mathbf{y}, \cdot)$. To overcome this difficulty, we provide the following additional analysis.

Consider the process Y as a solution to SDDE (4.6.8), and denote

$$\mathscr{E}_t = \exp\left(\int_0^t \beta_s \, dW_s - \frac{1}{2}\int_0^t \beta_s^2 \, ds\right),$$

Then the Girsanov theorem, see Ref. [98, Theorem 6.3], states that the process

$$\widetilde{W}_s, \quad s \le t$$

is a Wiener process w.r.t. the measure

$$d\mathsf{Q}^{(t)} = \mathscr{E}_t \, d\mathsf{P},$$

provided that

$$\mathsf{E}\mathscr{E}_t = 1. \tag{4.6.14}$$

We postpone verification of the condition (4.6.14) for a while, and perform the principal calculation first. Since the process $\widetilde{W}_s, s \le h$ is a Wiener process w.r.t. $\mathsf{Q}^{(h)}$, on the new probability space $(\Omega, \mathscr{F}, \mathsf{Q}^{(h)})$ the process Y solves, up to the time moment $t = h$, the SDDE (4.1.1), and thus the law of $\mathbf{Y}(h)$ w.r.t. $\mathsf{Q}^{(h)}$ equals $P^h(\mathbf{y}, \cdot)$. This means that

$$\|\mathrm{Law}\,(\mathbf{Y}(h)) - P^h(\mathbf{y}, \cdot)\|_{TV} \le \|\mathsf{P} - \mathsf{Q}^{(h)}\|_{TV}.$$

To estimate the latter distance, we use the Pinsker inequality:

$$\|\mathsf{P} - \mathsf{Q}^{(h)}\|_{TV} \le \sqrt{2\int_\Omega \log\left(\frac{d\mathsf{P}}{d\mathsf{Q}^{(h)}}\right) d\mathsf{P}},$$

for example, Ref. [130, Chapter 3]. Now we have

$$\int_\Omega \log\left(\frac{d\mathsf{P}}{d\mathsf{Q}^{(h)}}\right) d\mathsf{P} = -\mathsf{E}\log\mathscr{E}_h = \mathsf{E}\int_0^h \beta_s^2 \, ds.$$

Recall that the function $b^{[-1]}(x)$ is bounded, hence

$$\beta_s^2 \le C|X_t - Y_t|^2,$$

and using eq. (4.6.12) we easily get that, for some $C_3 > 0$,

$$\|\mathrm{Law}\,(\mathbf{Y}(h)) - P^h(\mathbf{y}, \cdot)\|_{TV} \le C_3\|\mathbf{x} - \mathbf{y}\|_C.$$

The similar calculation can be used to guarantee condition (4.6.14) with $t = h$. The localization argument here is quite standard, see Ref. [98, Chapter 7], for the reader's convenience we outline this argument. The pair of the processes X, Y satisfies

$$\begin{aligned} dX_t &= -a(X_t)\,dt + b(X_{t-r})\,d\widetilde{W}_t - \lambda b(X_{t-r})b^{[-1]}(Y_{t-r})(X_t - Y_t)\,dt,\\ dY_t &= -a(Y_t)\,dt + b(Y_{t-r})\,d\widetilde{W}_t. \end{aligned} \tag{4.6.15}$$

Let, w.r.t. some probability measure $\widetilde{\mathsf{P}}$ on $(\Omega, \mathscr{F})$, process $\widetilde{W}$ be a Wiener process, stopped at some stopping time τ. Then for any finite T we have

$$\widetilde{\mathsf{E}}\left\|\mathbf{X}(t\wedge\tau) - \mathbf{Y}(t\wedge\tau)\right\|_C^4 \le C\|\mathbf{x} - \mathbf{y}\|_C^4, \qquad t \in [0, T], \tag{4.6.16}$$

where $\widetilde{\mathsf{E}}$ denotes the expectation w.r.t. $\widetilde{\mathsf{P}}$. The proof of eq. (4.6.16) is similar to the proof of eq. (4.6.12), we leave the details for the reader. Define the sequence of stopping times

$$\tau_n = \inf\left\{t : \int_0^{\tau_n} \beta_s^2\,ds \ge n\right\} \wedge h, \quad n \ge 1,$$

then for each $n \ge 1$ the analogue of eq. (4.6.14) holds true:

$$\mathsf{E}\mathscr{E}_{\tau_n} = 1. \tag{4.6.17}$$

Then the Girsanov theorem applies to

$$d\mathsf{P}^{(\tau_n)} = \mathscr{E}_{\tau_n}\,d\mathsf{P},$$

and, in the previous notation, the process $\widetilde{W}$ stopped at τ_n is a Wiener process stopped at τ_n w.r.t. $\mathsf{P}^{(\tau_n)}$. Denote by $\mathsf{E}^{(\tau_n)}$ the expectation w.r.t. $\mathsf{P}^{(\tau_n)}$, then by eq. (4.6.16),

$$\mathsf{E}\mathscr{E}_{\tau_n}\left|\log \mathscr{E}_{\tau_n}\right| = \mathsf{E}^{(\tau_n)}\left|\int_0^{\tau_n} \beta_s\,d\widetilde{W}_s - \frac{1}{2}\int_0^{\tau_n} \beta_s^2\,ds\right| \le C\big(1 + \|\mathbf{x} - \mathbf{y}\|_C^2\big), \quad n \ge 1.$$

That is, the family $\{\mathscr{E}_{\tau_n}\}$ is uniformly integrable, and since $\tau_n \nearrow h$, we obtain eq. (4.6.14) from eq. (4.6.17) by passing to the limit.

Now we can finalize the entire proof. By the Coupling lemma, there exists a pair of $C([-r,0],\mathbb{R}^m)$-valued random elements η, ζ with the laws equal Law$(\mathbf{Y}(h))$ and $P^h(\mathbf{y},\cdot)$, and

$$\mathsf{P}(\eta \neq \zeta) \leq C_3\|\mathbf{x}-\mathbf{y}\|_C.$$

Then by Lemma 4.3.2, there exists a triple $\widehat{\xi}, \widehat{\eta}, \widehat{\zeta}$ such that the pair $\widehat{\xi}, \widehat{\eta}$ has the same law with the pair $\mathbf{X}(h), \mathbf{Y}(h)$, the law of $\widehat{\zeta}$ is $P^h(\mathbf{y},\cdot)$, and

$$\mathsf{P}(\widehat{\eta} \neq \widehat{\zeta}) \leq C_3\|\mathbf{x}-\mathbf{y}\|_C.$$

That is, $(\widehat{\xi}, \widehat{\zeta}) \in C\big(P^h(\mathbf{x},\cdot), P^h(\mathbf{y},\cdot)\big)$. This gives the bound

$$\begin{aligned} d\big(P^h(\mathbf{x},\cdot), P^h(\mathbf{y},\cdot)\big) &\leq \mathsf{E}d(\widehat{\xi},\widehat{\zeta}) \leq \mathsf{E}d(\widehat{\xi},\widehat{\eta}) + \mathsf{E}d(\widehat{\eta},\widehat{\zeta}) \\ &\leq \rho\mathsf{E}\big\|\mathbf{X}(h)-\mathbf{Y}(h)\big\|_C + \mathsf{P}(\widehat{\eta}\neq\widehat{\zeta}) \leq \left(\frac{\rho}{2}+C_3\right)\|\mathbf{x}-\mathbf{y}\|_C. \end{aligned}$$

That is, the required statement holds true if

$$\rho > 2C_3. \qquad \square$$

Since the metric $d(\cdot,\cdot)$ takes values ≤ 1 and equals 1 outside of $\widehat{B}$, it is a straightforward corollary of Proposition 4.6.3 that $d(\cdot,\cdot)$ is nonexpanding. However, Proposition 4.6.3 alone does not imply condition **I** of Theorem 4.5.2 because $\widehat{B}$ does not have the form (4.6.5). This motivates the second part of the construction, where we assume that the initial values $\mathbf{x}, \mathbf{y}$ have the difference $|\mathbf{x}_0-\mathbf{y}_0|$ bounded. Let h and ρ be the same as in the previous proposition.

Proposition 4.6.4. *For each $R > 0$, there exists $\delta > 0$ such that, for any $\mathbf{x},\mathbf{y} \in C([-r,0],\mathbb{R}^m)$ with*

$$|\mathbf{x}_0-\mathbf{y}_0| \leq R, \tag{4.6.18}$$

there exists a pair of processes $\{\widehat{X}_t, t\geq 0\}$, $\{\widehat{Y}_t, t\geq 0\}$ with the laws of the corresponding segment processes $\widehat{\mathbf{X}}, \widehat{\mathbf{Y}}$ equal $\mathsf{P}_\mathbf{x}, \mathsf{P}_\mathbf{y}$, respectively, and

$$P\Big(\|\widehat{\mathbf{X}}(h)-\widehat{\mathbf{Y}}(h)\|_C \leq \rho^{-1}\Big) > \delta. \tag{4.6.19}$$

Proof. Consider the pair X, Y defined by eq. (4.6.6), with the constant

$$\lambda > \|a\|_{\mathrm{Lip}},$$

which is yet to be specified. Take $p \geq 2$, and repeat the calculation from the beginning of the proof of Proposition 4.2.2, using instead of the Lipschitz condition on $b(x)$ just boundedness of $b(x)$. This will give

$$|\Delta_t|^p = |\mathbf{x}_0 - \mathbf{y}_0|^p + \int_0^t \left(\tilde{\eta}_s|\Delta_s|^p + C_{b,p}\right) ds + M_t^{(p)}$$

with a local martingale $M^{(p)}$, process $\{\tilde{\eta}_s\}$ satisfying

$$\tilde{\eta}_s \leq p(\|a\|_{\text{Lip}} - \lambda),$$

and constant $C_{b,p}$, which depends only on $b(x)$ and p. Then the argument, similar to the one used in the proof of Proposition 4.2.1 and based on the localization procedure and the Fatou lemma, gives

$$\begin{aligned} \mathsf{E}|\Delta_t|^p &\leq e^{p(\|a\|_{\text{Lip}}-\lambda)t}\left(|\mathbf{x}_0 - \mathbf{y}_0|^p + C_{b,p}\int_0^t e^{p(\|a\|_{\text{Lip}}-\lambda)(t-s)}\,ds\right) \\ &\leq e^{p(\|a\|_{\text{Lip}}-\lambda)t}|\mathbf{x}_0 - \mathbf{y}_0|^p + \frac{C_{b,p}}{p(\lambda - \|a\|_{\text{Lip}})}. \end{aligned} \tag{4.6.20}$$

On the other hand, we have similar to eq. (4.2.8),

$$|\Delta_t|^2 = |\Delta_r|^2 + m\|b\|_{\text{Lip}}^2 \int_0^r |\Delta_r|^2\,ds + \int_r^t \eta_s|\Delta_s|^2\,ds + M_t^{(2)} - M_r^{(2)}, \quad t \geq r \tag{4.6.21}$$

with

$$\eta_s \leq -2\lambda + \theta_2.$$

In what follows, we will take

$$\lambda > \frac{\theta_2}{2},$$

which will yield

$$\eta_s \leq 0.$$

Then,

$$\left\|\mathbf{X}(h) - \mathbf{Y}(h)\right\|_C^2 \leq |\Delta_{h-r}|^2 + m\|b\|_{\text{Lip}}^2 \int_{h-2r}^{h-r} |\Delta_s|^2\,ds + \sup_{t\in[h-r,h]}\left(M_t^{(2)} - M_{h-r}^{(2)}\right). \tag{4.6.22}$$

We have,

$$dM_t^{(2)} = \beta_t^{(2)}\, dW_t$$

with

$$|\beta_t^{(2)}| \le 2\|b\|_{\mathrm{Lip}}|\Delta_t||\Delta_{t-r}|.$$

Using eq. (4.6.20) with $p = 2$, eq. (4.6.22), and the Doob maximal inequality, we obtain similarly to the proof of Proposition 4.2.2

$$\mathsf{E}\left\|\mathbf{X}(h) - \mathbf{Y}(h)\right\|_C^2 \le \frac{C_{a,b,r,h}}{\lambda}|\mathbf{x}_0 - \mathbf{y}_0|^p$$

with a constant $C_{a,b,r,h}$ which does not depend on λ. Hence, we can fix λ large enough, such that for any $\mathbf{x}, \mathbf{y} \in C([-r, 0], \mathbb{R}^m)$ satisfying eq. (4.6.18), the corresponding pair of processes defined by eq. (4.6.6) satisfies

$$\mathsf{P}\Big(\|\mathbf{X}(h) - \mathbf{Y}(h)\|_C \le \rho^{-1}\Big) > \frac{1}{2}.$$

The pair X, Y by eq. (4.6.6) determines a generalized coupling for $\mathsf{P}_{\mathbf{x}}, \mathsf{P}_{\mathbf{y}}$, and in the second part of the proof, we construct a (true) coupling out of this pair using the Girsanov theorem. We will use respective notation from the previous proof. It follows from eq. (4.6.21) that there exists a constant C such that, for any $\mathbf{x}, \mathbf{y} \in C([-r, 0], \mathbb{R}^m)$ satisfying eq. (4.6.18), the corresponding Girsanov weight satisfies

$$\mathsf{E}\Big(\log \mathscr{E}_h\Big)^2 \le C.$$

Then there exists $\gamma > 0$ such that

$$\mathsf{P}\Big(\mathscr{E}_h < \gamma\Big) \le \frac{1}{4}.$$

Denote

$$\Omega^\gamma = \{\mathscr{E}_h \ge \gamma\}, \quad p^\gamma = \gamma\mathsf{P}(\Omega^\gamma) \ge \frac{3\gamma}{4} > 0.$$

Define

$$\mathsf{P}^\gamma = \gamma\mathsf{P}(\cdot \cap \Omega^\gamma),$$

then P^γ is a sub-probability measure with

$$\mathsf{P}^\gamma(\Omega) = p^\gamma.$$

The segment processes $\mathbf{X}, \mathbf{Y}$ have continuous trajectories in $\mathbb{C} = C([-r, 0], \mathbb{R}^m)$, and thus can be considered as random elements in the function space $C([0, \infty), \mathbb{C})$. For any measurable $A \subset C([0, \infty), \mathbb{C})$, we have by the construction,

$$\mathsf{P}^\gamma(\mathbf{Y} \in A) = \gamma \mathsf{E} 1_{\mathbf{Y} \in A} 1_{\Omega^\gamma} \le \mathsf{E} \mathscr{E}_h 1_{\mathbf{Y} \in A} 1_{\Omega^\gamma} \le \mathsf{P}_{\mathbf{y}}(A),$$

and if in addition we assume that $\gamma \le 1$, then the similar inequality holds true for $\mathbf{X}$. That is, the "laws" $\mathsf{P}^\gamma_{\mathbf{x}}, \mathsf{P}^\gamma_{\mathbf{y}}$ of the segment processes $\mathbf{X}, \mathbf{Y}$ w.r.t. P^γ are dominated by the laws $\mathsf{P}_{\mathbf{x}}, \mathsf{P}_{\mathbf{y}}$. Now we can finally specify the law of the required pair $\widetilde{\mathbf{X}}, \widetilde{\mathbf{Y}}$ as the sum

$$\mathsf{P}^\gamma((\mathbf{X}, \mathbf{Y}) \in \cdot) + (1 - p^\gamma)^{-1}(\mathsf{P}_{\mathbf{x}} - \mathsf{P}^\gamma_{\mathbf{x}}) \otimes (\mathsf{P}_{\mathbf{y}} - \mathsf{P}^\gamma_{\mathbf{y}}).$$

By the construction, this is a probability measure, and the laws of corresponding segment processes $\widetilde{\mathbf{X}}, \widetilde{\mathbf{Y}}$ are exactly equal $\mathsf{P}_{\mathbf{x}}, \mathsf{P}_{\mathbf{y}}$. On the other hand, we have

$$\begin{aligned} \mathsf{P}\Big(\|\widetilde{\mathbf{X}}(h) - \widetilde{\mathbf{Y}}(h)\|_C \le \rho^{-1}\Big) &\ge \mathsf{P}^\gamma\Big(\|\mathbf{X}(h) - \mathbf{Y}(h)\|_C \le \rho^{-1}\Big) \\ &\ge \gamma \mathsf{P}\Big(\|\mathbf{X}(h) - \mathbf{Y}(h)\|_C \le \rho^{-1}\Big) - \gamma \mathsf{P}(\Omega \setminus \Omega^\gamma) > \frac{\gamma}{4}, \end{aligned}$$

hence the required statement holds true with $\delta = \gamma/4$. □

Combining Propositions 4.6.3 and 4.6.4, we obtain the following

Corollary 4.6.5. *Let h, ρ be given by Proposition 4.6.3, and the metric $d(\cdot,\cdot)$ be defined by eq.* (4.6.11). *Then $d(\cdot,\cdot)$ is nonexpanding for the 2h-skeleton chain Markov process* $\mathbf{X}$ *defined by eq.* (4.1.1), *and it is contracting for* $\mathbf{X}$ *on each of the sets* (4.6.5).

Proof. By Proposition 4.6.3, metric $d(\cdot,\cdot)$ is contracting on the set $\widehat{B}$ and nonexpanding for the h-skeleton chain. Using the general Coupling lemma (Theorem 4.4.3) and the Markov property, we extend this property to the $2h$-skeleton chain; the argument here is the same as in the proof of Theorem 4.5.1. Hence, the only thing left to prove is that for any R metric $d(\cdot,\cdot)$ is contracting on the set

$$B_R = \{(\mathbf{x}, \mathbf{y}) : |\mathbf{x}_0 - \mathbf{y}_0| \le R\} \setminus \widehat{B}.$$

For given initial values $(\mathbf{x}, \mathbf{y}) \in B_R$, we define the pair of processes $\widetilde{X}, \widetilde{Y}$, a combining the constructions from Propositions 4.6.3 and 4.6.4. Namely, for given $\mathbf{x}, \mathbf{y}$, we take the pair $\widehat{X}, \widehat{Y}$ given by Proposition 4.6.4, and consider the values $\widehat{\mathbf{X}}(h), \widehat{\mathbf{X}}(h)$ of corresponding segment processes as the initial values $\hat{\mathbf{x}}, \hat{\mathbf{y}}$ of another pair of processes X, Y, which is chosen in such a way that their segment processes satisfy

$$\mathsf{E} d(\mathbf{X}(h), \mathbf{Y}(h)) = \inf_{(\mathbf{X}', \mathbf{Y}') \in C(\mathsf{P}_{\hat{\mathbf{x}}}, \mathsf{P}_{\hat{\mathbf{y}}})} d(\mathbf{X}'(h), \mathbf{Y}'(h)).$$

This construction is formally correct, since the pair X, Y can be chosen in a measurable way w.r.t. $\hat{\mathbf{x}}, \hat{\mathbf{y}}$, see Theorem 4.4.3. Then,

$$\begin{aligned}\mathsf{E}d(\widetilde{\mathbf{X}}(2h), \widetilde{\mathbf{Y}}(2h)) &= \mathsf{E}d(\mathbf{X}(h), \mathbf{Y}(h))1_{(\widehat{\mathbf{X}}(h),\widehat{\mathbf{X}}(h))\in\widehat{B}} \\ &\quad + \mathsf{E}d(\mathbf{X}(h), \mathbf{Y}(h))1_{(\widehat{\mathbf{X}}(h),\widehat{\mathbf{X}}(h))\notin\widehat{B}} \le 1 - \delta.\end{aligned}$$

Since $(\widetilde{\mathbf{X}}, \widetilde{\mathbf{Y}}) \in C(\mathsf{P}_{\mathbf{x}}, \mathsf{P}_{\mathbf{y}})$ by the construction, and $d(\mathbf{x}, \mathbf{y}) = 1$, this yields

$$d\big(\mathsf{P}^{2h}(\mathbf{x}, \cdot), \mathsf{P}^{2h}(\mathbf{y}, \cdot)\big) \le (1 - \delta)d(\mathbf{x}, \mathbf{y}),$$

which completes the proof. □

4.6.3 Summary

Likewise to Section 3.3.3, where ergodic rates for diffusion processes were established, we can summarize the above analysis and give sufficient conditions for exponential, subexponential, and polynomial weak ergodic rates solutions to SDDEs. Recall that we assume that coefficients $a(x)$, $b(x)$ are Lipschitz continuous, $a(x)$ satisfies the drift condition (3.3.4), coefficient $b(x)$ is bounded and $B(x) = b(x)b^*(x)$ satisfies the uniform ellipticity condition. Let the metric $d(\mathbf{x}, \mathbf{y})$ be defined by eq. (4.6.11), and

$$d_p(\mathbf{x}, \mathbf{y}) = \big(d(\mathbf{x}, \mathbf{y})\big)^{1/p}.$$

Theorem 4.6.6. *Let either $\kappa > 0$ and $\alpha > 0$ be arbitrary, or $\kappa = 0$ and $\alpha > 0$ satisfy*

$$\alpha < \frac{2A_0}{|||B|||}.$$

Then for every $p > 1$, there exists $c_1, c_2 > 0$ such that, for any $\mathbf{x}, \mathbf{y} \in C([-r, 0], \mathbb{R}^m)$, $t \ge 0$,

$$d_p\big(P^t(\mathbf{x}, \cdot), P^t(\mathbf{y}, \cdot)\big) \le c_1 e^{-c_2 t}\Big(e^{\alpha|\mathbf{x}_0|} + e^{\alpha|\mathbf{y}_0|}\Big)d_p(\mathbf{x}, \mathbf{y}). \tag{4.6.23}$$

In addition, there exists a unique IPM μ for $\mathbf{X}$, this measure satisfies

$$\int\limits_{C([-r,0],\mathbb{R}^m)} e^{\alpha|\mathbf{x}_0|}\,\mu(d\mathbf{x}) < \infty,$$

and, for every $\mathbf{x} \in C([-r, 0], \mathbb{R}^m)$, $t \ge 0$,

$$d_p\big(P^t(\mathbf{x}, \cdot), \mu\big) \le c_1 e^{-c_2 t}\left(e^{\alpha|\mathbf{x}_0|} + \int\limits_{C([-r,0],\mathbb{R}^m)} e^{\alpha|\mathbf{y}_0|}\,\mu(d\mathbf{y})\right). \tag{4.6.24}$$

Proof. Let h be given in Proposition 4.6.3. By statement I of Proposition 4.6.1, Theorem 3.2.3, Example 3.2.5, and Theorem 2.8.6, we have that condition **R** of Theorem 4.5.2 holds true for $2h$-skeleton chain for $\mathbf{X}$ on certain set of the form (4.6.5) with

$$W(\mathbf{x},\mathbf{y}) = V(\mathbf{x}_0) + V(\mathbf{y}_0), \quad \lambda(t) = e^{c_2 t},$$

where $V \in C^2$ is specified in Proposition 4.6.1 and $c_2 > 0$ is some constant. By Corollary 4.6.5, condition **I** of Theorem 4.5.2 holds true for $2h$-skeleton chain on any set of the form (4.6.5). Hence the required statement for the $2h$-skeleton chain follows by Theorem 4.5.2. Using Proposition 4.2.2 and the Markov property, we easily extend this statement to the entire process $\mathbf{X}$. □

The proofs of the following two statements are completely analogous, with statements II, III of Proposition 4.6.1 used instead of statement I.

Theorem 4.6.7. *Let $\kappa \in (-1,0)$. Then for any positive*

$$\alpha < \frac{2A_\kappa}{(1+\kappa)|||B|||}.$$

there exist $c_1, c_2 > 0$ such that, for any $\mathbf{x},\mathbf{y} \in C([-r,0],\mathbb{R}^m)$, $t \geq 0$

$$d_p\big(P^t(\mathbf{x},\cdot), P^t(\mathbf{y},\cdot)\big) \leq c_1 e^{-c_2 t^{(1+\kappa)/(1-\kappa)}}\Big(e^{\alpha|\mathbf{x}_0|^{1+\kappa}} + e^{\alpha|\mathbf{y}_0|^{1+\kappa}}\Big) d_p(\mathbf{x},\mathbf{y}). \tag{4.6.25}$$

In addition, there exists a unique IPM μ for $\mathbf{X}$, this measure satisfies

$$\int_{C([-r,0],\mathbb{R}^m)} e^{\alpha|\mathbf{x}_0|^{1+\kappa}}\,\mu(d\mathbf{x}) < \infty,$$

and, for every $\mathbf{x} \in C([-r,0],\mathbb{R}^m)$, $t \geq 0$,

$$d_p\big(P^t(\mathbf{x},\cdot),\mu\big) \leq c_1 e^{-c_2 t^{(1+\kappa)/(1-\kappa)}}\left(e^{\alpha|\mathbf{x}_0|^{1+\kappa}} + \int_{C([-r,0],\mathbb{R}^m)} e^{\alpha|\mathbf{y}_0|^{1+\kappa}}\,\mu(d\mathbf{y})\right). \tag{4.6.26}$$

Theorem 4.6.8. *Let $\kappa = -1$ and*

$$2A_{-1} > \sup_{x\in\mathbb{R}^m}\big(\operatorname{Trace} B(x)\big).$$

Then for every $p > 2$, $p_1 > 1$ such that

$$2A_{-1} > \sup_{x\in\mathbb{R}^m}\big(\operatorname{Trace} B(x) + (pp_1 - 2)\|B(x)\|\big),$$

there exist $c_1,\ c_2 > 0$ *such that, for any* $\mathbf{x}, \mathbf{y} \in C([-r, 0], \mathbb{R}^m)$, $t \geq 0$

$$d_{p_1}\big(P^t(\mathbf{x}, \cdot), P^t(\mathbf{y}, \cdot)\big) \leq c_1(1 + c_2 t)^{-p/2}\big(1 + |\mathbf{x}_0|^p + |\mathbf{y}_0|^p\big)d_{p_1}(\mathbf{x}, \mathbf{y}). \tag{4.6.27}$$

In addition, there exists a unique IPM μ *for* X*, this measure satisfies*

$$\int\limits_{C([-r,0],\mathbb{R}^m)} |\mathbf{x}_0|^{pp_1-2}\,\mu(d\mathbf{x}) < \infty,$$

and, for every $\mathbf{x} \in C([-r, 0], \mathbb{R}^m)$, $t \geq 0$,

$$\begin{aligned} d_{p_1}\big(P^t(\mathbf{x}, \cdot), \mu\big) &\leq c_1(1 + c_2 t)^{-(pp_1-2)/(2p_1)} \\ &\times \left(1 + |\mathbf{x}_0|^{pp_1-2} + \int\limits_{C([-r,0],\mathbb{R}^m)} |\mathbf{y}_0|^{pp_1-2}\,\mu(d\mathbf{y})\right). \end{aligned} \tag{4.6.28}$$

As a final remark, let us mention that the simple form of the SDDE (4.1.1) was adopted for clarity of exposition, only. Essentially the same results can be obtained for general SDDEs

$$\mathrm{d}X_t = -a(\mathbf{X}(t))\,\mathrm{d}t + b(\mathbf{X}(t))\,\mathrm{d}W_t$$

with the coefficients dependent on the entire segment. Once $a(\mathbf{x}), b(\mathbf{x})$ are Lipschitz continuous and $B(\mathbf{x}) = b(\mathbf{x})b^*(\mathbf{x})$ is uniformly elliptic, the construction from Section 4.6.2 can be repeated literally. The main novelty appears with the Lyapunov-type condition. For the drift coefficient of the form

$$a(\mathbf{x}) = a_0(\mathbf{x}_0) + \tilde{a}(\mathbf{x})$$

with bounded $\tilde{a}(\mathbf{x})$ and $a_0(x)$ satisfying the drift condition (3.3.4), the argument remains essentially the same. If such a decomposition is not available, and the argument in the drift coefficient is delayed substantially, the situation changes drastically, we refer to Ref. [14] for more details.

4.7 Comments

1. The general approach to weak ergodic rates, based on the use of a distance-like function $d(x, y)$, which is nonexpanding and is contracting on a certain set B, was introduced in Ref. [55]. This approach appears to be very convenient for dealing with Markov systems which neither are ergodic in total variation nor exhibit a dissipative behavior similar to the one discussed in Section 4.2. In Ref. [55] the general Harris-type theorem was obtained with the exponential rate of convergence, in Ref.

[13] it was extended to the general subexponential setting. Both in Refs. [55] and [13], the proofs were based on an analytical contraction-type argument. The probabilistic argument, based on the coupling construction and adopted in the current exposition, makes the proof of the main result of the chapter – Theorem 4.5.2 – quite short and transparent. For another proof of a general Harris-type theorem based on the coupling argument, we refer to Ref. [32].

2. The main difficulty in application of the general Harris-type theorem is a construction of a contracting distance-like function $d(x, y)$. We illustrate this topic by a particular example of an SDDE; actually the one treated in Ref. [55]. The proof of Ref. [55, Proposition 5.2] is rather involving, and in Section 4.6.2 we present another proof, based on the concept of the generalized coupling. This concept, under the name *asymptotic coupling*, was introduced in Ref. [55] in the spirit of Refs. [35, 50, 104]; see also Ref. [118]. Originally, this concept was designed as a tool for proving unique ergodicity; in Ref. [87], it was shown that, moreover, it can be used to guarantee weak stabilization of the transition probabilities. The argument developed in Section 4.6.2 looks quite promising because generalized couplings with required properties can be constructed quite easily in rather sophisticated settings; see Ref. [46]. Similar "stochastic control" argument was used in the proof of the local Dobrushin condition for degenerate diffusions in Ref. [1], and for solutions to Lévy-driven SDEs in Ref. [11].

Part II: **Limit Theorems**

5 The Law of Large Numbers and the Central Limit Theorem

5.1 Preliminaries

In this introductory section, we collect the facts and the simplest limit theorems, which can be derived easily, once ergodic rates for a Markov chain are specified. Throughout the entire chapter, we assume that a Markov chain X has unique IPM μ and satisfies

$$d\Big(P^n(x,\cdot),\mu\Big) \le V(x)r_n, \qquad n \ge 0. \tag{5.1.1}$$

Here d denotes the coupling distance on $\mathscr{P}(\mathbb{X})$ which corresponds to some distance-like function d on $\mathbb{X}$, sequence $r_n \to 0$ actually controls the rate of convergence of transition probabilities to the IPM, and function $V : \mathbb{X} \to [1,\infty)$ has the meaning of a "penalty term", see Remark 2.6.2 and discussion after Theorem 2.6.3.

We begin our consideration with the case $d(x,y) = 1_{x\neq y}$, where the corresponding coupling distance equals

$$d(\mu,\nu) = \frac{1}{2}\|\mu-\nu\|_{\mathrm{TV}}, \quad \mu,\nu \in \mathscr{P}(\mathbb{X}),$$

and thus eq. (5.1.1) actually gives an ergodic rate in the total variation distance. The following upper bounds for the α- and ϕ-mixing coefficients of the stationary version of the chain are similar to those given in Proposition 1.3.4 in the discrete state space setting.

Proposition 5.1.1. *Let X be a stationary Markov chain. Then,*

(i)

$$\alpha(n) \le \int_{\mathbb{X}} \|P^n(x,\cdot)-\mu\|_{TV}\mu(dx), \quad n \ge 1;$$

(ii)

$$\phi(n) \le \sup_{x\in\mathbb{X}} \|P^n(x,\cdot)-\mu\|_{TV}, \quad n \ge 1.$$

Proof. Fix n, take arbitrary $A \in \mathscr{F}_{n,\infty}$, and denote by f such a function that

$$\mathsf{E}[1_A|X_n] = f(X_n),$$

DOI 10.1515/9783110458930-006

then by the Markov property of X,

$$\mathsf{P}(A \cap B) = \mathsf{E}f(X_n)1_B = \mathsf{E}\left(1_B \int_{\mathbb{X}} f(y)P^n(X_0, dy)\right).$$

We also have by stationarity,

$$\mathsf{P}(A)\mathsf{P}(B) = \mathsf{E}f(X_n)\mathsf{E}1_B = \mathsf{E}\left(1_B \int_{\mathbb{X}} f(y)\mu(dy)\right).$$

Since $f(x) \in [0, 1]$, this gives,

$$|\mathsf{P}(A \cap B) - \mathsf{P}(A)\mathsf{P}(B)| \le \mathsf{E}\|P^n(X_0, \cdot) - \mu\|_{TV} 1_B,$$

which yields the required bounds. □

By Proposition 5.1.1, a Markov chain, which satisfies eq. (5.1.1) with $d(x, y) = 1_{x \neq y}$, is

- α-mixing, if the "penalty term" V is integrable w.r.t. μ;
- ϕ-mixing, if V is bounded.

In addition, the following bounds are available:

$$\alpha(n) \le 2r_n \int_{\mathbb{X}} V\, d\mu, \quad \phi(n) \le 2r_n \sup_{x \in \mathbb{X}} V(x), \quad n \ge 1. \tag{5.1.2}$$

The following statement extends Corollary 1.3.5, and shows that LLN and CLT for functionals of a Markov chain can be derived easily, once the chain possesses an ergodic rate in the total variation distance.

Theorem 5.1.2. *Let X be a stationary Markov chain, which satisfies eq.* (5.1.1) *with $d(x, y) = 1_{x \neq y}$ and*

$$\int_{\mathbb{X}} V\, d\mu < \infty.$$

Then the following statements hold true.

1. *(LLN). For any $f : \mathbb{X} \to \mathbb{R}$ such that $\mathsf{E}|f(X_0)| < \infty$*

$$\frac{1}{N} \sum_{k=1}^{N} f(X_k) \to \mathsf{E}f(X_0), \quad n \to \infty \tag{5.1.3}$$

   a.s. and in mean sense.

2. *(CLT). If the function V is bounded, then for any f such that* $Ef(X_0) = 0$, $Ef^2(X_0) < \infty$, *the series*

$$\sigma^2 = Ef^2(X_0) + 2\sum_{k=1}^{\infty} Ef(X_k)f(X_0) \tag{5.1.4}$$

converges, and if $\sigma^2 > 0$,

$$\frac{1}{\sqrt{N}}\sum_{k=1}^{N} f(X_k) \Rightarrow \mathcal{N}(0,\sigma^2), \quad n \to \infty. \tag{5.1.5}$$

For unbounded V, the same conclusion holds true under the following additional assumption: there exists $\delta > 0$ *such that*

$$E|f(X_0)|^{2+\delta} < \infty, \quad \sum_n r_n^{\delta/(2+\delta)} < \infty. \tag{5.1.6}$$

Proof. Under the assumptions of the theorem, the chain is α-mixing; see the discussion above. In particular, it is mixing, and thus ergodic; see Section 1.3. Hence LLN follows by the Birkhoff theorem.

If eq. (5.1.6) holds true, then by eq. (5.1.2) the required CLT follows immediately from Theorem 1.3.1. If V is bounded, then the chain X is actually uniformly ergodic w.r.t. the total variation distance, and then it is ergodic at exponential rate; see the discussion in Section 2.9. That is, an analogue of eq. (5.1.1) holds true with a new sequence $\tilde{r}_n = Ce^{-\theta n}$, where C, θ are positive constants. In that case the required CLT follows from Theorem 1.3.2. □

Our further aim is to extend the above results to the weak ergodic setting, where eq. (5.1.1) holds true with a coupling distance, weaker than the total variation distance. We begin the discussion with a simple observation that a weakly ergodic chain may fail to be α-mixing.

Example 5.1.3. Let $\mathbf{X} = \{\mathbf{X}(t), t \geq 0\}$ be the segment process in $\mathbb{X} = C(-r, 0)$, which corresponds to the solution to the SDDE (4.1.1); see Example 4.1.2. Assume that the process $\mathbf{X}$ is weakly ergodic with the IPM μ which is not degenerate, and consider the stationary version of $\mathbf{X}$ with Law $\mathbf{X}(0) = \mu$. We have seen that, given any $\mathbf{X}(t)$, the entire trajectory of the process before the time moment t can be reconstructed. That is, there exists a measurable mapping $F_t : \mathbb{X} \to \mathbb{X}$ such that

$$\mathbf{X}(0) = F_t(\mathbf{X}(t)) \quad \text{a.s.}$$

Since μ is not equal to a δ-measure, there exists a set $C \in \mathscr{X}$ such that $\mu(C) \in (0,1)$. We take

$$A = 1_{F_t^{-1}(C)}(\mathbf{X}(t)), \quad B = 1_C(\mathbf{X}(0)),$$

and get that for any $t \geq 0$

$$\alpha(t) \geq |\mathsf{P}(A \cap B) - \mathsf{P}(A)\mathsf{P}(B)| = \mu(C)\big(1 - \mu(C)\big) > 0. \quad t \geq 0.$$

This example clearly shows that the intrinsic memory effect, discussed in Section 4.1, prevents the system from having the mixing property on the level of mixing coefficients, which are defined *uniformly* over $A \in \mathscr{F}_{t,\infty}$, $B \in \mathscr{F}_{-\infty,0}$. However, the weak ergodic rate (5.1.1) is typically strong enough to produce "individual" mixing property (1.3.3). To prove this fact, we first introduce some notation and give a simple, but important auxiliary estimate.

Let a distance-like function d, a function $W : \mathbb{X} \to \mathbb{R}^+$, and $\gamma \in (0, 1]$ be fixed. Define the *weighted Hölder class* $H_{\gamma,W}(\mathbb{X}, d)$ *w.r.t.* d *with the index* γ *and the weight* W as the set of functions $f : \mathbb{X} \to \mathbb{R}$ such that

$$\|f\|_{d,\gamma,W} = \sup_{x\neq y} \frac{|f(x) - f(y)|}{d^\gamma(x, y)\big(W(x) + W(y)\big)^{1-\gamma}} < \infty.$$

Here and below we use the convention $a^0 = 1$, $a \in \mathbb{R}^+$; hence for $\gamma = 1$, the weight W is inessential, and $H_{1,W}(\mathbb{X}, d) = H_1(\mathbb{X}, d)$ is just the Lipschitz class w.r.t. d.

Proposition 5.1.4. *Let a function f belong to $H_{\gamma,W}(\mathbb{X}, d)$ for some $\gamma \in (0, 1]$. Then for any $\lambda, \nu \in \mathscr{P}(\mathbb{X})$,*

$$\left|\int_{\mathbb{X}} f\, d\lambda - \int_{\mathbb{X}} f\, d\nu\right| \leq \|f\|_{d,\gamma,W}\big(d(\lambda, \nu)\big)^\gamma \left(\int_{\mathbb{X}} W\, d\lambda + \int_{\mathbb{X}} W\, d\nu\right)^{1-\gamma}. \tag{5.1.7}$$

Proof. Take any pair $(\xi, \eta) \in C(\lambda, \nu)$ and observe that

$$\begin{aligned}\left|\int_{\mathbb{X}} f\, d\lambda - \int_{\mathbb{X}} f\, d\nu\right| &= \big|\mathsf{E}f(\xi) - \mathsf{E}f(\eta)\big| \leq \mathsf{E}\big|f(\xi) - f(\eta)\big| \\ &\leq \|f\|_{d,\gamma,W}\, \mathsf{E}\left(d^\gamma(\xi, \eta)\big(W(\xi) + W(\eta)\big)^{1-\gamma}\right).\end{aligned}$$

If $\gamma < 1$, we apply the Hölder inequality with $p = 1/\gamma$:

$$\begin{aligned}\left|\int_{\mathbb{X}} f\, d\lambda - \int_{\mathbb{X}} f\, d\nu\right| &\leq \|f\|_{d,\gamma,W}\big(\mathsf{E}d(\xi, \eta)\big)^\gamma\big(\mathsf{E}W(\xi) + \mathsf{E}W(\eta)\big)^{1-\gamma} \\ &= \|f\|_{d,\gamma,W}\big(\mathsf{E}d(\xi, \eta)\big)^\gamma \left(\int_{\mathbb{X}} W\, d\lambda + \int_{\mathbb{X}} W\, d\nu\right)^{1-\gamma}.\end{aligned}$$

For $\gamma = 1$ the same bound holds true directly. The left-hand side in the above inequality does not depend on the choice of $(\xi, \eta) \in C(\lambda, \nu)$, and

$$\inf_{(\xi,\eta)\in C(\lambda,\nu)} \left(\mathsf{E}d(\xi,\eta)\right)^{\gamma} = \left(\inf_{(\xi,\eta)\in C(\lambda,\nu)} \mathsf{E}d(\xi,\eta)\right)^{\gamma} = \left(d(\lambda,\nu)\right)^{\gamma}.$$

This gives the required statement. □

Combined with our principal assumption (5.1.1), the bound (5.1.7) provides the following "stabilization bound" for expected values of a function $f \in H_{\gamma,W}(\mathbb{X}, d)$ w.r.t. the transition probabilities of the chain X.

Corollary 5.1.5. *Let the chain X satisfy eq.* (5.1.1). *Then for each $f \in H_{\gamma,W}(\mathbb{X}, d) \cap L_1(\mathbb{X}, \mu)$, we have*

$$\left|\mathsf{E}_x f(X_n) - \int_{\mathbb{X}} f\, d\mu\right| \le r_n^{\gamma} \|f\|_{d,\gamma,W} U_n(x), \quad n \ge 0, \tag{5.1.8}$$

where

$$U_n(x) = V^{\gamma}(x)\left(\mathsf{E}_x W(X_n) + \int_{\mathbb{X}} W\, d\mu\right)^{1-\gamma}, \quad n \ge 0. \tag{5.1.9}$$

Proof. The proof follows immediately by eq. (5.1.1) and (5.1.7) applied to $\lambda(\mathrm{d}y) = P^n(x, \mathrm{d}y)$, $\nu(\mathrm{d}y) = \mu(\mathrm{d}y)$. □

Remark 5.1.6. The functions U_n, $n \ge 0$, likewise to the function V in eq. (5.1.1), play the role of "penalty terms" in the stabilization bound (5.1.8); the heuristics here is close to the one discussed in Remark 2.6.2. These functions do not depend on the choice of f, hence inequalities in eq. (5.1.8) actually provide uniform bounds within the class $H_{\gamma,W}(\mathbb{X}, d) \cap L_1(\mathbb{X}, \mu)$. Note that eq. (5.1.8) with $n = 0$ actually provides a bound for the deviation of $f \in H_{\gamma,W}(\mathbb{X}, d) \cap L_1(\mathbb{X}, \mu)$ from its mean value w.r.t. μ:

$$\left|f(x) - \int_{\mathbb{X}} f\, \mathrm{d}\mu\right| \le r_0^{\gamma} \|f\|_{d,\gamma,W} U_0(x), \quad U_0(x) = V^{\gamma}(x)\left(W(x) + \int_{\mathbb{X}} W\, \mathrm{d}\mu\right)^{1-\gamma}. \tag{5.1.10}$$

The following proposition provides L_p-bounds for the "penalty terms" U_n, $n \ge 0$.

Proposition 5.1.7. *For any $m, n \ge 0$ and $p \ge 1$,*

$$\mathsf{E}\big(U_n(X_m)\big)^p \le 2^{(p-1)(1-\gamma)}\left(\mathsf{E}V^p(X_m)\right)^{\gamma}\left(\mathsf{E}W^p(X_{n+m}) + \left(\int_{\mathbb{X}} W\, d\mu\right)^p\right)^{(1-\gamma)}. \tag{5.1.11}$$

In particular, if $V, W \in L_p(\mathbb{X}, \mu)$, *then*

$$\|U_n\|_{L_p(\mathbb{X},\mu)} \le 2^{(1-\gamma)} \|V\|^{\gamma}_{L_p(\mathbb{X},\mu)} \|W\|^{1-\gamma}_{L_p(\mathbb{X},\mu)}, \quad n \ge 0. \tag{5.1.12}$$

Proof. To prove eq. (5.1.11), we just apply the Hölder inequality, the elementary inequality

$$(a+b)^p \le 2^{p-1}(a^p + b^p), \quad a, b \ge 0,$$

the Hölder inequality again, and the Markov property:

$$\begin{aligned} \mathsf{E}\big(U_n(X_m)\big)^p &\le \big(\mathsf{E}V^p(X_m)\big)^{\gamma} \left(\mathsf{E}\left(\mathsf{E}_{X_m} W(X_n) + \int_{\mathbb{X}} W \, d\mu \right)^p \right)^{(1-\gamma)} \\ &\le 2^{1-\gamma}\big(\mathsf{E}V^p(X_m)\big)^{\gamma} \left(\mathsf{E}\mathsf{E}_{X_m} W^p(X_n) + \left(\int_{\mathbb{X}} W \, d\mu \right)^p \right)^{(1-\gamma)} \\ &= 2^{1-\gamma}\big(\mathsf{E}V^p(X_m)\big)^{\gamma} \left(\mathsf{E}W^p(X_{n+m}) + \left(\int_{\mathbb{X}} W \, d\mu \right)^p \right)^{(1-\gamma)}. \end{aligned}$$

To prove eq. (5.1.12), we consider stationary version of X, then $\mathrm{Law}(X_k) = \mu$, $k \ge 0$ and the required statement easily follows by eq. (5.1.11).

Combining eq. (5.1.8) and the moment bound (5.1.12), we derive the following.

Corollary 5.1.8. *Let* X *be a stationary Markov chain, which satisfies eq.* (5.1.1) *with* $V \in L_2(X, \mu)$. *Let* $f \in H_{\gamma,W}(\mathbb{X}, d)$ *with* $W \in L_2(X, \mu)$. *Then for* $g \in L_2(\mathbb{X}, \mu)$,

$$\Big|\mathrm{Cov}\big(f(X_n), g(X_0)\big)\Big| \le 2^{1-\gamma} r_n^{\gamma} \|f\|_{d,\gamma,W} \|g\|_{L_2(\mathbb{X},\mu)} \|V\|^{\gamma}_{L_2(\mathbb{X},\mu)} \|W\|^{1-\gamma}_{L_2(\mathbb{X},\mu)}, \quad n \ge 0. \tag{5.1.13}$$

Proof. By eq. (5.1.8) and stationarity of X,

$$\begin{aligned} \Big|\mathrm{Cov}\big(f(X_n), g(X_0)\big)\Big| &= \Big|\mathsf{E}\big(\mathsf{E}_{X_0} f(X_n) - \mathsf{E}f(X_0)\big) g(X_0)\Big| \\ &\le r_n^{\gamma} \|f\|_{d,\gamma,W} \|U_n\|_{L_2(\mathbb{X},\mu)} \|g\|_{L_2(\mathbb{X},\mu)}. \end{aligned}$$

Combining this with eq. (5.1.12), we complete the proof. □

This leads to the following important property.

Proposition 5.1.9. *Let* X *be a stationary Markov chain, which satisfies eq.* (5.1.1) *with* $V \in L_2(X, \mu)$. *Assume also that for some* $\gamma \in (0, 1]$, $W \in L_2(X, \mu)$ *the family of bounded* *functions from the class* $H_{\gamma,W}(\mathbb{X}, d)$ *is dense in* $L_2(\mathbb{X}, \mu)$*. Then eq.* (1.3.3) *holds true for any bounded* ξ, η*; that is, the sequence* X *is mixing.*

Proof. Take first $\xi = f(X_0)$ with bounded $f \in H_{\gamma,W}(\mathbb{X}, d)$, and $\eta = g(X_0)$ with bounded g. Then eq. (1.3.3) holds true by Corollary 5.1.8. Next, consider f just bounded, and take a sequence $\{f_k\}$ which approximates f in $L_2(\mathbb{X}, \mu)$. Denote $\xi_k = f(X_0)$. By stationarity, we have for any k

$$|\mathrm{Cov}(\theta_n\xi, \eta)| \leq |\mathrm{Cov}(\theta_n\xi - \theta_n\xi_k, \eta)| + \|f - f_k\|_{L_2(\mathbb{X},\mu)}\|g\|_{L_2(\mathbb{X},\mu)},$$

hence

$$\limsup_{n\to\infty} |\mathrm{Cov}(\theta_n\xi, \eta)| \leq \|f - f_k\|_{L_2(\mathbb{X},\mu)}\|g\|_{L_2(\mathbb{X},\mu)}.$$

Taking $k \to \infty$, we complete the proof of eq. (1.3.3) in this case. Next, let there exist $m \geq 1$ such that ξ, η are $\mathscr{F}_{0,m}$-measurable. Then, for $n > m$

$$\mathrm{Cov}(\theta_n\xi, \eta) = \mathrm{Cov}(\theta_{n-m}\xi^{(m)}, \eta),$$

where

$$\xi^{(m)} = \mathsf{E}[\xi|\mathscr{F}_0] = f^{(m)}(X_0),$$

and thus eq. (1.3.3) holds true in this case, as well. Since in the class such random variables are dense in $L_2(\Omega, \sigma(X), \mathsf{P})$, this completes the proof. □

Using the Birkhoff theorem, we can summarize the above results.

Theorem 5.1.10. *Let* X *be a stationary Markov chain, which satisfies eq.* (5.1.1)*, where* $V \in L_2(\mathbb{X}, \mu)$ *and the distance-like function* d *is such that, for some* $\gamma \in (0, 1]$ *and* $W \in L_2(\mathbb{X}, \mu)$*, the family of bounded functions from the class* $H_{\gamma,W}(\mathbb{X}, d)$ *is dense in* $L_2(\mathbb{X}, \mu)$*.*

Then for any $f : \mathbb{X} \to \mathbb{R}$ *such that* $\mathsf{E}|f(X_0)| < \infty$*,*

$$\frac{1}{N}\sum_{k=1}^{N} f(X_k) \to \mathsf{E}f(X_0), \quad n \to \infty \tag{5.1.14}$$

a.s. and in mean sense.

This Law of Large Numbers extends the one obtained in Theorem 5.1.2 to a general setting, where the chain X is weakly ergodic. The Central Limit Theorem can be extended, as well, but such an extension is far from being straightforward, because the chain X may fail to be α-mixing, and thus we cannot apply the ready-made Theorem 1.3.1. We give such an extension in the following section. Note that these results can be modified in order to hold true for non-stationary chains, as well; we will discuss this topic in Section 5.3.1.

5.2 The CLT for Weakly Ergodic Markov Chains

In this section, we explain one practical method to prove CLT (5.1.5) for a weakly ergodic Markov chain X. To make it easier to relate this result with other results in the field and to give further extensions, we introduce a general *auto-regressive (AR) model*

$$\zeta_n^\varepsilon = \zeta_0^\varepsilon + \sum_{k=1}^n \xi_k^\varepsilon, \quad n \geq 0. \tag{5.2.1}$$

Together with the model (5.2.1), we fix a family of filtrations

$$\mathbb{F}^\varepsilon = \{\mathscr{F}_k^\varepsilon, k \geq 0\}, \quad \varepsilon > 0,$$

assuming $\{\xi_k^\varepsilon\}$ to be adapted with $\mathbb{F}^\varepsilon$. Clearly, we can plug CLT (5.1.5) into the AR model (5.2.1), simply taking

$$\zeta_0^\varepsilon = 0, \quad \xi_k^\varepsilon = \sqrt{\varepsilon} f(X_k), \quad \varepsilon = n^{-1}, \quad \mathscr{F}_k^\varepsilon = \mathscr{F}_k = \sigma(X_j, j \leq k).$$

5.2.1 The Martingale CLT

We start our exposition by CLT with the principal assumption that the family $\{\xi_k^\varepsilon\}$, up to asymptotically negligible terms, forms a *martingale difference* w.r.t. $\mathbb{F}^\varepsilon$. Extension of CLT to dependent random sequences, which have the martingale structure, was initiated by P. Lévy, Refs. [92–94]; see also a survey paper Ref. [66]. Now this theory is deeply developed; see Refs. [56] and [64]. For the reader's convenience, we formulate explicitly the version of the martingale CLT which will be used below, and give its complete proof.

Theorem 5.2.1. *(The martingale CLT) Let the family $\{\xi_k^\varepsilon\}$ satisfy the following:*

(i) *(the asymptotic martingale property):*

$$E[\xi_k^\varepsilon | \mathscr{F}_{k-1}^\varepsilon] = \varepsilon \gamma_k^\varepsilon, \quad k \geq 1$$

and

$$\sup_k E|\gamma_k^\varepsilon|^2 \to 0, \quad \varepsilon \to 0;$$

(ii) *(LLN for conditional variances): the random variables*

$$\beta_k^\varepsilon = E[(\xi_k^\varepsilon)^2 | \mathscr{F}_{k-1}^\varepsilon], \quad k \geq 1$$

are well defined, and for some $\sigma \geq 0$

$$\sum_{k \leq \varepsilon^{-1} t} \beta_k^\varepsilon \to t\sigma^2, \quad \varepsilon \to 0, \quad t \geq 0,$$

in mean sense;

(iii) *(the Lyapunov condition): for some* $p > 2$,

$$\mathsf{E}|\xi_k^\varepsilon|^p \leq C\varepsilon^{p/2}, \quad k \geq 1.$$

Then,

$$\sum_{k \leq \varepsilon^{-1}} \xi_k^\varepsilon \Rightarrow \mathcal{N}(0, \sigma^2), \quad \varepsilon \to 0.$$

Proof. Denote

$$Y^\varepsilon(t) = \zeta_{[\varepsilon^{-1} t]}^\varepsilon = \sum_{k \leq \varepsilon^{-1} t} \xi_k^\varepsilon, \quad t \geq 0.$$

We have

$$\mathsf{E}(Y^\varepsilon(t) - Y^\varepsilon(s))^2 = \mathsf{E} \sum_{\varepsilon^{-1} s < k \leq \varepsilon^{-1} t} \left(\beta_k^\varepsilon + (\gamma_k^\varepsilon)^2\right), \quad s < t.$$

The Lyapunov condition and the Hölder inequality guarantee that

$$\mathsf{E}\beta_k^\varepsilon = \mathsf{E}(\xi_k^\varepsilon)^2 \leq C\varepsilon,$$

hence we have

$$\mathsf{E}(Y^\varepsilon(t) - Y^\varepsilon(s))^2 \leq C\left(t - s + \varepsilon\right), \quad 0 \leq s \leq t. \tag{5.2.2}$$

Then it is easy to prove that the family $\{Y^\varepsilon\}$ is *weakly compact* in the sense of convergence of finite-dimensional distributions; see Ref. [124]. That is, for any sequence $\{\varepsilon_m\}$ with $\varepsilon_m \to 0$, there exists a subsequence $\{\varepsilon_{m_k}\}$ such that, for any $t_1, \dots t_m \in [0, 1]$,

$$\left(Y^{\varepsilon_{m_k}}(t_1), \dots, Y^{\varepsilon_{m_k}}(t_m)\right) \Rightarrow \left(Y(t_1), \dots, Y(t_m)\right).$$

In the rest of the proof, we show that the limiting process Y equals σW with a Wiener process W; this will immediately yield the required weak convergence for

$$\sum_{k \leq \varepsilon^{-1}} \xi_k^\varepsilon = Y^\varepsilon(1).$$

We fix $\phi \in C_b^3$ and write

$$\phi(\zeta_k^\varepsilon) - \phi(\zeta_{k-1}^\varepsilon) = \phi'(\zeta_{k-1}^\varepsilon)\xi_k^\varepsilon + \frac{1}{2}\phi''(\zeta_{k-1}^\varepsilon)(\xi_k^\varepsilon)^2 + \chi_k^\varepsilon, \tag{5.2.3}$$

where χ_k^ε is just the notation; that is,

$$\chi_k^\varepsilon = \phi(\zeta_k^\varepsilon) - \phi(\zeta_{k-1}^\varepsilon) - \phi'(\zeta_{k-1}^\varepsilon)\xi_k^\varepsilon - \frac{1}{2}\phi''(\zeta_{k-1}^\varepsilon)(\xi_k^\varepsilon)^2.$$

We have

$$\mathsf{E}\phi(Y^\varepsilon(t)) = \phi(0) + \frac{1}{2}\varepsilon\sigma^2 \sum_{k\le\varepsilon^{-1}t} \mathsf{E}\phi''(Y^\varepsilon(k\varepsilon)) + r_\phi^\varepsilon(t), \tag{5.2.4}$$

where

$$r_\phi^\varepsilon(t) = \mathsf{E}\sum_{k\le\varepsilon^{-1}t} \phi'(\zeta_{k-1}^\varepsilon)\left(\varepsilon\gamma_k^\varepsilon\right) + \frac{1}{2}\mathsf{E}\sum_{k\le\varepsilon^{-1}t}\phi''(\zeta_{k-1}^\varepsilon)\left(\beta_k^\varepsilon - \varepsilon\sigma^2\right) + \mathsf{E}\sum_{k\le\varepsilon^{-1}t}\chi_k^\varepsilon = \sum_{i=1}^3 r_\phi^{i,\varepsilon}(t).$$

We aim is to show that

$$r_\phi^\varepsilon(t) = \sum_{i=1}^3 r_\phi^{i,\varepsilon}(t) \to 0, \quad \varepsilon \to 0, \tag{5.2.5}$$

and we analyze the terms $r_\phi^{i,\varepsilon}(t), i = 1, 2, 3$ separately. The first term is easily estimated by

$$|r_\phi^{1,\varepsilon}(t)| \le t\sup_\zeta |\phi'(\zeta)| \sup_k \mathsf{E}\gamma_k^\varepsilon \to 0, \quad \varepsilon \to 0.$$

The third term is also simple. Namely, we apply, if necessary, the Hölder inequality, and observe that the Lyapunov condition is satisfied with some $p \in (2, 3]$. Since ϕ'', ϕ''' are bounded, we have that ϕ'' is Hölder continuous with the index $p - 2 \le 1$. Then,

$$|\phi(y+z) - \phi(y) - \phi'(y)z - \frac{1}{2}z^2\phi''(z)| = \left| z^2 \int_0^1 \int_0^t (\phi''(y+sz) - \phi''(y))\, ds dt \right| \le C|z|^p. \tag{5.2.6}$$

That is, $|\chi_k^\varepsilon| \le C|\xi_k^\varepsilon|^p$, and the Lyapunov condition gives

$$|r_\phi^{3,\varepsilon}(t)| \le Ct\varepsilon^{-1+p/2} \to 0, \quad \varepsilon \to 0.$$

To analyze the second term, we introduce an auxiliary construction. Fix a partition $0 = t_0 < \cdots < t_m = t$ and denote

$$\tilde{r}_{\phi}^{2,\varepsilon}(t) = \frac{1}{2}\mathsf{E}\sum_{j=1}^{m} \sum_{k:\varepsilon k\in(t_{j-1},t_j]} \phi''(Y^{\varepsilon}(t_{j-1}))\left(\beta_k^{\varepsilon} - \varepsilon\sigma^2\right).$$

For each j,

$$\sum_{k:\varepsilon k\in(t_{j-1},t_j]} \left(\beta_k^{\varepsilon} - \varepsilon\sigma^2\right) \to 0, \quad \varepsilon \to 0$$

in the mean sense, and because f'' is bounded, this gives

$$|\tilde{r}_{\phi}^{2,\varepsilon}(t)| \to 0, \quad \varepsilon \to 0.$$

On the other hand, ϕ'' is Hölder continuous with the index $2(p-2)/p < 1$. Combining eq. (5.2.2) and the inequality

$$\mathsf{E}\left|\beta_k^{\varepsilon} - \varepsilon\sigma^2\right|^{p/2} \le C\varepsilon^{p/2}$$

(which holds true by the Lyapunov condition), we get

$$\begin{aligned}
&\mathsf{E}\left|\phi''(Y^{\varepsilon}(\varepsilon(k-1)) - \phi''(Y^{\varepsilon}(t_{j-1}))\right|\left|\beta_k^{\varepsilon} - \varepsilon\sigma^2\right| \\
&\quad \le C\mathsf{E}\left|Y^{\varepsilon}(\varepsilon(k-1) - Y^{\varepsilon}(t_{j-1})\right|^{2(p-2)/p)}\left|\beta_k^{\varepsilon} - \varepsilon\sigma^2\right| \\
&\quad \le C\left(\mathsf{E}\left(Y^{\varepsilon}(\varepsilon(k-1)) - Y^{\varepsilon}(t_{j-1})\right)^2\right)^{(p-2)/p}\left(\mathsf{E}\left|\beta_k^{\varepsilon} - \varepsilon\sigma^2\right|^{p/2}\right)^{2/p} \\
&\quad \le C\varepsilon\left(|\varepsilon(k-1) - t_{j-1}| + \varepsilon\right)^{(p-2)/p}.
\end{aligned}$$

That is,

$$|r_{\phi}^{2,\varepsilon}(t) - \tilde{r}_{\phi}^{2,\varepsilon}(t)| \le Ct\max_j\left(|t_j - t_{j-1}| + \varepsilon\right)^{(p-2)/p},$$

and thus

$$\limsup_{\varepsilon\to 0}|r_{\phi}^{2,\varepsilon}(t)| \le Ct\max_j|t_j - t_{j-1}| + \varepsilon^{(p-2)/p}.$$

Since the auxiliary partition $\{t_j\}$ can be taken arbitrarily fine, this completes the proof of eq. (5.2.5).

Now we can complete the proof. Passing to the limit in eq. (5.2.4) by the subsequence $\{\varepsilon_{m_k}\}$, we get for the corresponding limit process Y the relation

$$\mathsf{E}\phi(Y(t)) = \phi(0) + \frac{1}{2}\sigma^2 \int_0^t \mathsf{E}\phi''(Y(s))\,ds, \tag{5.2.7}$$

valid for arbitrary $\phi \in C_b^3(\mathbb{R})$. Take $\phi(y) = e^{i\lambda y}$ with some $\lambda \in \mathbb{R}$, then $\phi'' = -\lambda^2\phi$ and the function $m(t) = \mathsf{E}\phi(Y(t))$ satisfies the linear ODE

$$m'(t) = -\frac{1}{2}\sigma^2\lambda^2 m(t), \quad m(0) = 1.$$

Thus,

$$\mathsf{E}e^{i\lambda Y(t)} = e^{-\sigma^2\lambda^2 t/2}, \quad \lambda \in \mathbb{R},$$

which proves that $Y(t) \sim \mathcal{N}(0, t\sigma^2)$. Similarly, one can prove that the joint distribution for $\Big(Y(t_1), \dots, Y(t_m)\Big)$ is the same with the one for σW; we omit the details. This completes the proof. □

Remark 5.2.2. We actually have proved a stronger statement than the "individual" CLT (5.1.5), claimed initially: for the family of processes

$$Y^\varepsilon(t) = \sum_{k \le \varepsilon^{-1}t} \xi_k^\varepsilon, \quad t \ge 0, \tag{5.2.8}$$

all finite-dimensional distributions weakly converge to those for the process σW. This result has a further extension, and Y^ε actually converges weakly to σW in a proper functional sense; see Remark 6.2.5.

5.2.2 The CLT for a Weakly Ergodic Markov Chain

In this section, we fix a chain X, which satisfies eq. (5.1.1), and a function $f \in H_{\gamma,W}(\mathbb{X}, d)$. We consider the AR model (5.2.1) with

$$\zeta_0^\varepsilon = 0, \quad \xi_k^\varepsilon = \sqrt{\varepsilon} f(X_k), \quad \varepsilon = n^{-1}, \quad \mathcal{F}_k^\varepsilon = \mathcal{F}_k = \sigma(X_j, j \le k),$$

and study the limit behavior of

$$\zeta_n^\varepsilon = \frac{1}{\sqrt{n}} \sum_{k=1}^n f(X_k), \quad \varepsilon = n^{-1} \to 0.$$

We say that the function f is *centered*, if $f \in L_1(\mathbb{X}, \mu)$ and

$$\int_{\mathbb{X}} f(x)\mu(\mathrm{d}x) = 0. \tag{5.2.9}$$

Condition that f is centered is a natural pre-requisite to CLT (5.1.5), and it is substantially weaker than the assumption that

$$\xi_k^\varepsilon = \sqrt{\varepsilon} f(X_k)$$

is a martingale difference. Namely, we have

$$\mathsf{E}[\xi_k^\varepsilon | \mathscr{F}_{k-1}] = \sqrt{\varepsilon} \int_{\mathbb{X}} f(y) P(X_{k-1}, \mathrm{d}y),$$

and the martingale property requires

$$\int_{\mathbb{X}} f(y) P(x, dy) = 0, \quad x \in \mathbb{X}, \tag{5.2.10}$$

which is substantially more restrictive than eq. (5.2.9). Below, we explain the *corrector term* trick, which makes it possible to transform the model in such a way that the martingale CLT becomes applicable.

For a given function $f : \mathbb{X} \to \mathbb{R}$, define the *potential* as the sum of the series

$$\mathscr{R} f(x) = \sum_{k=0}^{\infty} \mathsf{E}_x f(X_k), \tag{5.2.11}$$

provided that this sum is well defined.

Proposition 5.2.3. *Let a stationary Markov chain X satisfy eq.* (5.1.1) *and function $f \in H_{\gamma,W}(\mathbb{X}, d)$ be centered. Assume also that*

$$\sum_k r_k^\gamma < \infty$$

and for some $p \geq 1$

$$V \in L_p(\mathbb{X}, \mu), \quad W \in L_p(\mathbb{X}, \mu). \tag{5.2.12}$$

Then the series (5.2.11) *converges in $L_p(\mathbb{X}, \mu)$ sense, and the sequence*

$$M_k = \sum_{j=0}^{k-1} f(X_j) + \mathscr{R} f(X_k), \qquad k \geq 0,$$

is a martingale w.r.t. the natural filtration $\{\mathscr{F}_k\}$ *of the process* X. *If, in addition,*

$$\sup_k E_x W(X_k) < \infty, \quad x \in \mathbb{X}, \tag{5.2.13}$$

then the series (5.2.11) *converges for every* $x \in \mathbb{X}$.

Proof. Write

$$P^k f(x) = \mathsf{E}_x f(X_k) = \int_{\mathbb{X}} f(y) P^k(x, dy).$$

Because f is centered, we have by eq. (5.1.8)

$$|P^k f(x)| = \left| \mathsf{E}_x f(X_k) - \int_{\mathbb{X}} f \, d\mu \right| \le \|f\|_{d,\gamma,W} r_k^\gamma U_k(x). \tag{5.2.14}$$

This immediately yields that the series in eq. (5.2.11) converges for every $x \in \mathbb{X}$ under the additional assumption eq. (5.2.13). On the other hand, under eq. (5.2.12) the sequence $\{U_k\}$ is bounded in $L_p(\mathbb{X}, \mu)$; see eq. (5.1.12). This proves that eq. (5.2.11) converges in $L_p(\mathbb{X}, \mu)$.

Because X is stationary and $f \in L_p(\mathbb{X}, \mu)$, we have that the series

$$S^{(\rho)} := \sum_{j=0}^{\infty} \rho^j f(X_j)$$

converges in $L_p(\Omega, \mathscr{F}, \mathsf{P})$ for every $\rho \in (0, 1)$. Then,

$$M_k^{(\rho)} := \mathsf{E}\big(S^{(\rho)} \big| \mathscr{F}_k\big) = \sum_{j=0}^{\infty} \rho^j \mathsf{E}\big(f(X_j) \big| \mathscr{F}_k\big), \quad k \ge 0$$

is a martingale. Denote

$$\mathscr{R}^{(\rho)} f(x) = \sum_{k=0}^{\infty} \rho^k \mathsf{E}_x f(X_k) = \sum_{k=0}^{\infty} \rho^k P^k f(x),$$

then we get,

$$M_k^{(\rho)} = \sum_{j=0}^{k-1} \rho^j f(X_j) + \rho^k \mathscr{R} f^{(\rho)}(X_k), \qquad k \ge 0.$$

It follows from eqs. (5.2.14) and (5.1.12) that

$$\mathscr{R}^{(\rho)} f \to \mathscr{R} f, \qquad \rho \to 1$$

in $L_p(\mathbb{X}, \mu)$. Hence, because every X_k obeys law μ, we deduce

$$M_k^{(\rho)} \to M_k, \qquad \rho \to 1$$

in $L_p(\Omega, \mathcal{F}, \mathsf{P})$ for every $k \geq 0$. Therefore $\{M_k\}$ is a martingale. □

Denote

$$\widetilde{\mathcal{R}}f(x) = \mathcal{R}f(x) - f(x) = \sum_{k=1}^{\infty} \mathsf{E}_x f(X_k),$$

then we have the following representation for the AR sequence ζ^ε:

$$\zeta_n^\varepsilon = \sqrt{\varepsilon} \sum_{k \leq n} f(X_k) = \sqrt{\varepsilon} M_n - \sqrt{\varepsilon} \widetilde{\mathcal{R}} f(X_n), \quad n \geq 0,$$

where M is the martingale sequence from the proposition above. Denote

$$\tilde{\zeta}_n^\varepsilon = \sqrt{\varepsilon} M_n - \sqrt{\varepsilon} \widetilde{\mathcal{R}} f(X_0) = \sum_{k=1}^{n} \tilde{\xi}_k^\varepsilon,$$

$$\tilde{\xi}_k^\varepsilon = \sqrt{\varepsilon}\Big(f(X_k) + \widetilde{\mathcal{R}}f(X_k) - \widetilde{\mathcal{R}}f(X_{k-1})\Big) = \sqrt{\varepsilon}\Big(\mathcal{R}f(X_k) - \widetilde{\mathcal{R}}f(X_{k-1})\Big).$$

Then $\{\tilde{\xi}_k^\varepsilon\}$ is a martingale difference sequence. We have

$$\zeta_n^\varepsilon = \tilde{\zeta}_n^\varepsilon + \eta_n^\varepsilon, \quad \eta_n^\varepsilon = -\sqrt{\varepsilon}\widetilde{\mathcal{R}}f(X_n) + \sqrt{\varepsilon}\widetilde{\mathcal{R}}f(X_0),$$

and since $\widetilde{\mathcal{R}}f \in L_p(\mathbb{X}, \mu)$ and X is stationary, it is clear that

$$\eta_n^\varepsilon \to 0, \quad \varepsilon \to 0$$

in $L_p(\Omega, \mathcal{F}, \mathsf{P})$. This is the essence of the *corrector term* method: by adding an asymptotically negligible term $\{\eta_n^\varepsilon\}$, we transform the AR sequence $\{\zeta_n^\varepsilon\}$ into the form, where the martingale CLT becomes applicable. This particular form of the method, where the AR sequence itself is transformed, dates back to Gordin [47], see also Refs. [48] and [12]; another version of the corrector term method will be discussed in the next section.

Now we are ready to give the main result of this section.

Theorem 5.2.4. *Let a stationary Markov chain X satisfy eq.* (5.1.1), *and a function $f \in H_{\gamma,W}(\mathbb{X}, d)$ be centered; see eq.* (5.2.9). *Assume that*

$$\sum_k r_k^\gamma < \infty,$$

and for some $p > 2$

$$V \in L_p(\mathbb{X}, \mu), \quad W \in L_p(\mathbb{X}, \mu).$$

Then CLT (5.1.5) *holds true with σ^2 given by eq.* (5.1.4).

Proof. We will apply Theorem 5.2.1 to the AR sequence $\{\widetilde{\zeta}_n^\varepsilon\}$ constructed above. Assumption (i) of this theorem holds true by the construction. To verify the assumption (ii), observe that

$$\mathsf{E}[(\widetilde{\xi}_k^\varepsilon)^2 | \mathscr{F}_{k-1}] = \varepsilon g(X_{k-1})$$

with

$$g(x) = \int_{\mathbb{X}} (\mathscr{R}f(y))^2 P(x, dy) - (\widetilde{\mathscr{R}}f(x))^2.$$

Since $\mathscr{R}f, \widetilde{\mathscr{R}}f \in L_p(\mathbb{X}, \mu)$ and $g(x) \geq 0$, we have

$$\begin{aligned}\int_{\mathbb{X}} |g(x)| \mu(dx) = \int_{\mathbb{X}} g(x) \mu(dx) &= \int_{\mathbb{X}} \left(\int_{\mathbb{X}} (\mathscr{R}f(y))^2 P(x, dy) - (\widetilde{\mathscr{R}}f(x))^2 \right) \mu(dx) \\ &= \int_{\mathbb{X}} \left((\mathscr{R}f(x))^2 - (\widetilde{\mathscr{R}}f(x))^2 \right) \mu(dx) < \infty.\end{aligned}$$

Then, applying Theorem 5.1.10 to the function $g \in L_1(\mathbb{X}, \mu)$, we get assumption (ii) of Theorem 5.2.1 with

$$\begin{aligned}\sigma^2 = \int_{\mathbb{X}} \left((\mathscr{R}f(x))^2 - (\widetilde{\mathscr{R}}f(x))^2 \right) \mu(dx) &= \int_{\mathbb{X}} \left(f^2(x) + 2f(x)\widetilde{\mathscr{R}}f(x) \right) \mu(dx) \\ &= \mathsf{E}f^2(X_0) + 2\sum_{k=1}^\infty \mathsf{E}f(X_k)f(X_0).\end{aligned}$$

Finally, by the stationarity we have

$$\mathsf{E}|\widetilde{\xi}_k^\varepsilon|^p = C\varepsilon^{p/2},$$

where

$$C = \mathsf{E}\left|(\mathscr{R}f(X_1) - \widetilde{\mathscr{R}}f(X_0)\right|^p$$

is finite because $\mathscr{R}f, \widetilde{\mathscr{R}}f \in L_p(\mathbb{X}, \mu)$. This verifies assumption (iii) of Theorem 5.2.1, and completes the proof. □

5.3 Extensions

In previous sections, for clarity of presentation, we considered LLN and CLT in the stationary setting and for a discrete-time chain. Of course, these limitations are not crucial, and in this section we briefly explain how the previous results can be extended to nonstationary or/and continuous-time settings.

5.3.1 Non-Stationary Setting

Let us explain a simple trick, which allow one to remove the assumption on the chain X to be stationary, made in Theorems 5.1.2, 5.1.10, and 5.2.4. This trick apparently dates back to paper [8]; see also Ref. [97]. Another possibility to treat nonstationary systems will be explored in Section 6.3.2.

Consider first the case where the chain X is ergodic in total variation distance. Let all the assumptions of Theorem 5.1.2 hold true except the stationarity one; denote $\mathrm{Law}(X_0) = \nu$. By eq. (5.1.1), the laws

$$\nu^M = \int_{\mathbb{X}} P^M(x, \cdot)\nu(dx), \quad M \geq 0$$

of the respective values of X satisfy

$$\delta_M = \|\nu^M - \mu\|_{TV} \to 0, \quad M \to \infty.$$

By the Coupling lemma, for a given M there exists a pair $\xi^M, \eta^M \in C(\nu^M, \mu)$ such that

$$\mathsf{P}(\xi \neq \eta) = \frac{1}{2}\delta_M.$$

Then, applying Lemma 4.3.2, we can construct the families of random variables

$$\{\xi_n^M, n = \overline{0, M}\}, \quad \{\eta_n^M, n = \overline{0, M}\}$$

such that the law of $\{\xi_n^M, n = \overline{0, M}\}$ coincides with the joint law of $X_0, \dots X_M$ w.r.t. P_ν, the law of $\{\xi_n^M, n = \overline{0, M}\}$ coincides with the joint law of $X_0, \dots X_M$ w.r.t. P_μ, and

$\mathrm{Law}(\xi_M^M, \eta_M^M) = \mathrm{Law}(\xi^M, \eta^M)$. We also can take (ξ_M^M, η_M^M) as a new initial value of an (independent) greedy coupling, which will give finally the pair of sequences $\{\xi_n\}, \{\eta_n\}$ such that their laws are equal $\mathsf{P}_\nu, \mathsf{P}_\mu$, respectively, and

$$\mathsf{P}(\xi_n \neq \eta_n, n \geq M) = \frac{1}{2}\|\nu^M - \mu\|_{TV} = \frac{1}{2}\delta_M. \tag{5.3.1}$$

By Theorem 5.1.2, we have

$$\frac{1}{N}\sum_{k=1}^N f(\eta_k) \to \int f \, \mathrm{d}\mu, \quad N \to \infty$$

in probability and, provided that $f \in L_2(\mathbb{X}, \mu)$ satisfies the centering condition

$$\int_{\mathbb{X}} f \, \mathrm{d}\mu = 0,$$

$$\frac{1}{\sqrt{N}}\sum_{k=1}^N f(\eta_k) \Rightarrow \mathcal{N}(0, \sigma^2), \quad N \to \infty.$$

By eq. (5.3.1), these statements remain true with $\{\eta_k\}$ replaced by $\{\xi_k\}$. That is, Theorem 5.1.2 remains true without the stationarity assumption, with the minor difference that LLN (5.1.3) now holds true in P_ν-probability. Note that, in nonstationary setting, the right-hand side in eq. (5.1.3) and the variance σ^2 of the weak limit in eq. (5.1.5) are calculated w.r.t. the stationary measure P_μ.

To consider the general case, we introduce a new condition: we assume that, for any $x \in \mathbb{X}$, there exists a sequence $Z = (X, Y)$ such that its components X, Y have the laws $\mathsf{P}_x, \mathsf{P}_\mu$ and

$$\mathsf{E}d(X_n, Y_n) \leq V(x)r_n, \quad n \geq 0. \tag{5.3.2}$$

This condition is formally stronger than eq. (5.1.1): the latter one requires that, *for any given n*, there exists a coupling (X_n, Y_n) for $P^n(x, \cdot), \mu$ such that the inequality from eq. (5.3.2) holds true, while eq. (5.3.2) actually requires that there exists *one* coupling for the measures $\mathsf{P}_x, \mathsf{P}_\mu$ on the path space, which provides the inequality *for all n* simultaneously. We will call eq. (5.3.2) a *path coupling* condition, to emphasize the difference with the *individual coupling* condition eq. (5.1.1). Note, however, that there is almost no practical difference between these two types of conditions. For $d(x, y) = 1_{x\neq y}$, these two conditions simply coincide; see eq. (5.3.1). For general d, all the theorems from Chapter 4, which provide individual coupling conditions, actually give a bound of the form (5.3.2) for the d-greedy Markov coupling, and thus provide corresponding path coupling conditions, as well.

Proposition 5.3.1. *Assume the path coupling condition* (5.3.2) *holds true,* $\nu \in \mathscr{P}(\mathbb{X})$ *be such that*

$$E_\nu V(X_0) = \int_{\mathbb{X}} V\, d\nu < \infty,$$

and the function f belong to $H_{\gamma,W}(\mathbb{X}, d)$ *with*

$$\int_{\mathbb{X}} W\, d\mu < \infty.$$

I. *If all conditions of Theorem 5.1.10, except the stationarity of X, hold true, and*

$$\frac{1}{N}\sum_{k=1}^{N} r_n^\gamma \Big(E_\nu W(X_k)\Big)^{1-\gamma} \to 0, \quad N \to \infty, \tag{5.3.3}$$

then eq. (5.1.3) *holds true in the mean sense w.r.t.* P_ν.

II. *If all conditions of Theorem 5.2.4, except the stationarity of X, hold true, and*

$$\frac{1}{\sqrt{N}}\sum_{k=1}^{N} r_k^\gamma \Big(E_\nu W(X_k)\Big)^{1-\gamma} \to 0, \quad N \to \infty, \tag{5.3.4}$$

then eq. (5.1.5) *holds true w.r.t.* P_ν.

Remark 5.3.2. Note that

$$\frac{1}{N}\sum_{k=1}^{N} r_k^\gamma \Big(E_\nu W(X_k)\Big)^{1-\gamma} \le \left(\frac{1}{N}\sum_{k=1}^{N} r_k\right)^\gamma \left(\frac{1}{N}E_\nu \sum_{k=1}^{N} W(X_k)\right)^{1-\gamma}$$

and

$$\frac{1}{N}\sum_{k=1}^{N} r_k \to 0, \quad N \to \infty.$$

Hence eq. (5.3.3) holds true if

$$\sup_N \frac{1}{N} E_\nu \sum_{k=1}^{N} W(X_k) < \infty, \tag{5.3.5}$$

which is easy to verify using a Lyapunov-type condition; see Proposition 2.8.5. Similarly, eq. (5.3.4) is provided by eq. (5.3.5) and

$$N^{-1+1/(2\gamma)} \sum_{k=1}^{N} r_k \to 0, \quad N \to \infty.$$

Proof. We have

$$\mathsf{E}|f(X_k) - f(Y_k)| \le \mathsf{E}d(X_k, Y_k)^\gamma W(X_k, Y_k)^{1-\gamma}$$
$$\le r_n^\gamma \Big(\mathsf{E}V(X_0)\Big)^\gamma \Big(\mathsf{E}W(X_k) + \mathsf{E}W(Y_k)\Big)^{1-\gamma}$$

and

$$\Big(\mathsf{E}W(X_k) + \mathsf{E}W(Y_k)\Big)^{1-\gamma} \le \Big(\mathsf{E}W(X_k)\Big)^{1-\gamma} + \Big(\mathsf{E}W(Y_k)\Big)^{1-\gamma}.$$

If (X, Y) is a path coupling for $\mathsf{P}_\nu, \mathsf{P}_\mu$, we have

$$\mathsf{E}W(Y_k) = \int_{\mathbb{X}} W\, d\mu < \infty.$$

Taking a coupling such that eq. (5.3.2) holds, and using eq. (5.3.3), we easily get the required statements. □

5.3.2 Continuous-Time Setting

Let $X = \{X_t\}$ be a Markov process, which satisfies

$$d\Big(P^t(x, \cdot), \mu\Big) \le V(x) r_t, \qquad t \ge 0, \tag{5.3.6}$$

a continuous-time analogue of eq. (5.1.1). We begin with the following continuous-time version of LLN.

Theorem 5.3.3. *Let X be a stationary Markov process, which satisfies eq. (5.3.6) with $V \in L_2(\mathbb{X}, \mu)$ and the distance-like function d such that, for some $\gamma \in (0, 1]$ and $W \in L_2(\mathbb{X}, \mu)$, the family of bounded functions from the class $H_{\gamma,W}(\mathbb{X}, d)$ is dense in $L_2(\mathbb{X}, \mu)$.*

Then for any $f : \mathbb{X} \to \mathbb{R}$ such that $\mathsf{E}|f(X_0)| < \infty$

$$\frac{1}{T}\int_0^T f(X_t)\, dt \to \mathsf{E}f(X_0), \quad T \to \infty \tag{5.3.7}$$

in the mean sense.

Proof. The argument here is essentially the same as the one which leads to Theorem 5.1.10, hence we just outline the proof. First, observe that by conditions (5.3.6) and $V, W \in L_2(\mathbb{X}, \mu)$, we have

$$\mathrm{Cov}\Big(f(X_t), f(X_0)\Big) \to 0, \quad t \to \infty$$

for any $f \in L_2(\mathbb{X}, \mu)$ such that $f \in H_{\gamma,W}(\mathbb{X}, d)$; see Corollary 5.1.8. Then using the L_2-isometry caused by stationarity, we extend this statement to any $f \in L_2(\mathbb{X}, \mu)$; the argument here is the same as in Proposition 5.1.9. This proves (5.3.7) for $f \in L_2(\mathbb{X}, \mu)$. Using the L_1-isometry, we complete the proof for arbitrary $f \in L_1(\mathbb{X}, \mu)$. □

Next, we give a continuos-time version of CLT.

Theorem 5.3.4. *Let a stationary Markov chain X satisfy eq.* (5.3.6), *and a function* $f \in H_{\gamma,W}(\mathbb{X}, d)$ *be centered; see eq.* (5.2.9). *Assume that*

$$\int_0^\infty r_t^\gamma \, dt < \infty,$$

and for some $p > 2$

$$V \in L_p(\mathbb{X}, \mu), \quad W \in L_p(\mathbb{X}, \mu).$$

Then the integral

$$\sigma^2 = 2 \int_0^\infty Ef(X_t)f(X_0)\, dt \tag{5.3.8}$$

converges, and

$$\frac{1}{\sqrt{T}} \int_0^T f(X_t)\, dt \Rightarrow \mathcal{N}(0, \sigma^2), \quad T \to \infty. \tag{5.3.9}$$

Proof. By stationarity, for each $S \le T$

$$\mathsf{E}\left| \frac{1}{\sqrt{T}} \int_S^T f(X_t)\, dt \right| \le \frac{1}{\sqrt{T}} \int_S^T \mathsf{E}|f(X_t)|\, dt = \frac{T-S}{\sqrt{T}} \mathsf{E}|f(X_0)|\, dt.$$

Hence it is enough to prove eq. (5.3.9) for $T = n$ which take values in $\mathbb{N}$. As in the proof of Theorem 5.2.4, we reduce this statement to the martingale setting by using a proper corrector term. Namely, we put

$$\mathscr{R}f(x) = \int_0^\infty \mathsf{E}_x f(X_t)\, dt.$$

This function is well defined and belongs to $L_p(\mathbb{X}, \mu)$; in addition,

$$M_n = \int_0^n f(X_s)\, ds + \mathscr{R}f(X_n), \quad n \ge 0$$

is a martingale. The proof of these statements repeats the proof of Proposition 5.2.3 and is omitted.

Similarly to the proof of Theorem 5.2.4, now we apply the martingale CLT (Theorem 5.2.1) to the family

$$\zeta_n^\varepsilon = \sqrt{\varepsilon}\int_0^n f(X_s)\,ds + \sqrt{\varepsilon}\mathscr{R}f(X_n), \quad \varepsilon = n^{-1}.$$

The martingale assumption (i) of this theorem holds true by the construction. We have

$$\xi_k^\varepsilon = \sqrt{\varepsilon}\int_{k-1}^k f(X_s)\,ds + \sqrt{\varepsilon}(\mathscr{R}f(X_k) - \mathscr{R}f(X_{k-1})),$$

which immediately yield the Lyapunov condition (iii) because $f, \mathscr{R}f$ belong to $L_p(\mathbb{X}, \mu)$ and X is stationary. We also have

$$\beta_k^\varepsilon = \varepsilon g(X_{k-1}),$$
$$g(x) = \mathsf{E}_x\left(\int_0^1 f(X_s)\,ds + \mathscr{R}f(X_1) - \mathscr{R}f(x)\right)^2.$$

Since $f, \mathscr{R}f$ belong to $L_2(\mathbb{X}, \mu)$, we have by stationarity

$$\int_{\mathbb{X}} g(x)\mu(dx) = \mathsf{E}\left(\int_0^1 f(X_s)\,ds + \mathscr{R}f(X_1) - \mathscr{R}f(X_0)\right)^2 < \infty.$$

Then by Theorem 5.1.10, condition (ii) holds true with

$$\begin{aligned}
\sigma^2 &= \int_{\mathbb{X}} g(x)\mu(dx) = \mathsf{E}\,(M_1 - M_0)^2 = \mathsf{E}M_1^2 - \mathsf{E}M_0^2 \\
&= \mathsf{E}\left(\int_0^1 f(X_s)\,ds\right)^2 + 2\mathsf{E}\int_0^1 \mathscr{R}f(X_1)f(X_s)\,ds \\
&= 2\int_0^1\int_0^t \mathsf{E}f(X_t)f(X_s)\,ds\,dt + 2\int_0^\infty\int_0^1 \mathsf{E}f(X_{1+t})f(X_s)\,ds\,dt \\
&= 2\int_0^\infty\int_0^{t\wedge 1} \mathsf{E}f(X_t)f(X_s)\,ds\,dt = 2\int_0^\infty\int_0^1 \mathsf{E}f(X_{s+v})f(X_s)\,ds\,dv \\
&= 2\int_0^\infty \mathsf{E}f(X_v)f(X_0)\,dv.
\end{aligned}$$

The required statement now follows directly from Theorem 5.2.1. □

The continuous-time LLN and CLT, given above, can be easily extended to a non-stationary setting, provided the following continuous-time path coupling condition holds true: for each x, there exists a pair of processes (X, Y) such that the laws of X, Y are $\mathsf{P}_x, \mathsf{P}_\mu$ respectively, and

$$\mathsf{E}d(X_t, Y_t) \le V(x)r_t, \quad t \ge 0. \tag{5.3.10}$$

Proposition 5.3.5. *Assume the path coupling condition* (5.3.10) *hold true,* $\nu \in \mathscr{P}(\mathbb{X})$ *be such that*

$$E_\nu V(X_0) = \int_{\mathbb{X}} V\, d\nu < \infty,$$

and the function f *belong to* $H_{\gamma,W}(\mathbb{X}, d)$ *with*

$$\int_{\mathbb{X}} W\, d\mu < \infty.$$

I. *If all conditions of Theorem 5.3.3, except the stationarity of* X, *hold true, and*

$$\frac{1}{T}\int_0^T r_t^\gamma \Big(E_\nu W(X_t)\Big)^{1-\gamma}\, dt \to 0, \quad T \to \infty,$$

then eq. (5.3.7) *holds true in the mean sense w.r.t.* P_ν.

II. *If all conditions of Theorem 5.3.4, except the stationarity of* X, *hold true, and*

$$\frac{1}{\sqrt{T}}\int_0^T r_t^\gamma \Big(E_\nu W(X_t)\Big)^{1-\gamma}\, dt \to 0, \quad T \to \infty,$$

then eq. (5.3.9) *holds true w.r.t.* P_ν.

The proof, with obvious changes, repeats the proof of Proposition 5.3.1, and is omitted.

5.4 Comments

1. The list of references concerning CLT for Markov chains, ergodic in total variation, is vast. Apart from the approach, outlined in Section 5.1 here, based on evaluation of the mixing coefficients and CLT for strictly stationary sequences, the combination of the martingale argument and the correction term trick is widely used. In particular, the latter argument was used in the proof of the well known Kipnis-Varadhan CLT, see Ref. [75]. Within this general streamline, numerous variations had been developed, according to different possible ways to construct the correction term, or, in the equivalent and commonly used terminology, to solve the corresponding Poisson equation, see Refs. [16, 23, 24, 58, 80, 105] and references therein. We also refer to Ref. [41] (and references therein) for a detailed exposition of the weak spectral method, based on the Keller–Liverani perturbation theorem, well applicable in the case where the L_2-semigroup of the process possesses L_2-spectral gap.

2. The difference between the principal assumptions in the first and the second parts of Theorem 5.1.2 is that LLN for a Markov system requires just the stabilization property, while for CLT the ergodic rate should be specified. In general, the erodic rate required for CLT should be related with the moment bounds for the sequence subject to CLT. Theorem 5.1.2 gives just one such a set of conditions, for the survey of similar results, available in the literature, we refer to Ref. [68]. It is a kind of a "common knowledge" that CLT for dependent sequence requires ergodic rates or semigroup bounds in the Markov setting, bounds for mixing coefficients in the strictly stationary setting, etc. A notable exception to this heuristic principle is given by CLT for strictly stationary martingale differences, see Ref. [59], valid just under the basic ergodicity assumption.

3. The list of references concerning CLT for weakly ergodic Markov chains is surprisingly short, when compared with the one discussed in comment **1.** In CLT established in Ref. [78], the underlying Markov process was assumed to have the contraction property, similar to eq. (4.5.1), w.r.t. the Wasserstein distance. In Ref. [89], CLT was obtained as a particular case of a diffusion approximation type theorem. The contraction property was not included explicitly in the list of assumptions therein; however, it was still present, in a hidden form, as the condition which guarantees Hölder continuity of the generalized potentials. The contraction property is a strong structural restriction, which one should not expect to have in a hypothetically most general version of CLT. Principally, this restriction was removed in Ref. [85, Section 3.3] thanks to a more precise analysis, relied on Hölder continuity of the partial sums for the potential (5.2.11), rather than of the potential itself. In Theorem 5.2.4, this argument is further refined, and an auxiliary (though nonrestrictive) stabilization assumption (3.2.12) from Ref. [85] is replaced just by the basic assumption (5.1.1): this becomes possible thanks to Theorem 5.1.10, which is free from any Hölder-type assumptions (unlike, for instance, Ref. [85, Theorem 3.10]).

6 Functional Limit Theorems

6.1 Preliminaries

The limit theorems obtained in the previous section actually guarantee weak convergence of the corresponding family of *processes* rather than just individual weak convergence. This was observed for the martingale CLT (see Remark 5.2.2), and thus holds true for the CLTs obtained in Theorems 5.2.4 and 5.3.4, since they were derived directly from Theorem 5.2.1. In this section, we will make this observation more precise; namely, we will specify a proper form of the weak convergence, stronger than just the weak convergence of finite-dimensional distributions, which actually holds true in the setting, considered in Theorems 5.2.1, 5.2.4, and 5.3.4. We also extend the model to a substantially more general one, which we explain now.

In the continuous-time setting of Section 5.3.2, denote

$$X^\varepsilon(t) = X_{\varepsilon^{-1}t}, \quad t \geq 0,$$

take $\varepsilon = T^{-1}$, and consider the process Y^ε defined by

$$\mathrm{d}Y^\varepsilon(t) = a^\varepsilon\big(X^\varepsilon(t)\big)\,\mathrm{d}t, \quad Y^\varepsilon(0) = y_0. \tag{6.1.1}$$

If $a^\varepsilon(x) = f(x)$, convergence of corresponding $Y^\varepsilon(1)$ is actually equivalent to LLN (5.3.7); to get CLT (5.3.9) one has to take $a^\varepsilon(x) = \varepsilon^{-1/2}f(x)$. Equation (6.1.1) describes the behavior of a (continuous-time) dynamical system, influenced by an external random perturbation X^ε. This system is fairly simple because its right-hand side does not involve the process Y^ε; that is, it is *additive* w.r.t. the random perturbations. More general (non-additive) setting should include dependence on Y^ε, and possibly a stochastic term:

$$\mathrm{d}Y^\varepsilon(t) = a^\varepsilon\big(Y^\varepsilon(t), X^\varepsilon(t)\big)\,\mathrm{d}t + \sigma^\varepsilon\big(Y^\varepsilon(t), X^\varepsilon(t)\big)\,\mathrm{d}W_t. \tag{6.1.2}$$

One particular further aim for us will be to establish the limit behavior for such systems. Processes $X^\varepsilon, Y^\varepsilon$ are called *the fast* and *the slow* components of the system, respectively. Likewise to the additive equation (6.1.1), we have a natural distinction between the case where the drift coefficient $a^\varepsilon(x, y)$ is bounded, and the one where $a^\varepsilon(x, y)$ contain a term of the order $\varepsilon^{-1/2}$. These two cases extend LLN and CLT, respectively; in the literature, such limit theorems are called *the averaging principle* and *the diffusion approximation theorem* correspondingly.

Let us give a brief outline of the basic notions and constructions, which will be used to formulate and to prove functional limit theorems below; for a detailed exposition of this topic, we refer to Refs. [44, Chapter IX], [10] and [38, Chapter 3].

DOI 10.1515/9783110458930-007

We will consider families of processes

$$Y^\varepsilon = \{Y^\varepsilon(t), t \in [0,T]\}, \quad \varepsilon > 0,$$

defined on a finite time interval; the terminal time moment T will be arbitrary, but fixed. The processes will be assumed real-valued. An extension to multivariate case is straightforward, but requires a more cumbersome notation, hence for transparency of the exposition we restrict ourselves to one-dimensional case.

Assuming processes $Y^\varepsilon, \varepsilon > 0$ to have either continuous or càdlàg trajectories, we will respectively consider these processes as random elements taking values in the space $\mathbb{C}(0,T)$ of continuous functions, or the Shorokhod space $\mathbb{D}(0,T)$ of càdlàg functions. Both these spaces are Polish under a proper choice of the metric. For $\mathbb{C}(0,T)$ this is just the *uniform metric*

$$\|x - y\| = \sup_{t\in[0,T]} |x(t) - y(t)|,$$

for $\mathbb{D}(0,T)$ the corresponding *Skorokhod metric* is given by

$$\rho(x,y) = \inf\left\{\delta : \exists\lambda \in \Lambda \text{ s.t. } \|x(\lambda(\cdot)) - y\| < \delta, \sup_{s\neq t}\left|\log\left(\frac{\lambda(t)-\lambda(s)}{t-s}\right)\right| < \delta\right\},$$

where Λ denotes the class of strictly increasing functions $\lambda : [0,T] \to [0,T]$ with $\lambda(0) = 0, \lambda(T) = T$.

To prove weak convergence of a family of processes $\{Y^\varepsilon\}$, considered as elements of a proper functional space, we will follow the standard plan, which consists of two principal steps:

(I) to prove that the family of the laws of $\{Y^\varepsilon\}$ is *weakly relatively compact*, that is, each of its subsequence contains a weakly convergent subsequence;

(II) to identify the weak limit points of this sequence, and prove that, actually, such a limit point is unique.

By the technical reasons, typically it is much simpler to prove weak relative compactness of the laws of a family $\{Y^\varepsilon\}$ in $\mathbb{D}(0,T)$ than in $\mathbb{C}(0,T)$, even if $\{Y^\varepsilon\}$ have continuous trajectories and thus can be considered as random elements in $\mathbb{C}(0,T)$. Fortunately, we have the following simple observation.

Proposition 6.1.1. *Let the family of the laws of $\{Y^\varepsilon\}$ in $\mathbb{D}(0,T)$ weakly converge to the law of certain Y, and both Y and each Y^ε have continuous trajectories. Then the laws of $\{Y^\varepsilon\}$ in $\mathbb{C}(0,T)$ weakly converge to the law of Y, as well.*

Proof. It follows from the definition of the Skorokhod metric ρ, that any sequence $\{x_n\}$, which converge to a *continuous* limit x w.r.t. ρ, actually converges w.r.t. the uniform metric. The required statement follows from this observation and the "common probability space" principle, which now is well applicable because $\mathbb{D}(0, T)$ is a Polish space. □

Our strategy will be as follows: even if $\{Y^\varepsilon\}$ have continuous trajectories, we will first prove weak relative compactness of their laws in $\mathbb{D}(0, T)$, which together with the identification of weak limit points will prove weak convergence in $\mathbb{D}(0, T)$. Then we will make the argument more precise, applying the above proposition and proving weak convergence in $\mathbb{C}(0, T)$.

The following sufficient condition will be strong enough for our needs; for a general criteria for a weak relative compactness in $\mathbb{D}(0, T)$, we refer to Ref. [38, Chapter 3.8]. Denote

$$Q(x) = |x| \wedge 1.$$

Proposition 6.1.2. *Let a family $\{Y^\varepsilon\}$ be such that there exist $C > 0, \alpha > 0, \beta > 0$ such that for any $t_1 < t_2 < t_3, t_i \in [0, T], i = 1, 2, 3$*

$$EQ\big(Y^\varepsilon(t_2) - Y^\varepsilon(t_1)\big)^\alpha Q\big(Y^\varepsilon(t_3) - Y^\varepsilon(t_2)\big)^\alpha \le C|t_3 - t_1|^{1+\beta}. \tag{6.1.3}$$

Then the family of the laws of $\{Y^\varepsilon\}$ in $\mathbb{D}(0, T)$ is weakly relatively compact.

To identify the weak limit points, we will use the *martingale approach*, which dates back to Refs. [126, Chapter 11] and [110], and is based on the concept of *martingale problem*. Let us briefly recall this concept and related results; for a detailed exposition, we refer to Refs. [126] and [38].

Let $\mathscr{L}$ be an operator defined on some set $\mathscr{D}$ of functions $\phi : \mathbb{R} \to \mathbb{R}$. A process $Y = \{Y(t), t \in [0, T]\}$ is called a *solution to the martingale problem* $(\mathscr{L}, \mathscr{D})$, if for any $\phi \in \mathscr{D}$ the process

$$\phi\big(Y(t)\big) - \int_0^t \mathscr{L}\phi\big(Y(s)\big)\, ds, \qquad t \in [0, 1]$$

is a martingale w.r.t. the natural filtration of Y. This definition includes, as a natural pre-requisite, the claim that the above integral is well defined; typically, it is assumed that process Y is *measurable.*

A martingale problem $(\mathscr{L}, \mathscr{D})$ is said to be *well posed*, if for every $y \in \mathbb{R}$ there exists a solution $(\mathscr{L}, \mathscr{D})$ with the initial value $Y(0) = y$, and for any two such solutions their finite-dimensional distributions agree. We will use the following classical statement, see Ref. [126, Chapter 6].

Theorem 6.1.3. *Let $A : \mathbb{R} \to \mathbb{R}$ be a measurable locally bounded function and $B \in C(\mathbb{R})$ take positive values. Then the martingale problem $(\mathscr{L}, \mathscr{D})$ with $\mathscr{D} = C_0^\infty$ (the class of compactly supported C^∞-functions) and*

$$\mathscr{L}\phi = A\phi' + \frac{1}{2}B\phi''$$

is well posed.

6.2 Autoregressive Models with Markov Regime

Let a Markov chain $X = \{X_k\}$ and an independent i.i.d. sequence $\{\eta_k\}$ with

$$\mathsf{E}\eta_k = 0, \quad \mathsf{E}\eta_k^2 = 1$$

be given. Consider the AR sequence (5.2.1) with $\zeta_0^\varepsilon = y \in \mathbb{R}$ and

$$\xi_k^\varepsilon = \varepsilon f(\zeta_{k-1}^\varepsilon, x) + \sqrt{\varepsilon} g(\zeta_{k-1}^\varepsilon, x) + \sqrt{\varepsilon} h(\zeta_{k-1}^\varepsilon, x)\eta_k, \quad k \geq 1, \tag{6.2.1}$$

where

$$f(\zeta, x), \quad g(\zeta, x), \quad h(\zeta, x)$$

are certain functions. This system can be considered as a discrete-time version of system (6.1.2) with

$$a^\varepsilon(x, y) = f(y, x) + \varepsilon^{-1/2} g(y, x), \quad \sigma^\varepsilon(x, y) = h(x, y). \tag{6.2.2}$$

We use the discrete-time setting in order to explain the essence of the *corrector term method*, avoiding technical complications which arise in the continuous-time case; see discussion at the end of this section.

Because the increments of the AR model (6.2.1) are perturbed by the values of a Markov chain, we call it an *autoregressive model with Markov regime*. Note that, in the available literature, when an autoregressive model with Markov regime is discussed, it is typically assumed that the increments additionally depend on a sequence of i.i.d. random variables $\{\varepsilon_k\}$. Introducing an additional "random seed" $\{\varepsilon_k\}$ allows one to extend the model (6.2.1), and to consider sequences with ζ_k^ε, not being defined by values ζ_{k-1}, X_k, η_k through a functional relation, but having its conditional distribution defined by these values. At least formally, there is no substantial difference between this general settings and our current one, because we can consider the pair (X_k, ε_k) as a new Markov "driving noise".

Theorem 6.2.1. *Let stationary X satisfy condition* (5.1.1), *and the AR family $\{\zeta_k^\varepsilon\}$ be defined by eq.* (6.2.1). *Assume the following.*

1. *There exists a derivative $\partial_\zeta g(\zeta, x)$ and, for some $\gamma \in (0,1]$ and function W, the functions*

$$f(\zeta, \cdot), \quad g(\zeta, \cdot), \quad \partial_\zeta g(\zeta, \cdot), \quad h(\zeta, \cdot)$$

belong to $H_{\gamma,W}(\mathbb{X}, d)$. In addition,

$$\sup_{\zeta \in \mathbb{R}} \left(\|f(\zeta, \cdot)\|_{d,\gamma,W} + \|g(\zeta, \cdot)\|_{d,\gamma,W} + \|\partial_\zeta g(\zeta, \cdot)\|_{d,\gamma,W} + \|h(\zeta, \cdot)\|_{d,\gamma,W} \right) < \infty,$$

and there exists $\delta > 0$ such that

$$\sup_{\zeta_1 \neq \zeta_2} |\zeta_1 - \zeta_2|^{-\delta} \|f(\zeta_1, \cdot) - f(\zeta_2, \cdot)\|_{d,\gamma,W} < \infty,$$

$$\sup_{\zeta_1 \neq \zeta_2} |\zeta_1 - \zeta_2|^{-\delta} \left(\|g(\zeta_1, \cdot) - g(\zeta_2, \cdot)\|_{d,\gamma,W} + \|\partial_\zeta g(\zeta, \cdot) - \partial_\zeta g(\zeta_2, \cdot)\|_{d,\gamma,W} \right) < \infty,$$

$$\sup_{\zeta_1 \neq \zeta_2} |\zeta_1 - \zeta_2|^{-\delta} \|h(\zeta_1, \cdot) - h(\zeta_2, \cdot)\|_{d,\gamma,W} < \infty.$$

2. *For every $\zeta \in \mathbb{R}$, the functions $g(\zeta, \cdot), \partial_\zeta g(\zeta, \cdot) \in L_1(\mathbb{X}, \mu)$ are centered; that is,*

$$\int_{\mathbb{X}} g(\zeta, x)\, \mu(dx) = 0, \qquad \int_{\mathbb{X}} \partial_\zeta g(\zeta, x)\, \mu(dx) = 0.$$

The functions $f(\zeta, \cdot), h(\zeta, \cdot) \in L_1(\mathbb{X}, \mu)$ satisfy

$$\sup_{\zeta \in \mathbb{R}} \left| \int_{\mathbb{X}} f(\zeta, x)\, \mu(dx) \right| < \infty, \quad \sup_{\zeta \in \mathbb{R}} \left| \int_{\mathbb{X}} h(\zeta, x)\, \mu(dx) \right| < \infty,$$

$$\sup_{\zeta_1 \neq \zeta_2} |\zeta_1 - \zeta_2|^{-\delta} \left| \int_{\mathbb{X}} f(\zeta_1, x)\, \mu(dx) - \int_{\mathbb{X}} f(\zeta_2, x)\, \mu(dx) \right| < \infty,$$

$$\sup_{\zeta_1 \neq \zeta_2} |\zeta_1 - \zeta_2|^{-\delta} \left| \int_{\mathbb{X}} h(\zeta_1, x)\, \mu(dx) - \int_{\mathbb{X}} h(\zeta_2, x)\, \mu(dx) \right| < \infty.$$

3.

$$\sum_n r_n^\gamma < \infty, \quad \int_{\mathbb{X}} V^{2+\delta}\, d\mu < \infty, \quad \int_{\mathbb{X}} W^{2+\delta}\, d\mu < \infty,$$

and there exists $\delta' \in (0, \delta)$ such that

$$\mathsf{E}|\eta_k|^{(2+\delta)(2+\delta')/(\delta-\delta')} < \infty.$$

Then for the family

$$Y^\varepsilon(t) = \zeta^\varepsilon_{[\varepsilon^{-1}t]}, \quad t \in [0, T], \quad \varepsilon > 0,$$

the following statements hold true:

I. *The family $\{Y^\varepsilon\}$ is weakly compact in $\mathbb{D}(0, T)$.*
II. *Any weak limit point of $Y^\varepsilon, \varepsilon \to 0$ is a solution to the martingale problem $(\mathscr{L}, \mathscr{D})$ with*

$$\mathscr{L}\phi(\zeta) = A(\zeta)\phi'(\zeta) + \frac{1}{2}B(\zeta)\phi''(\zeta), \qquad \mathscr{D} = C_0^\infty(\mathbb{R}^d), \tag{6.2.3}$$

where

$$A(\zeta) = \int_{\mathbb{X}} f(\zeta, x)\,\mu(dx) + \sum_{k=1}^\infty \iint_{\mathbb{X}^2} g(\zeta, x)\partial_\zeta g(\zeta, y)\,\mu(dx)P^k(x, dy),$$

$$B(\zeta) = \int_{\mathbb{X}} g^2(\zeta, x)\,\mu(dx) + 2\sum_{k=1}^\infty \iint_{\mathbb{X}^2} g(\zeta, x)g(\zeta, y)\,\mu(dx)P^k(x, dy)$$
$$+ \int_{\mathbb{X}} h^2(\zeta, x)\,\mu(dx).$$

III. *If the martingale problem (6.2.3) is well posed, then the family Y^ε weakly converges in $\mathbb{D}(0, T)$ to a diffusion process Y, which is the unique solution of this martingale problem. In this case, the family*

$$\widetilde{Y}^\varepsilon(t) = Y^\varepsilon(t) + \Big(\varepsilon[\varepsilon^{-1}t] + \varepsilon - t\Big)\xi^\varepsilon_{[\varepsilon^{-1}t]+1}, \quad t \in [0, T]$$

weakly converges in $\mathbb{C}(0, T)$ to the diffusion process Y.

Before proceeding with the formal proof of Theorem 6.2.1, we outline the *corrector term* construction this proof is based on. We would like to repeat the argument used in the proof of Theorem 5.2.1. Namely, we take $\phi \in C^3$ with bounded derivatives, and express the increments of the sequence $\phi(\zeta_k^\varepsilon), k \ge 0$ using the Taylor formula; see eq. (5.2.3). Then we would like to take an expectation (actually, a conditional expectation), which should lead to an analogue of eq. (5.2.4), with $(1/2)\sigma^2\phi''$ replaced by $\mathscr{L}\phi$. However, it is not possible to do that immediately, because the first-order term

$$\phi'(\zeta^\varepsilon_{k-1})\xi^\varepsilon_k = \varepsilon\phi'(\zeta^\varepsilon_{k-1})f(\zeta^\varepsilon_{k-1}, X_k) + \sqrt{\varepsilon}\phi'(\zeta^\varepsilon_{k-1})g(\zeta^\varepsilon_{k-1}, X_k)$$
$$+ \sqrt{\varepsilon}\phi'(\zeta^\varepsilon_{k-1})h(\zeta^\varepsilon_{k-1}, X_k)\eta_k$$

is no longer negligible. The second summand

$$\sqrt{\varepsilon}\phi'(\zeta^\varepsilon_{k-1})g(\zeta^\varepsilon_{k-1}, X_k) \tag{6.2.4}$$

in the above decomposition is most "dangerous": it is not a martingale difference (unlike the third one), and has an upper bound $\sqrt{\varepsilon}$, which is not sufficient to dominate the sum of $\sim \varepsilon^{-1}$ such summands (unlike the first one). To negate this "danger", we introduce the corrector term

$$\Upsilon_k^\varepsilon = \phi'(\zeta_k^\varepsilon)\widetilde{\mathscr{R}}g(\zeta_k^\varepsilon, X_k).$$

Here,

$$\widetilde{\mathscr{R}}g(\zeta, x) = \mathscr{R}g(\zeta, x) - g(\zeta, x),$$

and $\mathscr{R}g(\zeta, x)$ is the potential of the function g, similar to the one introduced in Section 5.2.2. Now the function g depends on an additional variable ζ, and in the rigorous definition of the potential this variable is treated as a "frozen" parameter:

$$\mathscr{R}g(\zeta, x) = \sum_{k=0}^{\infty} \mathsf{E}_x g(\zeta, X_k), \qquad x \in \mathbb{X}. \tag{6.2.5}$$

The following properties of the potential can be proved analogously to Proposition 5.2.3; we omit the proof. Denote by $\mathbb{F} = \{\mathscr{F}_k\}$ the natural filtration for the pair (X, η):

$$\mathscr{F}_k = \sigma(X_j, \eta_j, j \le k), \quad k \ge 0.$$

Proposition 6.2.2. *Let conditions of Theorem 6.2.1 hold true. Then for every ζ the series (6.2.5) converges in $L_{2+\delta}(\mathbb{X}, \mu)$ and for every $k \ge 1$, $\zeta \in \mathbb{R}$,*

$$E[\mathscr{R}g(\zeta, X_k)|\mathscr{F}_{k-1}] = \widetilde{\mathscr{R}}g(\zeta, X_{k-1}). \tag{6.2.6}$$

As a corollary, the sequence

$$\sqrt{\varepsilon}\Upsilon_k^\varepsilon - \sqrt{\varepsilon}\Upsilon_{k-1}^\varepsilon + \sqrt{\varepsilon}\phi'(\zeta_{k-1}^\varepsilon)g(\zeta_{k-1}^\varepsilon, X_k)$$

is a martingale difference w.r.t. $\mathbb{F}$. This explains the essence of the method: by adding a small corrector term $\sqrt{\varepsilon}\Upsilon_k^\varepsilon$, we negate the "most dangerous" term (6.2.4) in the Taylor expansion of $\phi(\zeta_k^\varepsilon)$, and transform this sequence to a more feasible one

$$\psi_k^\varepsilon = \phi(\zeta_k^\varepsilon) + \sqrt{\varepsilon}\Upsilon_k^\varepsilon, \quad k \ge 0. \tag{6.2.7}$$

Note that here we do not correct the sequence $\{\zeta_k^\varepsilon\}$ itself, but do that for its image under a "test function" ϕ; this is the difference between the current form of the corrector term method, which dates back to ref. [110], and the one used in Section 5.2.2.

Below we give two technical results, which are required in order to apply rigorously the corrector term method outlined above. In these results, conditions of Theorem 6.2.1 are assumed to hold true.

Proposition 6.2.3. *There exists a function $H \in L_{2+\delta}(\mathbb{X}, \mu)$ such that*

$$
\begin{aligned}
|f(\zeta, x)| + |g(\zeta, x)| + |\mathscr{R}g(\zeta, x)| + |h(\zeta, x)| &\leq H(x), \\
|f(\zeta_1, x) - f(\zeta_2, x)| + |g(\zeta_1, x) - g(\zeta_2, x)| + |\mathscr{R}g(\zeta_1, x) - \mathscr{R}g(\zeta_2, x)| & \\
+ |h(\zeta_1, x) - h(\zeta_2, x)| &\leq |\zeta_1 - \zeta_2|^{\delta} H(x).
\end{aligned}
$$

Proof. Denote

$$U(x) = \sum_{n=0}^{\infty} r_n^{\gamma} U_n(x)$$

with $U_n, n \geq 0$ defined by eq. (5.1.9). Note that by condition 3 of Theorem 6.2.1 and eq. (5.1.12) this series converges in $L_{2+\delta}(\mathbb{X}, \mu)$. By eq. (5.1.10), we have

$$|f(\zeta, x)| \leq C_f^1 + C_f^2 U(x), \quad |f(\zeta_1, x) - f(\zeta_2, x)| \leq |\zeta_1 - \zeta_2|^{\delta} \Big(C_f^3 + C_f^4 U(x) \Big),$$

where

$$
\begin{aligned}
C_f^1 &= \sup_{\zeta} \left| \int_{\mathbb{X}} f(\zeta, x) \mu(dx) \right|, \quad C_f^2 = \sup_{\zeta} \|f(\zeta, \cdot)\|_{d,\gamma,W}, \\
C_f^3 &= \sup_{\zeta_1 \neq \zeta_2} |\zeta_1 - \zeta_2|^{-\delta} \left| \int_{\mathbb{X}} f(\zeta_1, x) - f(\zeta_2, x) \mu(dx) \right|, \\
C_f^4 &= \sup_{\zeta_1 \neq \zeta_2} |\zeta_1 - \zeta_2|^{-\delta} \|f(\zeta_1, \cdot) - f(\zeta_2, \cdot)\|_{d,\gamma,W}.
\end{aligned}
$$

Similar bounds are also available for g, h, and $\mathscr{R}g$; the calculation is analogous and thus is omitted. Denote by

$$C_g^i, \quad C_h^i, \quad C_{\mathscr{R}g}^i, \quad i = 1, \ldots, 4$$

corresponding constants, then the required statement holds true with

$$H(x) = C_1 + C_2 H(x),$$

where

$$
\begin{aligned}
C_1 &= \max \left\{ C_f^1, C_g^1, C_h^1, C_{\mathscr{R}g}^1, C_f^3, C_g^3, C_h^3, C_{\mathscr{R}g}^3 \right\}, \\
C_2 &= \max \left\{ C_f^2, C_g^2, C_h^2, C_{\mathscr{R}g}^2, C_f^4, C_g^4, C_h^4, C_{\mathscr{R}g}^4 \right\}.
\end{aligned}
$$

□

Proposition 6.2.4. *The function $\mathscr{R}g$, considered as a map*

$$\mathbb{R} \ni \zeta \mapsto \mathscr{R}g(\zeta, \cdot) \in L_{2+\delta}(\mathbb{X}, \mu),$$

has a continuous derivative equal to

$$\partial_\zeta \mathscr{R}g(\zeta, x) = \sum_{k=0}^{\infty} \mathsf{E}_x \partial_\zeta g(\zeta, X_k).$$

In addition, there exists a function $\widetilde{H} \in L_{2+\delta}(\mathbb{X}, \mu)$ such that

$$|\partial_\zeta g(\zeta, x)| \le \widetilde{H}(x), \qquad |\partial_\zeta g(\zeta_1, x) - \partial_\zeta g(\zeta_2, x)| \le |\zeta_1 - \zeta_2|^\delta \widetilde{H}(x),$$
$$|\partial_\zeta \mathscr{R}g(\zeta, x)| \le H(x), \qquad |\partial_\zeta \mathscr{R}g(\zeta_1, x) - \partial_\zeta \mathscr{R}g(\zeta_2, x)| \le |\zeta_1 - \zeta_2|^\delta \widetilde{H}(x).$$

Proof. Consider the potential of the function $\partial_\zeta g(\zeta, x)$:

$$\mathscr{R}\big(\partial_\zeta g\big)(\zeta, x) = \sum_{k=0}^{\infty} \mathsf{E}_x \partial_\zeta g(\zeta, X_k),$$

which is well defined by condition 1, because $\partial_\zeta g(\zeta, x)$ is centered. Denote also

$$\mathscr{R}^L\big(\partial_\zeta g\big)(\zeta, x) = \sum_{k=0}^{L} \mathsf{E}_x \partial_\zeta g(\zeta, X_k), \qquad L \ge 0;$$

note that $\mathscr{R}^0(\partial_\zeta g) = \partial_\zeta g$. Similarly to the proof of the previous Proposition 6.2.3, one constructs a function $\widetilde{H} \in L_{2+\delta}(\mathbb{X}, \mu)$ such that, for any $L \ge 0$,

$$\left|\mathscr{R}^L\big(\partial_\zeta g\big)(\zeta, x)\right| \le \widetilde{H}(x),$$
$$\left|\mathscr{R}^L\big(\partial_\zeta g\big)(\zeta_1, x) - \mathscr{R}^L\big(\partial_\zeta g\big)(\zeta_2, x)\right| \le |\zeta_1 - \zeta_2|^\delta \widetilde{H}(x).$$

Hence, for every $L \ge 0$, the function $\mathscr{R}^L g(\partial_\zeta g)$, considered as a map $\mathbb{R} \to L_{2+\delta}(\mathbb{X}, \mu)$, is continuous. Then

$$\mathscr{R}^L g(\zeta_2, x) - \mathscr{R}^L g(\zeta_1, x) = \int_{\zeta_1}^{\zeta_2} \mathscr{R}^L\big(\partial_\zeta g\big)(\zeta, x)\, d\zeta$$

where the integral in the right-hand side is well defined as the Bochner integral of an $L_{2+\delta}(\mathbb{X}, \mu)$-valued function. Because $\mathscr{R}^L g$, $\mathscr{R}^L(\partial_\zeta g)$ respectively converge to $\mathscr{R}g$, $\mathscr{R}(\partial_\zeta g)$

in $L_{2+\delta}(\mathbb{X},\mu)$ as $L \to \infty$ for every fixed ζ, we have the same identity and the same bounds for $\mathscr{R}g$ and $\mathscr{R}(\partial_\zeta g)$. □

Now we are ready to proceed with the proof of Theorem 6.2.1.

Proof of Theorem 6.2.1. Step I: Semi-martingale representation of the corrected sequence (6.2.7). We fix a function $\phi \in C^3(\mathbb{R})$ with bounded derivatives, and analyze the structure of the corresponding sequence (6.2.7). We have

$$\begin{aligned}\phi(\zeta_k^\varepsilon) &= \phi(\zeta_{k-1}^\varepsilon)\\ &\quad + \phi'(\zeta_{k-1}^\varepsilon)\left(\varepsilon f(\zeta_{k-1}^\varepsilon, X_k) + \sqrt{\varepsilon} g(\zeta_{k-1}^\varepsilon, X_k) + \sqrt{\varepsilon} h(\zeta_{k-1}^\varepsilon, X_k)\eta_k\right)\\ &\quad + \frac{\varepsilon}{2}\phi''(\zeta_{k-1}^\varepsilon)\Big(g(\zeta_{k-1}^\varepsilon, X_k) + h(\zeta_{k-1}^\varepsilon, X_k)\eta_k\Big)^2 + \vartheta_k^{\varepsilon,1},\end{aligned} \tag{6.2.8}$$

where the residual term equals

$$\vartheta_k^{\varepsilon,1} = \phi(\zeta_k^\varepsilon) - \phi(\zeta_{k-1}^\varepsilon) - \phi'(\zeta_{k-1}^\varepsilon)\xi_k^\varepsilon - \frac{\varepsilon}{2}\phi''(\zeta_{k-1}^\varepsilon)\Big(g(\zeta_{k-1}^\varepsilon, X_k) + h(\zeta_{k-1}^\varepsilon, X_k)\eta_k\Big)^2.$$

Since ζ_{k-1}^ε is $\mathscr{F}_{k-1}$-measurable, by the property (6.2.6) of the potential $\mathscr{R}g$, we have

$$\begin{aligned}\mathsf{E}[\mathscr{R}g(\zeta_{k-1}^\varepsilon, X_k)|\mathscr{F}_{k-1}] &= \mathsf{E}[\mathscr{R}g(\zeta, X_k)|\mathscr{F}_{k-1}]_{\zeta=\zeta_{k-1}^\varepsilon}\\ &= \widetilde{\mathscr{R}}g(\zeta, X_{k-1})\Big|_{\zeta=\zeta_{k-1}^\varepsilon} = \widetilde{\mathscr{R}}g(\zeta_{k-1}^\varepsilon, X_{k-1}).\end{aligned}$$

That is,

$$M_k^\varepsilon = \sum_{j=1}^{k}\Big(\mathscr{R}g(\zeta_{j-1}^\varepsilon, X_j) - \mathscr{R}g(\zeta_{j-1}^\varepsilon, X_{j-1})\Big), \quad k \geq 0$$

is a martingale, and we have

$$\widetilde{\mathscr{R}}g(\zeta_{k-1}^\varepsilon, X_k) - \widetilde{\mathscr{R}}g(\zeta_{k-1}^\varepsilon, X_{k-1}) = -g(\zeta_{k-1}^\varepsilon, X_k) + M_k^\varepsilon - M_{k-1}^\varepsilon. \tag{6.2.9}$$

By the definition of Υ^ε and eq. (6.2.9), we have

$$\begin{aligned}\Upsilon_k^\varepsilon - \Upsilon_{k-1}^\varepsilon &= \big(\phi'(\zeta_k^\varepsilon) - \phi'(\zeta_{k-1}^\varepsilon)\big)\widetilde{\mathscr{R}}g(\zeta_k^\varepsilon, X_k)\\ &\quad + \phi'(\zeta_{k-1}^\varepsilon)\big(\widetilde{\mathscr{R}}g(\zeta_k^\varepsilon, X_k) - \widetilde{\mathscr{R}}g(\zeta_{k-1}^\varepsilon, X_k)\big)\\ &\quad + \phi'(\zeta_{k-1}^\varepsilon)\big(-g(\zeta_{k-1}^\varepsilon, X_k) + M_k^\varepsilon - M_{k-1}^\varepsilon\big).\end{aligned} \tag{6.2.10}$$

Denote

$$\vartheta_k^{\varepsilon,2} = \sqrt{\varepsilon}\big(\phi'(\zeta_k^\varepsilon) - \phi'(\zeta_{k-1}^\varepsilon)\big)\big(\mathscr{R}g(\zeta_k^\varepsilon, X_k) - \mathscr{R}g(\zeta_{k-1}^\varepsilon, X_k)\big),$$
$$\vartheta_k^{\varepsilon,3} = \sqrt{\varepsilon}\left(\phi'(\zeta_k^\varepsilon) - \phi'(\zeta_{k-1}^\varepsilon) - \sqrt{\varepsilon}g(\zeta_k^\varepsilon, X_k)\phi''(\zeta_{k-1}^\varepsilon)\right)\mathscr{R}g(\zeta_{k-1}^\varepsilon, X_k),$$
$$\vartheta_k^{\varepsilon,4} = \sqrt{\varepsilon}\phi'(\zeta_{k-1}^\varepsilon)\left(\mathscr{R}g(\zeta_k^\varepsilon, X_k) - \mathscr{R}g(\zeta_{k-1}^\varepsilon, X_k) - \sqrt{\varepsilon}g(\zeta_k^\varepsilon, X_k)\partial_\zeta\mathscr{R}g(\zeta_{k-1}^\varepsilon, X_k)\right),$$

then we have from eqs. (6.2.8) and (6.2.10)

$$\begin{aligned}\psi_k^\varepsilon = \psi_{k-1}^\varepsilon &+ \sqrt{\varepsilon}\phi'(\zeta_{k-1}^\varepsilon)(M_k^\varepsilon - M_{k-1}^\varepsilon)\\ &+ \phi'(\zeta_{k-1}^\varepsilon)\left(\varepsilon f(\zeta_{k-1}^\varepsilon, X_k) + \sqrt{\varepsilon}h(\zeta_{k-1}^\varepsilon, X_k)\eta_k\right)\\ &+ \frac{\varepsilon}{2}\phi''(\zeta_{k-1}^\varepsilon)\Big(g(\zeta_{k-1}^\varepsilon, X_k) + h(\zeta_{k-1}^\varepsilon, X_k)\eta_k\Big)^2\\ &+ \varepsilon\phi'(\zeta_{k-1}^\varepsilon)g(\zeta_{k-1}^\varepsilon, X_k)\big(\partial_\zeta\mathscr{R}g\big)(\zeta_{k-1}^\varepsilon, X_k)\\ &+ \varepsilon\phi''(\zeta_{k-1}^\varepsilon)g(\zeta_{k-1}^\varepsilon, X_k)\mathscr{R}g(\zeta_{k-1}^\varepsilon, X_k) + \vartheta_k^\varepsilon\end{aligned}\tag{6.2.11}$$

with

$$\vartheta_k^\varepsilon = \vartheta_k^{\varepsilon,1} + \vartheta_k^{\varepsilon,2} + \vartheta_k^{\varepsilon,3} + \vartheta_k^{\varepsilon,4}.$$

Let us show that each of the residual terms $\vartheta_k^{\varepsilon,i}$, $i = 1, \dots, 4$ satisfies

$$\mathsf{E}|\vartheta_{k,i}^\varepsilon| \le C\left(\varepsilon^{1+\delta'/2} + \varepsilon^{3/2}\right),\tag{6.2.12}$$

without loss of generality we can assume that $\delta' < 1$ and $2 + \delta/2 < 3$. Because ϕ has bounded derivatives ϕ', ϕ'', ϕ''' and $2 + \delta' \le 3$, this implies

$$\left|\phi(y) - \phi(x) - (y - x)\phi'(x) - \frac{1}{2}(y - x)^2\phi''(x)\right| \le C|y - x|^{2+\delta'}.$$

Then,

$$\begin{aligned}|\vartheta_k^{\varepsilon,1}| &\le \frac{1}{2}\phi''(\zeta_{k-1}^\varepsilon)\left|(\xi_k^\varepsilon)^2 - \varepsilon\Big(g(\zeta_{k-1}^\varepsilon, X_k) + h(\zeta_{k-1}^\varepsilon, X_k)\eta_k\Big)^2\right| + C\left|\xi_k^\varepsilon\right|^{2+\delta'}\\ &\le C\varepsilon^2 f^2(\zeta_{k-1}^\varepsilon, X_k)\\ &\quad + C\varepsilon^{3/2}\left|f(\zeta_{k-1}^\varepsilon, X_k)\Big(g(\zeta_{k-1}^\varepsilon, X_k) + h(\zeta_{k-1}^\varepsilon, X_k)\eta_k\Big)\right| + C\left|\xi_k^\varepsilon\right|^{2+\delta'}\end{aligned}$$

Using Proposition 6.2.3, we get after straightforward rearrangements (which are somewhat cumbersome, and therefore omitted):

$$|\vartheta_k^{\varepsilon,1}| \leq C\varepsilon^{3/2}H^2(X_k) + C\varepsilon^{3/2}H^2(X_k)|\eta_k| + C\varepsilon^{1+\delta'/2}H^{2+\delta'}(X_k)|\eta_k|^{2+\delta'}.$$

By stationarity of X and the moment condition on η_k (see condition 3 of the theorem),

$$\mathsf{E}H^{2+\delta'}(X_k)|\eta_k|^{2+\delta'} \leq \left(\int_{\mathbb{X}} \widehat{H}^{2+\delta}\, \mathrm{d}\mu\right)^{(2+\delta')/(2+\delta)} \left(\mathsf{E}|\eta_k|^{(2+\delta)(2+\delta')/(\delta-\delta')}\right)^{(\delta-\delta')/(2+\delta)} = C < \infty. \tag{6.2.13}$$

Similarly, we have the bounds

$$\mathsf{E}H^2(X_k) \leq C, \quad \mathsf{E}H^2(X_k)|\eta_k| \leq C, \tag{6.2.14}$$

which complete the proof of eq. (6.2.12) for $i = 1$.

The proof of eq. (6.2.12) for $i = 3$ is similar and simpler, and we leave it for the reader; we only note that Proposition 6.2.3 yields $|\widetilde{\mathscr{R}}g(\zeta, x)| \leq 2H(x)$ because $\widetilde{\mathscr{R}}g = \mathscr{R}g - g$.

To prove eq. (6.2.12) for $i = 2$, observe that by Proposition 6.2.3

$$|\vartheta_k^{\varepsilon,2}| \leq C\sqrt{\varepsilon}|\zeta_k^{\varepsilon} - \zeta_k^{\varepsilon}|^{1+\delta'}H(X_k),$$

and

$$|\zeta_k^{\varepsilon} - \zeta_k^{\varepsilon}| \leq \varepsilon|f(\zeta_{k-1}^{\varepsilon}, X_k)| + \sqrt{\varepsilon}|g(\zeta_{k-1}^{\varepsilon}, X_k)| + \sqrt{\varepsilon}|h(\zeta_{k-1}^{\varepsilon}, X_k)\eta_k|.$$

One has

$$\mathsf{E}H^{2+\delta'}(X_k) \leq C$$

which combined with eq. (6.2.13) gives

$$\mathsf{E}H^{2+\delta'}(X_k)|\eta_k|^{1+\delta'} \leq C.$$

Combined with Proposition 6.2.3, this proves eq. (6.2.12) for $i = 2$.

The bound (6.2.12) with $i = 4$ can be obtained similarly, using the integral identity

$$\widetilde{\mathscr{R}}g(\zeta_k^{\varepsilon}, X_k) - \widetilde{\mathscr{R}}g(\zeta_{k-1}^{\varepsilon}, X_k) = \int_{\zeta_{k-1}^{\varepsilon}}^{\zeta_k^{\varepsilon}} \partial_\zeta \widetilde{\mathscr{R}}g(\zeta, X_k)\, \mathrm{d}\zeta$$

and the Hölder continuity of $\partial_\zeta \widetilde{\mathscr{R}}g(\zeta, x)$ w.r.t. ζ; we leave the details for the reader.

Now we can complete the rearrangement of representation (6.2.11). Denote

$$M_k^{\varepsilon,1} = \sum_{j=1}^{k} \phi'(\zeta_{j-1}^{\varepsilon})\Big((M_j^{\varepsilon} - M_{j-1}^{\varepsilon}) + h(\zeta_{j-1}^{\varepsilon}, X_j)\eta_j\Big),$$

$$M_k^{\varepsilon,2} = \frac{1}{2}\sum_{j=1}^{k} \phi''(\zeta_{j-1}^{\varepsilon})\Big(2g(\zeta_{j-1}^{\varepsilon}, X_j)h(\zeta_{j-1}^{\varepsilon}, X_j)\eta_j + h^2(\zeta_{j-1}^{\varepsilon}, X_j)(\eta_j^2 - 1)\Big).$$

These sequences are martingales: this follows from the martingale property for M^{ε} and an easy observation that for each k the (centered) random variables

$$\eta_k, \quad \eta_k^2 - 1$$

are independent with the σ-algebra

$$\widetilde{\mathscr{F}}_{k-1} = \mathscr{F}_{k-1} \bigvee \sigma(X_k).$$

This finally gives representation

$$\begin{aligned}\psi_k^{\varepsilon} &= \psi_{k-1}^{\varepsilon} + \varepsilon\phi'(\zeta_{k-1}^{\varepsilon})\Big(f(\zeta_{k-1}^{\varepsilon}, X_k) + g(\zeta_{k-1}^{\varepsilon}, X_k)\big(\partial_{\zeta}\widetilde{\mathscr{R}}g\big)(\zeta_{k-1}^{\varepsilon}, X_k)\Big)\\ &\quad + \frac{\varepsilon}{2}\phi''(\zeta_{k-1}^{\varepsilon})\Big(g^2(\zeta_{k-1}^{\varepsilon}, X_k) + 2g(\zeta_{k-1}^{\varepsilon}, X_k)\widetilde{\mathscr{R}}g(\zeta_{k-1}^{\varepsilon}, X_k) + h^2(\zeta_{k-1}^{\varepsilon}, X_k)\Big)\\ &\quad + \sqrt{\varepsilon}M_k^{\varepsilon,1} + \varepsilon M_k^{\varepsilon,2} + \vartheta_k^{\varepsilon}.\end{aligned} \tag{6.2.15}$$

The residual term satisfies

$$\mathsf{E}|\vartheta_k^{\varepsilon}| \le C\big(\varepsilon^{1+\delta'/2} + \varepsilon^{3/2}\big). \tag{6.2.16}$$

Repeating the argument which led to eq. (6.2.16), we get the following moment bounds for the increments of the martingales $M_k^{\varepsilon,1}, M_k^{\varepsilon,2}$:

$$\mathsf{E}|M_k^{\varepsilon,1} - M_{k-1}^{\varepsilon,1}|^{2+\delta'} \le C, \quad \mathsf{E}|M_k^{\varepsilon,2} - M_{k-1}^{\varepsilon,2}|^{1+\delta'/2}. \tag{6.2.17}$$

Step II: Weak relative compactness. Take $\phi(x) = x$; clearly, this is a C^3-function with bounded derivatives, and thus eq. (6.2.15) holds true. Now we have $\phi'(x) \equiv 1, \phi''(x) \equiv 0$, hence by eq. (6.2.15)

$$Y^{\varepsilon}(t) = Y^{\varepsilon,1}(t) + Y^{\varepsilon,2}(t) + Y^{\varepsilon,3}(t)$$

with

$$Y^{\varepsilon,1}(t) = \varepsilon \sum_{k\le\varepsilon^{-1}t} \Big(f(\zeta^\varepsilon_{k-1}, X_k) + g(\zeta^\varepsilon_{k-1}, X_k)\big(\partial_\zeta \widetilde{\mathcal{R}} g\big)(\zeta^\varepsilon_{k-1}, X_k)\Big),$$
$$Y^{\varepsilon,2}(t) = \sqrt{\varepsilon} M^{\varepsilon,1}_{[\varepsilon^{-1}t]} + \sqrt{\varepsilon}\Big(\Upsilon^\varepsilon_0 - \Upsilon^\varepsilon_{[\varepsilon^{-1}t]}\Big),$$
$$Y^{\varepsilon,3}(t) = \sum_{k\le\varepsilon^{-1}t} \vartheta^\varepsilon_k.$$

By eq. (6.2.16), we have

$$\mathsf{E}\sup_{t\le T} |Y^{\varepsilon,3}(t)| \le \mathsf{E} \sum_{k\le\varepsilon^{-1}t} |\vartheta^\varepsilon_k| \le C\big(\varepsilon^{\delta'/2} + \varepsilon^{1/2}\big) \to 0, \tag{6.2.18}$$

hence weak relative compactness for the family $\{Y^\varepsilon\}$ is equivalent to that for the family

$$\widehat{Y}^\varepsilon = Y^{\varepsilon,1} + Y^{\varepsilon,2}.$$

To prove weak relative compactness for $\{\widehat{Y}^\varepsilon\}$, we will use Proposition 6.1.2. Observe that

$$\Big|f(\zeta^\varepsilon_{k-1}, X_k) + g(\zeta^\varepsilon_{k-1}, X_k)\big(\partial_\zeta \widetilde{\mathcal{R}} g\big)(\zeta^\varepsilon_{k-1}, X_k)\Big| \le H(X_k) + 2H(X_k)\widetilde{H}(X_k),$$

and

$$\mathsf{E}\big(H(X_k) + 2H(X_k)\widetilde{H}(X_k)\big)^{1+\delta'/2} \le C.$$

Thus for any $s < t$

$$\begin{aligned}\mathsf{E}(Y^{\varepsilon,1}(t) - Y^{\varepsilon,1}(s))^{1+\delta'/2} &\le \mathsf{E}\left(\varepsilon \sum_{\varepsilon^{-1}s<k\le\varepsilon^{-1}t} \big(H(X_k) + 2H(X_k)\widetilde{H}(X_k)\big)\right)^{1+\delta'/2} \\ &\le C\big(|t-s| + \varepsilon\big)^{1+\delta'/2}.\end{aligned} \tag{6.2.19}$$

Next, by the Burkholder–Davis–Gundy inequality we have

$$\mathsf{E}\Big(\sqrt{\varepsilon} M^{\varepsilon,1}_{[\varepsilon^{-1}t]} - \sqrt{\varepsilon} M^{\varepsilon,1}_{[\varepsilon^{-1}s]}\Big)^{2+\delta'} \le \mathsf{E}\left(\varepsilon \sum_{\varepsilon^{-1}s<k\le\varepsilon^{-1}t} \Big(M^{\varepsilon,1}_{[\varepsilon^{-1}t]} - M^{\varepsilon,1}_{[\varepsilon^{-1}s]}\Big)^2\right)^{1+\delta'/2}.$$

Then, using the first inequality in eq. (6.2.17) and the same calculation as above, we get

$$\mathsf{E}\Big|\sqrt{\varepsilon} M^{\varepsilon,1}_{[\varepsilon^{-1}t]} - \sqrt{\varepsilon} M^{\varepsilon,1}_{[\varepsilon^{-1}s]}\Big|^{2+\delta'} \le C\big(|t-s| + \varepsilon\big)^{1+\delta'/2}.$$

In addition, we have

$$|\Upsilon_k^\varepsilon| = |\mathscr{R}g(\zeta_k^\varepsilon, X_k)| \le 2\widetilde{H}(X_k),$$

and therefore

$$\mathsf{E}\left|\sqrt{\varepsilon}\left(\Upsilon^\varepsilon_{[\varepsilon^{-1}t]} - \Upsilon^\varepsilon_{[\varepsilon^{-1}s]}\right)\right|^{2+\delta'} \le C\varepsilon^{1+\delta'/2}.$$

Combined with the previous inequality, this gives

$$\mathsf{E}(Y^{\varepsilon,2}(t) - Y^{\varepsilon,2}(s))^{2+\delta'} \le C\big(|t-s| + \varepsilon\big)^{1+\delta'/2}. \tag{6.2.20}$$

Now we can finalize the argument. We will prove that eq. (6.1.3) holds true for $\{\widehat{Y}^\varepsilon\}$ with $\alpha = 1 + \delta'/2$, $\beta = \delta'/2$. Observe that the function $Q(x) = |x| \wedge 1$ satisfies the triangle inequality

$$Q(x+y) \le Q(x) + Q(y),$$

hence

$$\begin{aligned}
&\mathsf{E}Q\big(\widehat{Y}^\varepsilon(t_2) - \widehat{Y}^\varepsilon(t_1)\big)^\alpha Q\big(\widehat{Y}^\varepsilon(t_3) - \widehat{Y}^\varepsilon(t_2)\big)^\alpha \\
&\qquad \le \sum_{i,j=1}^{2} \mathsf{E}Q\big(Y^{\varepsilon,i}(t_2) - Y^{\varepsilon,i}(t_1)\big)^\alpha Q\big(Y^{\varepsilon,j}(t_3) - Y^{\varepsilon,j}(t_2)\big)^\alpha.
\end{aligned}$$

If $i = 1$, we use the inequalities $Q(x) \le |x|$, $Q(x) \le 1$ and easily get by eq. (6.2.19)

$$\begin{aligned}
&\mathsf{E}Q\big(Y^{\varepsilon,i}(t_2) - Y^{\varepsilon,i}(t_1)\big)^\alpha Q\big(Y^{\varepsilon,j}(t_3) - Y^{\varepsilon,j}(t_2)\big)^\alpha \\
&\qquad \le \mathsf{E}\big|Y^{\varepsilon,i}(t_2) - Y^{\varepsilon,i}(t_1)\big|^\alpha \le C\big(|t_3 - t_1| + \varepsilon\big)^{1+\delta'/2}.
\end{aligned}$$

The same argument applies for $j = 1$. If $i = j = 2$, the we have by eq. (6.2.20) and the Cauchy inequality

$$\begin{aligned}
&\mathsf{E}Q\big(Y^{\varepsilon,i}(t_2) - Y^{\varepsilon,i}(t_1)\big)^\alpha Q\big(Y^{\varepsilon,j}(t_3) - Y^{\varepsilon,j}(t_2)\big)^\alpha \\
&\qquad \le \left(\mathsf{E}\big|Y^{\varepsilon,i}(t_2) - Y^{\varepsilon,i}(t_1)\big|^{2+\delta'}\right)^{1/2} \left(\mathsf{E}\big|Y^{\varepsilon,j}(t_3) - Y^{\varepsilon,j}(t_2)\big|^{2+\delta'}\right)^{1/2} \\
&\qquad \le C\big(|t_3 - t_1| + \varepsilon\big)^{1+\delta'/2}.
\end{aligned}$$

This proves that eq. (6.1.3) holds true for $\{\widehat{Y}^\varepsilon\}$ for any t_1, t_2, t_3 with $t_3 - t_1 \ge \varepsilon$. On the other hand, for $t_3 - t_1 < \varepsilon$ one of the intervals

$$[\varepsilon^{-1}t_1, \varepsilon^{-1}t_2], \quad [\varepsilon^{-1}t_2, \varepsilon^{-1}t_3]$$

does not contain integer points, and thus the corresponding increment of $\widehat{Y}^\varepsilon$ equals 0. Hence eq. (6.1.3) holds true for any $t_1 < t_2 < t_3$. This completes the proof of the weak compactness.

Step III: Identification of a limit point. Let Y be a weak limit point for the family $\{Y^\varepsilon\}$ as $\varepsilon \to 0$; that is, some subsequence $\{Y^{\varepsilon_n}\}$ weakly converges in $D(0,T)$ to Y. In order to simplify the notation, we assume that Y is a weak limit of the entire family Y^ε.

Recall that at the previous step we had decomposed

$$Y^\varepsilon = \widehat{Y}^\varepsilon + Y^{\varepsilon,3},$$

where $\{Y^{\varepsilon,3}\}$ is negligible, see eq. (6.2.16). Hence Y is a weak limit point for $\{\widehat{Y}^\varepsilon\}$, as well. By eqs. (6.2.19) and (6.2.20), we have

$$\mathsf{E}(\widehat{Y}^\varepsilon(t) - \widehat{Y}^\varepsilon(s))^{1+\delta'/2} \le C\big(|t-s| + \varepsilon\big)^{1/2+\delta'/4}. \tag{6.2.21}$$

Hence by the Fatou lemma, similar bound holds true for the process Y:

$$\mathsf{E}(Y(t) - Y(s))^{1+\delta'/2} \le C|t-s|^{1/2+\delta'/4}. \tag{6.2.22}$$

That is, Y is continuous in $L_{1+\delta/2}$, and thus there exists its measurable modification. In what follows, we assume that Y itself is measurable.

For an arbitrary but fixed $\phi \in C_0^\infty(\mathbb{R})$, we will prove that for every $t > s$, $m \ge 1$, $s_1, \dots, s_m \in [0,s]$, $F \in C_b(\mathbb{R}^m)$

$$\mathsf{E}\left(\phi\big(Y(t)\big) - \int_s^t \mathscr{L}\phi\big(Y(r)\big)\,dr\right)F\big(Y(s_1),\dots,Y(s_m)\big) = 0. \tag{6.2.23}$$

This will imply that

$$\phi\big(Y(t)\big) - \int_0^t \mathscr{L}\phi\big(Y(s)\big)\,ds, \qquad t \in [0,1]$$

is a martingale w.r.t. the natural filtration of Y, and prove the second statement of the theorem. Denote

$$\begin{aligned}\mathscr{Q}\phi(\zeta,x) &= \phi'(\zeta)\Big(f(\zeta,x) + g(\zeta,x)\big(\partial_\zeta \overline{\mathscr{R}}g\big)(\zeta,x)\Big)\\ &\quad + \frac{\varepsilon}{2}\phi''(\zeta)\Big(g^2(\zeta,x) + 2g(\zeta,x)\overline{\mathscr{R}}g(\zeta,x) + h^2(\zeta,x)\Big),\end{aligned} \tag{6.2.24}$$

and observe that

$$\int_{\mathbb{X}} \mathscr{Q}\phi(\zeta,x)\,\mu(dx) = \mathscr{L}\phi(x). \tag{6.2.25}$$

By eq. (6.2.15), the process

$$\phi\big(Y^\varepsilon(t)\big) + \sqrt{\varepsilon}\Upsilon^\varepsilon_{[\varepsilon^{-1}t]} - \int_0^{\varepsilon[\varepsilon^{-1}t]} \Big(\mathcal{Q}\phi\big(Y^\varepsilon(r), X_{[\varepsilon^{-1}r]}\big) + \varepsilon^{-1}\vartheta^\varepsilon_{[\varepsilon^{-1}r]}\Big)\,\mathrm{d}r$$

is a martingale w.r.t. the filtration

$$\Big\{\mathcal{G}^\varepsilon_t = \sigma\big(X_k, \eta_k, k \le \varepsilon^{-1}t\big), t \in [0, T]\Big\}.$$

Recall that $\{\vartheta^\varepsilon_k\}$ satisfies eq. (6.2.16) and

$$\mathsf{E}|\Upsilon^\varepsilon_k| \le C, \qquad \mathsf{E}|\mathcal{Q}\phi(Y^\varepsilon(r), X_{[\varepsilon^{-1}r]})| \le C;$$

see the moment bounds in Step I of the proof. Hence

$$\mathsf{E}\left(\phi\big(Y^\varepsilon(t)\big) - \int_0^t \mathcal{Q}\phi\big(Y^\varepsilon(r), X_{[\varepsilon^{-1}r]}\big)\,\mathrm{d}r\right)F\big(Y^\varepsilon(s_1), \dots, Y^\varepsilon(s_m)\big) \to 0 \tag{6.2.26}$$

as $\varepsilon \to 0$. If $t, s, s_1, \dots, s_m$ are such that, with probability 1, Y does not have a discontinuity in these points, then

$$\big(Y^\varepsilon(t), Y^\varepsilon(s), \dots, Y^\varepsilon(s_m)\big) \Rightarrow \big(Y(t), Y(s_1), \dots, Y(s_m)\big), \quad \varepsilon \to 0,$$

see Ref. [10, Chapter 3]. Combined with eq. (6.2.22), this yields the same weak convergence for arbitrary $t, s, s_1, \dots, s_m$. Since ϕ, F are continuous and bounded, we deduce from that

$$\mathsf{E}\phi\big(Y^\varepsilon(t)\big)F\big(Y^\varepsilon(s_1), \dots, Y^\varepsilon(s_m)\big) \to \mathsf{E}\phi\big(Y(t)\big)F\big(Y(s_1), \dots, Y(s_m)\big), \qquad \varepsilon \to 0.$$

Hence, to obtain eq. (6.2.23), we have to prove that

$$\begin{aligned}\mathsf{E}&\left(\int_s^t \mathcal{Q}\phi\big(Y^\varepsilon(r), X_{[\varepsilon^{-1}r]}\big)\,\mathrm{d}r\right)F\big(Y^\varepsilon(s_1), \dots, Y^\varepsilon(s_m)\big)\\ &\to \mathsf{E}\left(\int_s^t \mathcal{L}\phi\big(Y(r)\big)\,\mathrm{d}r\right)F\big(Y(s_1), \dots, Y(s_m)\big), \quad \varepsilon \to 0.\end{aligned} \tag{6.2.27}$$

To do that, we first replace Y^ε in this relation by $\widehat{Y}^\varepsilon$. Using Propositions 6.2.3 and 6.2.4, we construct a function $G \in L_{1+\delta/2}(\mathbb{X}, \mu)$ such that

$$|\mathcal{Q}\phi(\zeta, x)| \le G(x), \qquad |\mathcal{Q}\phi(\zeta_1, x) - \mathcal{Q}\phi(\zeta_2, x)| \le |\zeta_1 - \zeta_2|^\delta G(x). \tag{6.2.28}$$

The difference $Y^{\varepsilon,3} = Y^{\varepsilon} - \widehat{Y}^{\varepsilon}$ is L_1-negligible, see eq. (6.2.16), hence

$$\mathsf{E}\left|\int_s^t \mathscr{Q}\phi\big(Y^{\varepsilon}(r), X_{[\varepsilon^{-1}r]}\big)\,dr - \int_s^t \mathscr{Q}\phi\big(\widehat{Y}^{\varepsilon}(r), X_{[\varepsilon^{-1}r]}\big)\,dr\right|$$
$$\leq \mathsf{E}\int_s^t \big(|Y^{\varepsilon,3}(r)|^{\delta}\wedge 2\big)G(X_{[\varepsilon^{-1}r]})\,dr$$
$$\leq \left(\int_s^t \left(\mathsf{E}\big(|Y^{\varepsilon,3}(r)|^{\delta}\wedge 2\big)^{(2+\delta)/\delta}\right)^{\delta/(2+\delta)} dr\right)\|G\|_{L_{1+\delta/2}(\mathbb{X},\mu)} \to 0$$

as $\varepsilon \to 0$ by the dominated convergence theorem. That is, eq. (6.2.27) holds true provided the same relation, with Y^{ε} under the integral replaced by $\widehat{Y}^{\varepsilon}$, holds true. To prove the latter relation, we follow essentially the same idea we used in the proof of Theorem 5.2.1: with a controlled accuracy, we replace the integral by an integral sum, and then apply (a kind of) LLN.

Let $s = t_0 < \cdots < t_m = t$ be a given partition. By eq. (6.2.28), we have

$$|\mathscr{Q}\phi(\zeta_1, x) - \mathscr{Q}\phi(\zeta_2, x)| \leq |\zeta_1 - \zeta_2|^{\delta'/2}G(x).$$

Hence by eq. (6.2.22),

$$\mathsf{E}\left|\int_{t_{j-1}}^{t_j} \Big(\mathscr{Q}\phi\big(\widehat{Y}^{\varepsilon}(r), X_{[\varepsilon^{-1}r]}\big) - \mathscr{Q}\phi\big(\widehat{Y}^{\varepsilon}(t_{j-1}), X_{[\varepsilon^{-1}r]}\big)\Big)\,dr\right|$$
$$\leq \int_{t_{j-1}}^{t_j} \mathsf{E}|\widehat{Y}^{\varepsilon}(r) - \widehat{Y}^{\varepsilon}(t_{j-1})|^{\delta'/2}G\big(X_{[\varepsilon^{-1}r]}\big)\,dr$$
$$\leq (t_j - t_{j-1})\Big(C(t_j - t_{j-1})^{1/2+\delta'/4}\Big)^{\delta'/(2+\delta')}\|G\|^{2/\delta'}_{L_{1+\delta'/2}(\mathbb{X},\mu)}.$$

That is,

$$\mathsf{E}\left|\int_s^t \mathscr{Q}\phi\big(\widehat{Y}^{\varepsilon}(r), X_{[\varepsilon^{-1}r]}\big)\,dr - \sum_{j=1}^m \int_{t_{j-1}}^{t_j} \mathscr{Q}\phi\big(\widehat{Y}^{\varepsilon}(t_{j-1}), X_{[\varepsilon^{-1}r]}\big)\,dr\right| \leq C\Big(\max_j(t_j - t_{j-1})\Big)^{\delta'/4}.$$

We also have

$$\mathsf{E}\left|\int_s^t \mathscr{L}\phi(Y(r))\,dr - \sum_{j=1}^m \mathscr{L}\phi(Y(t_{j-1}))(t_j - t_{j-1})\right| \leq C\Big(\max_j(t_j - t_{j-1})\Big)^{\delta'/4},$$

the proof is similar and simpler, and thus is omitted. Hence we will have eq. (6.2.27) once we prove that for any $t_1 < t_2$, $t_1, t_2 \in [s, t]$,

$$\mathsf{E}\left|\int_{t_1}^{t_2} \mathscr{Q}\phi\big(\widehat{Y}^\varepsilon(t_1), X_{[\varepsilon^{-1}r]}\big)\,dr - (t_2 - t_1)\mathscr{L}\phi\big(\widehat{Y}^\varepsilon(t_1)\big)\right| \to 0, \quad \varepsilon \to 0. \tag{6.2.29}$$

To prove eq. (6.2.29), consider the uniform partitions $\{\zeta_{n,N} = n/N, n \in \mathbb{Z}\}$, $N \ge 1$ of the real line, denote $I_{n,N} = (\zeta_{n-1,N}, \zeta_{n,N})$, and put

$$\widehat{Y}^{\varepsilon,N}(t_{j-1}) = \sum_{n=-\infty}^{\infty} \zeta_{n,N} 1_{\widehat{Y}^\varepsilon(t_{j-1}) \in I_{n,N}}.$$

Denote

$$f_{n,N}(x) = \mathscr{Q}(\zeta_{n,N}, x), \quad n \in \mathbb{Z}, \quad x \in \mathbb{X},$$

then by Theorem 5.2.1 and eq. (6.2.25), we have

$$\int_{t_1}^{t_2} f_{n,N}\big(X_{[\varepsilon^{-1}r]}\big)\,dr \to (t_2 - t_1)\mathscr{L}\phi(\zeta_{n,N})$$

in mean sense. Hence for each $N, R \in \mathbb{N}$

$$\begin{aligned}
&1_{|\widehat{Y}^\varepsilon(t_{j-1})| \le R}\left(\int_{t_1}^{t_2} \mathscr{Q}\phi\big(\widehat{Y}^{\varepsilon,N}(t_1), X_{[\varepsilon^{-1}r]}\big)\,dr - (t_2 - t_1)\mathscr{L}\phi\big(\widehat{Y}^{\varepsilon,N}(t_1)\big)\right) \\
&= \sum_{n=-RN+1}^{RN} 1_{\widehat{Y}^\varepsilon(t_{j-1}) \in I_{n,N}}\left(\int_{t_1}^{t_2} f_{n,N}\big(X_{[\varepsilon^{-1}r]}\big)\,dr - (t_2 - t_1)\mathscr{L}\phi(\zeta_{n,N})\right) \to 0
\end{aligned}$$

as $\varepsilon \to 0$ in mean sense. That is,

$$\begin{aligned}
&\limsup_{\varepsilon \to 0} \mathsf{E}\left|\int_{t_1}^{t_2} \mathscr{Q}\phi\big(\widehat{Y}^\varepsilon(t_1), X_{[\varepsilon^{-1}r]}\big)\,dr - (t_2 - t_1)\mathscr{L}\phi\big(\widehat{Y}^\varepsilon(t_1)\big)\right| \\
&\le \limsup_{\varepsilon \to 0} \mathsf{E}1_{|\widehat{Y}^\varepsilon(t_{j-1})| > R}\left|\int_{t_1}^{t_2} \mathscr{Q}\phi\big(\widehat{Y}^\varepsilon(t_1), X_{[\varepsilon^{-1}r]}\big)\,dr - (t_2 - t_1)\mathscr{L}\phi\big(\widehat{Y}^\varepsilon(t_1)\big)\right| \\
&+ \limsup_{\varepsilon \to 0} \mathsf{E}\int_{t_1}^{t_2} \left|\mathscr{Q}\phi\big(\widehat{Y}^\varepsilon(t_1), X_{[\varepsilon^{-1}r]}\big) - \mathscr{Q}\phi\big(\widehat{Y}^{\varepsilon,N}(t_1), X_{[\varepsilon^{-1}r]}\big)\right|\,dr \\
&+ \limsup_{\varepsilon \to 0}(t_2 - t_1)\mathsf{E}\left|\mathscr{L}\phi\big(\widehat{Y}^\varepsilon(t_1) - \mathscr{L}\phi\big(\widehat{Y}^{\varepsilon,N}(t_1)\right|
\end{aligned}$$

By eq. (6.2.25), inequality (6.2.28) yields similar bounds for the function $\mathscr{L}\phi$:

$$|\mathscr{L}\phi(\zeta)| \le C, \quad |\mathscr{L}\phi(\zeta_1) - \mathscr{L}\phi(\zeta_2)| \le C|\zeta_1 - \zeta_2|^{\delta'}.$$

Then by eq. (6.2.21),

$$\begin{aligned}
&\mathsf{E}1_{|\widehat{Y}^\varepsilon(t_{j-1})|>R}\left|\int_{t_1}^{t_2} \mathscr{Q}\phi\big(\widehat{Y}^\varepsilon(t_1), X_{[\varepsilon^{-1}r]}\big)\,\mathrm{d}r - (t_2-t_1)\mathscr{L}\phi\big(\widehat{Y}^\varepsilon(t_1)\big)\right| \\
&\quad \le C\int_{t_1}^{t_2} \mathsf{E}1_{|\widehat{Y}^\varepsilon(t_{j-1})|>R}\Big(G\big(X_{[\varepsilon^{-1}r]}\big)+1\Big)\,\mathrm{d}r \\
&\quad \le CR^{-\delta'/2}\int_{t_1}^{t_2} \mathsf{E}|\widehat{Y}^\varepsilon(t_{j-1})|^{\delta'/2}\Big(G\big(X_{[\varepsilon^{-1}r]}\big)+1\Big)\,\mathrm{d}r \le CR^{-\delta'/2},
\end{aligned}$$

$$\begin{aligned}
&\mathsf{E}\int_{t_1}^{t_2} |\mathscr{Q}\phi\big(\widehat{Y}^\varepsilon(t_1), X_{[\varepsilon^{-1}r]}\big) - \mathscr{Q}\phi\big(\widehat{Y}^{\varepsilon,N}(t_1), X_{[\varepsilon^{-1}r]}\big)|\,\mathrm{d}r \\
&\quad \le C\int_{t_1}^{t_2} \mathsf{E}|\widehat{Y}^\varepsilon(t_1) - \widehat{Y}^{\varepsilon,N}|^{\delta'/2}\Big(G\big(X_{[\varepsilon^{-1}r]}\big)+1\Big)\,\mathrm{d}r \\
&\quad \le C\Big(\mathsf{E}|\widehat{Y}^\varepsilon(t_1) - \widehat{Y}^{\varepsilon,N}|^{1+\delta'/2}\Big)\delta'/(2+\delta') \le CN^{-\delta'/4},
\end{aligned}$$

$$(t_2-t_1)\mathsf{E}|\mathscr{L}\phi\big(\widehat{Y}^\varepsilon(t_1)\big) - \mathscr{L}\phi\big(\widehat{Y}^{\varepsilon,N}(t_1)\big)| \le CN^{-\delta'/4}.$$

That is,

$$\limsup_{\varepsilon\to 0}\mathsf{E}\left|\int_{t_1}^{t_2} \mathscr{Q}\phi\big(\widehat{Y}^\varepsilon(t_1), X_{[\varepsilon^{-1}r]}\big)\,\mathrm{d}r - (t_2-t_1)\mathscr{L}\phi\big(\widehat{Y}^\varepsilon(t_1)\big)\right| \le CR^{-\delta'/2} + CN^{-\delta'/4}.$$

Since N, R here are arbitrary, this proves eq. (6.2.29), and completes the proof of statement II of the theorem.

Step IV: Completion of the proof. If the martingale problem (6.2.3) is well posed, then the weak convergence in $\mathbb{D}(0,T)$ of the family $\{Y^\varepsilon\}$ follows by the statements I, II of the theorem. Note that

$$\sup_{t\in[0,T]} |Y^\varepsilon(t) - \widetilde{Y}^\varepsilon(t)| \le \max_{k\le\varepsilon^{-1}T} |\xi_k^\varepsilon|.$$

By Proposition 6.2.3 and eq. (6.2.13), we have

$$\mathsf{E}|\xi_k^\varepsilon|^{2+\delta'} \le C\varepsilon^{1+\delta'/2},$$

and thus

$$\mathsf{E}\sup_{t\in[0,T]}|Y^\varepsilon(t)-\widetilde{Y}^\varepsilon(t)|^{2+\delta'}\le\sum_{k\le\varepsilon^{-1}T}\mathsf{E}|\xi_k^\varepsilon|^{2+\delta'}\le C\varepsilon^{\delta'/2}\to 0.$$

That is, for any $\eta>0$

$$\mathsf{P}\left(\sup_{t\in[0,T]}|Y^\varepsilon(t)-\widetilde{Y}^\varepsilon(t)|>\eta\right)\to 0,\quad \varepsilon\to 0,$$

which means that the family $\{\widetilde{Y}^\varepsilon\}$ weakly converges in $\mathbb{D}(0,T)$ to the same limit. This family has continuous trajectories by the construction, and the limiting diffusion process has continuous trajectories, as well; see Ref. [126, Chapter 7]. Applying Proposition 6.1.1, we complete the proof. □

Remark 6.2.5. The CLT provided by Theorem 5.2.4 can be considered just as a particular case of Theorem 6.2.1, because in the case $f\equiv 0, h\equiv 0$, and $g(\zeta,x)=g(x)$, the assumptions of Theorem 6.2.1 actually repeat those of Theorem 5.2.4, with $f(x)$ renamed to $g(x)$ and $p=2+\delta$. It is well known that the martingale problem

$$\mathscr{L}\phi(\zeta)=\frac{1}{2}\sigma^2\phi''(\zeta),\quad \mathscr{D}=C_0^\infty(\mathbb{R})$$

is well posed, and its unique solution is σW; hence applying Theorem 6.2.1 we get a more detailed functional version of the CLT stated in Theorem 5.2.4: the processes

$$Y^\varepsilon(t)=\sum_{k\le\varepsilon^{-1}t}\xi_k^\varepsilon,\quad \widetilde{Y}^\varepsilon(t)=Y^\varepsilon(t)+\left(\varepsilon[\varepsilon^{-1}t]+\varepsilon-t\right)\xi^\varepsilon_{[\varepsilon^{-1}t]+1}$$

weakly converge to σW in $\mathbb{D}(0,T)$ and $\mathbb{C}(0,T)$, respectively. Similar functional version is available for the martingale CLT given by Theorem 5.2.1.

Let us discuss two possibilities to extend the results of this section to continuous-time setting, for example, for the family $\{Y^\varepsilon\}$ given by eq. (6.1.2). First, one can try to repeat the entire argument, with proper changes, in the continuous-time setting. The key step would be to introduce, for a given "test function" $\phi\in C^2(\mathbb{R}^m)$, a corrected process

$$\psi_t^\varepsilon=\phi(Y_t^\varepsilon)+\sqrt{\varepsilon}\Upsilon_t^\varepsilon,\quad t\ge 0 \tag{6.2.30}$$

with properly chosen corrector term Υ_t^ε, which will make it possible to use the martingale methods in the further analysis. This is the essence of the corrector term method developed in Ref. [110]; however, implementation of this method meets substantial technical difficulties which we explain below. The corrector term now should have the form

$$\Upsilon_t^\varepsilon=\phi'(\zeta_t^\varepsilon)\mathscr{R}^{(c)}g(Y_t^\varepsilon,X_t^\varepsilon),$$

where the (continuous-time) potential $\mathscr{R}^{(c)}g$ is well defined by the formula

$$\mathscr{R}^{(c)}g(y,x) = \int_0^\infty \mathsf{E}_x g(y, X_s)\, ds, \qquad x \in \mathbb{X} \tag{6.2.31}$$

provided that X satisfies eq. (5.3.6) and g satisfies the assumptions of Theorem 6.2.1. To clarify the semi-martingale structure of the corrected process (6.2.30), one should use a continuous-time analogue of the Taylor formula used before. A natural substitute for the Taylor formula in this framework is the Itô formula, and this is exactly the way the corrector term method was applied in Ref. [110]. However, even in the comparatively simple case of X being a diffusion process, this brings serious difficulties. Namely, to apply the Itô formula when X is a diffusion process, one should require that $\mathscr{R}^{(c)}g$ is C^2-smooth w.r.t. the variable x. This is a strong additional assumption, which can not be verified unless strong structural assumptions on the process X are imposed; the situation here is similar to the one mentioned in Section 5.4, comment **3**. In the diffusion setting, this problem is resolved in Ref. [112], where the Itô–Krylov formula is used instead of the classical Itô one, which requires less regularity from $\mathscr{R}^{(c)}g$, and thus can be applied under the principal assumptions, formulated only in the terms of ergodic properties of the process X. This line of arguments can be extended even more, and instead of the Itô formula an analogue of the Dynkin formula can be used; see Refs. [88, 89]. This version of the corrector term method is well applicable when X is ergodic w.r.t. total variation distance, but for the weakly ergodic X it still requires $\mathscr{R}^{(c)}g$ to belong to certain weighted Hölder class w.r.t. x, which is a hidden structural limitation on the process X; see discussion in Section 5.4, comment **3**. It is not clear, if it is possible to develop a continuous-time version of the corrector term method, where all the regularity assumptions on $\mathscr{R}^{(c)}g$ are removed, like we did that in the discrete-time setting in Theorem 6.2.1.

Another possibility, which is apparently simpler than the previous one, is to embed the continuous-time model into a discrete-time one, like it was done, for example, in the proof of continuous-time CLT; Theorem 5.3.4. For instance, if Y^ε is given by eq. (6.1.2) with $\sigma^\varepsilon \equiv 0$, then

$$Y^\varepsilon_{k\varepsilon} = \zeta^\varepsilon_k = \sum_{j=1}^k \xi^\varepsilon_j, \quad \xi^\varepsilon_k = \int_{(k-1)\varepsilon}^{k\varepsilon} a^\varepsilon(Y^\varepsilon_s, X^\varepsilon_s)\, ds.$$

However, while trying to embed the corresponding AR sequence into the model given by eq. (6.2.1), we encounter new technical complications, for example, we need to consider *segment chain*

$$\mathbf{X}(k) = \{X_{k+s}, s \in [0,1]\}, \quad k \ge 0$$

instead of the initial process X, to track the dependence of the trajectory of Y^ε on its initial value, and so on. This approach looks realistic, but overcomplicated, especially when compared with an alternative and simpler way to study systems even more general than eq. (6.1.2), which we develop in the following section.

6.3 The Time Delay Method for General Autoregressive Models

In this section, we develop an alternative method for the study of asymptotic behavior of autoregressive models, which will not use the corrector term in any form. This method is based on the *time delay* trick, which will be explained later on.

Unlike we did that in eq. (6.2.1), now we do not specify the increments of the AR sequence (5.2.1) in a functional form, but impose the conditions on the dependence of ξ_k^ε on $\mathscr{F}_j^\varepsilon$ for $k > j$. Our principal assumption is the following.

$\mathbf{H}_0$. For some sequence $\{R_l, l \geq 0\} \subset \mathbb{R}^+$, family $\{L^\varepsilon, \varepsilon > 0\} \subset \mathbb{N}$, and families of random variables $\{\gamma_{k,l}^\varepsilon\}$, $\{\upsilon_{k,l}^\varepsilon\}$,

$$\mathsf{E}\left[\xi_{k+l}^\varepsilon \middle| \mathscr{F}_k^\varepsilon\right] = \varepsilon\gamma_{k,l}^\varepsilon + \varepsilon^{1/2}\upsilon_{k,l}^\varepsilon R_l, \quad k \geq 0, \quad l = 1, \ldots, L^\varepsilon. \tag{6.3.1}$$

Later on we will specify the conditions imposed on the families $\{R_l\}$, $\{L^\varepsilon\}$, $\{\gamma_{k,l}^\varepsilon\}$, $\{\upsilon_{k,l}^\varepsilon\}$, and provide examples where these conditions are verified. Now we shortly discuss two simplest models of that kind, which will explain the heuristics of the model specified above.

Example 6.3.1. Let $\mathbf{H}_0$ hold true with $L^\varepsilon = 1$ and $R_1 = 0$. This just means that the sequence $\{\xi_k^\varepsilon\}$ admits a representation

$$\xi_{k+1}^\varepsilon = \eta_{k+1}^\varepsilon + \varepsilon\gamma_k^\varepsilon \tag{6.3.2}$$

with $\{\eta_k^\varepsilon\}$ being a martingale difference. In this case the required limit behavior is well understood: under proper bondedness/convergence conditions imposed on $\{\gamma_k^\varepsilon\}$ and the quadratic variation of $\{\eta_k^\varepsilon\}$, respective family $\{\zeta_n^\varepsilon, \varepsilon = n^{-1}\}$ is weakly compact/converges weakly; see Theorem 5.2.1 for a particular result of such a type. We note that, in order the quadratic variation of $\{\eta_k^\varepsilon\}$ to be asymptotically nontrivial, it should be typically assumed that the first term in eq. (6.3.2) has the "absolute order" $\varepsilon^{1/2} \gg \varepsilon = n^{-1}$, but this does not cause a "blowup" because of the martingale structure of this term.

Example 6.3.2. Let $\{\xi_k^\varepsilon\}$ have the form

$$\xi_k^\varepsilon = \varepsilon^{1/2} f(X_k)$$

where X is a Markov chain; that is, now we are in the framework studied in Section 5.2.2. If X satisfies eq. (5.1.1) and $f \in H_{\gamma,W}(\mathbb{X}, d)$ is centered, then by Proposition 5.1.4 we have that eq. (6.3.1) holds true for

$$L_\varepsilon = \infty, \quad \gamma^\varepsilon_{k,l} = 0, \quad R_l = r_l^\gamma,$$

$$\upsilon^\varepsilon_{k,l} = \|f\|_{d,\gamma,W}\big(V(X_k)\big)^\gamma \left(W(X_k) + \int_{\mathbb{X}} W\, d\mu \right)^{1-\gamma}.$$

Now $\{\xi^\varepsilon_k\}$ cannot be treated as a martingale difference, but its conditional expectations exhibit a stabilization property "in a long run", since $R_l \to 0, l \to \infty$.

Our principal assumption $\mathbf{H}_0$ combines these two basic models: the term $\varepsilon\gamma^\varepsilon_{k,l}$ in eq. (6.3.1) has the similar meaning with the term γ^ε_k in eq. (6.3.2), and the term $\varepsilon^{1/2}\upsilon^\varepsilon_{k,l}R_l$ corresponds to the part of the model which exhibits a stabilization "in a long run". From this point of view, the sequence $\{R_l\}$ has a natural interpretation of a "stabilization rate", and the family $\{\upsilon^\varepsilon_{k,l}\}$ has the meaning of a "penalty", similarly to the term $V(x)$ in the nonuniform ergodic rate (5.1.1).

6.3.1 General Statements

We begin our study with a weak relative compactness result for the family of processes

$$Y^\varepsilon(t) = \sum_{k \le \varepsilon^{-1}t} \xi^\varepsilon_k.$$

Theorem 6.3.3. *Let assumption $\mathbf{H}_0$ hold true with $L^\varepsilon = [\varepsilon^{-\rho}] + 1$ and*

$$R_l \le Cl^{-r}, \quad l \le L^\varepsilon \tag{6.3.3}$$

for some

$$\rho \in \left(0, \frac{1}{2}\right], \quad r > \frac{1}{2\rho}.$$

Assume in addition that there exist $C > 0, p > 2$ such that

$$\mathsf{E}|\gamma^\varepsilon_{k,l}|^p \le C, \quad \mathsf{E}|\upsilon^\varepsilon_{k,l}|^p \le C, \quad k \ge 0, \quad l \le L^\varepsilon \tag{6.3.4}$$

and

$$\mathsf{E}|\xi^\varepsilon_k|^p \le C\varepsilon^{p/2}, \quad k \ge 0. \tag{6.3.5}$$

Then for every $T > 0$, there exists a constant C_T such that

$$\mathsf{E}\big|Y^\varepsilon(t) - Y^\varepsilon(s)\big|^p \le C_T\big(|t-s| + \varepsilon\big)^{p/2}, \quad |s-t| \le T, \tag{6.3.6}$$

and thus the family $\{Y^\varepsilon\}$ is weakly relatively compact in $\mathbb{D}(0, T)$.

Remark 6.3.4. Weak relative compactness of $\{Y^\varepsilon\}$ follows from eq. (6.3.6) by Proposition 6.1.2.

Proof. Since Y^ε is piece-wise constant, it is sufficient to prove eq. (6.3.6) only for s, t of the form $i\varepsilon$, $k\varepsilon$. To simplify the notation, we prove the required inequality only for $i = 0$ assuming $\zeta_0^\varepsilon = 0$. That is, our aim is to prove the inequality

$$\mathsf{E}|\zeta_k^\varepsilon|^p \le C_T(\varepsilon k)^{p/2}, \quad \varepsilon k \le T. \tag{6.3.7}$$

Applying, if necessary, the Hölder inequality, we can assume that $p \in (2, 3]$. Denote

$$f(x) = |x|^p, \quad g(x) = f'(x) = p|x|^{p-1}\operatorname{sign} x, \quad h(x) = f''(x) = p(p-1)|x|^{p-2}.$$

It follows from the elementary inequality

$$|1+z|^p \le 1 + pz + \frac{p(p-1)}{2}z^2 + C|z|^p, \quad z \in \mathbb{R}$$

that

$$f(\zeta_k^\varepsilon) \le f(\zeta_{k-1}^\varepsilon) + g(\zeta_{k-1}^\varepsilon)\xi_k^\varepsilon + \frac{1}{2}h(\zeta_{k-1}^\varepsilon)(\xi_k^\varepsilon)^2 + C|\xi_k^\varepsilon|^p.$$

Therefore,

$$\mathsf{E}f(\zeta_k^\varepsilon) \le \sum_{j=1}^k \mathsf{E}g(\zeta_{j-1}^\varepsilon)\xi_j^\varepsilon + \frac{1}{2}\sum_{j=1}^k \mathsf{E}h(\zeta_{j-1}^\varepsilon)(\xi_j^\varepsilon)^2 + C\sum_{j=1}^k \mathsf{E}|\xi_j^\varepsilon|^p. \tag{6.3.8}$$

We perform further transformation of the first sum in the right hand side of eq. (6.3.8) using the "time delay" trick, which we explain below. A direct estimation of the expectations involved into the first sum is hardly possible, because under our principal assumption $\mathbf{H}_0$ the decomposition of the conditional expectation for ξ_j^ε w.r.t. $\mathcal{F}_{j-1}^\varepsilon$ contains the term $\varepsilon^{1/2}R_1 v_{j-1,1}^\varepsilon$, which has the order $\varepsilon^{1/2} \gg \varepsilon$. In order to resolve this difficulty, we will change ζ_{j-1}^ε to $\zeta_{(j-L^\varepsilon)_+}^\varepsilon$; that is, we will perform a delay of the time variable in the sequence ζ^ε. Because

$$\varepsilon^{1/2}R_{L^\varepsilon} \le C\varepsilon^{1/2+\rho r} \ll \varepsilon, \tag{6.3.9}$$

the conditional expectation of ξ_j^ε w.r.t. $\mathcal{F}_{(j-L^\varepsilon)_+}^\varepsilon$ will be well manageable; this is actually our main tool which makes it possible to use the "long run stabilization" property of the model.

Let us develop in details the procedure outlined above. We have

$$g(x+\delta) - g(x) = h(x)\delta + \varrho(x, \delta),$$

where

$$\varrho(x,\delta) = \delta(h(x+\theta) - h(x))$$

with θ being an intermediate point between x and $x+\delta$. Recall that $h(x) = C|x|^{p-2}$ and $p-2 \in (0,1]$. Then,

$$|h(x+\theta) - h(x)| \le C|\theta|^{p-2},$$

which yields $|\varrho(x,\delta)| \le C|\delta|^{p-1}$. That is,

$$g(\zeta_i^\varepsilon) = g(\zeta_{i-1}^\varepsilon) + h(\zeta_{i-1}^\varepsilon)\xi_i^\varepsilon + \varrho_i^\varepsilon, \quad |\varrho_i^\varepsilon| \le C|\xi_i^\varepsilon|^{p-1},$$

and thus,

$$g(\zeta_{j-1}^\varepsilon) = g(\zeta_{(j-L^\varepsilon)_+}^\varepsilon) + \sum_{i=(j-L^\varepsilon)_+ +1}^{j-1} \left(h(\zeta_{i-1}^\varepsilon)\xi_i^\varepsilon + \varrho_i^\varepsilon\right).$$

We rearrange eq. (6.3.8) and then analyze separately each sum which appear in the right hand side. We have

$$\begin{aligned} \mathsf{E}f(\zeta_k^\varepsilon) \le{}& \sum_{j=1}^{k} \mathsf{E}g(\zeta_{(j-L^\varepsilon)_+}^\varepsilon)\xi_j^\varepsilon + \sum_{j=1}^{k}\sum_{i=(j-L^\varepsilon)_+ +1}^{j-1} \mathsf{E}h(\zeta_{i-1}^\varepsilon)\xi_i^\varepsilon\xi_j^\varepsilon \\ &+ \sum_{j=1}^{k}\sum_{i=(j-L^\varepsilon)_+ +1}^{j-1} \mathsf{E}\varrho_i^\varepsilon\xi_j^\varepsilon + \frac{1}{2}\sum_{j=1}^{k} \mathsf{E}h(\zeta_{j-1}^\varepsilon)(\xi_j^\varepsilon)^2 + C\sum_{j=1}^{k} \mathsf{E}|\xi_j^\varepsilon|^p =: \sum_{l=1}^{5} S_k^{l,\varepsilon}. \end{aligned} \tag{6.3.10}$$

Recall that $f(0) = 0$ and ζ_0^ε is assumed to be equal 0. That is, each term in the sum $S_k^{1,\varepsilon}$ with $j \le L^\varepsilon$ equals 0. For all the other terms, we have $\zeta_{(j-L^\varepsilon)_+}^\varepsilon = \zeta_{j-L^\varepsilon}^\varepsilon$ and thus by $\mathbf{H}_0$

$$S_k^{1,\varepsilon} = \sum_{j=L^\varepsilon+1}^{k} \mathsf{E}g(\zeta_{j-L^\varepsilon}^\varepsilon)\left(\varepsilon^{1/2}\upsilon_{j-L^\varepsilon,L^\varepsilon}^\varepsilon R_{L^\varepsilon} + \varepsilon\gamma_{j-L^\varepsilon,L^\varepsilon}^\varepsilon\right).$$

Recall that $|g(x)|^{p/(p-1)} = Cf(x)$, and denote $\mu_j^\varepsilon = \mathsf{E}f(\zeta_j^\varepsilon)$. Then by eq. (6.3.4) and the Hölder inequality,

$$S_k^{1,\varepsilon} \le C\sum_{j=L^\varepsilon+1}^{k} \left(\mu_{j-L^\varepsilon}^\varepsilon\right)^{(p-1)/p}\left(\varepsilon^{1/2}R_{L^\varepsilon} + \varepsilon\right) \le C\varepsilon \sum_{j=L^\varepsilon+1}^{k} \left(\mu_{j-L^\varepsilon}^\varepsilon\right)^{(p-1)/p};$$

the second inequality holds true because of eq. (6.3.9).

Next, by $\mathbf{H}_0$

$$\mathsf{E}h(\zeta_{i-1}^\varepsilon)\xi_i^\varepsilon\xi_j^\varepsilon = \mathsf{E}h(\zeta_{i-1}^\varepsilon)\xi_i^\varepsilon\left(\varepsilon^{1/2}\upsilon_{i,j-i}^\varepsilon R_{j-i} + \varepsilon\gamma_{i,j-i}^\varepsilon\right).$$

We have $|h(x)|^{p/(p-2)} = Cf(x)$, hence by eqs. (6.3.4) and (6.3.5), we get

$$\mathsf{E}h(\zeta_{i-1}^{\varepsilon})\xi_i^{\varepsilon}\upsilon_{i,j-i}^{\varepsilon} \le C\varepsilon^{1/2}\Big(\mu_{i-1}^{\varepsilon}\Big)^{(p-2)/p}, \quad \mathsf{E}h(\zeta_{i-1}^{\varepsilon})\xi_i^{\varepsilon}\gamma_{i,j-i}^{\varepsilon} \le C\varepsilon^{1/2}\Big(\mu_{i-1}^{\varepsilon}\Big)^{(p-2)/p}.$$

Taking into account that $\sum_l R_l < \infty$ and $L^{\varepsilon} \le C\varepsilon^{-1/2}$, we deduce that

$$S_k^{2,\varepsilon} \le C\varepsilon \sum_{i=1}^{k} \Big(\mu_{i-1}^{\varepsilon}\Big)^{(p-2)/p}.$$

Similarly,

$$\mathsf{E}\varrho_i^{\varepsilon}\xi_j^{\varepsilon} = \mathsf{E}\varrho_i^{\varepsilon}\Big(\varepsilon^{1/2}\upsilon_{i,j-i}^{\varepsilon}R_{j-i} + \varepsilon\gamma_{i,j-i}^{\varepsilon}\Big),$$

and because $|\varrho_i^{\varepsilon}| \le C|\xi_i^{\varepsilon}|^{p-1}$, we have

$$\mathsf{E}\varrho_i^{\varepsilon}\upsilon_{i,j-i}^{\varepsilon} \le \Big(\mathsf{E}|\varrho_i^{\varepsilon}|^{p/(p-1)}\Big)^{(p-1)/p}\Big(\mathsf{E}|\upsilon_{i,j-i}^{\varepsilon}|^{p}\Big)^{1/p} \le C\varepsilon^{(p-1)/2}$$

$$\mathsf{E}\varrho_i^{\varepsilon}\gamma_{i,j-i}^{\varepsilon} \le C\varepsilon^{(p-1)/2}.$$

Taking into account that $\sum_l R_l < \infty$ and $L^{\varepsilon} \le C\varepsilon^{-1/2}$, we deduce

$$S_k^{3,\varepsilon} \le C\sum_{i=1}^{k} \varepsilon^{p/2}.$$

We also have

$$S_k^{4,\varepsilon} \le C\varepsilon \sum_{i=1}^{k} \Big(\mu_{i-1}^{\varepsilon}\Big)^{(p-2)/p}$$

(the proof is omitted since it is similar and simpler to the proof of the inequality for $S_k^{2,\varepsilon}$), and

$$S_k^{5,\varepsilon} \le C\sum_{i=1}^{k} \varepsilon^{p/2}$$

(which follows directly from eq. (6.3.5)). We denote $M_k^{\varepsilon} = \max_{i\le k}\mu_i^{\varepsilon}$ and summarize the above estimates as follows:

$$M_k^{\varepsilon} \le C_* \sum_{j=1}^{k} \Big(\varepsilon(M_{j-1}^{\varepsilon})^{(p-1)/p} + \varepsilon(M_{j-1}^{\varepsilon})^{(p-2)/p} + \varepsilon^{p/2}\Big)$$

(we fix the constant C_* for the further reference needs). Now we use the induction to prove that, under a proper choice of $C_T \geq 1$, eq. (6.3.7) holds true. Assume eq. (6.3.7) to hold for all $k' \leq k$ ($\leq \varepsilon^{-1}T$), then by the above inequality,

$$\begin{aligned} M^{\varepsilon}_{k+1} &\leq C_* \sum_{j=1}^{k} \left(\varepsilon C_T^{(p-1)/p} (\varepsilon j)^{(p-1)/2} + \varepsilon C_T^{p-2/p} (\varepsilon j)^{(p-2)/2} + \varepsilon^{p/2} \right) \\ &\leq C_* \sum_{j=1}^{k} \left(\varepsilon C_T^{(p-1)/p} T^{1/2} (\varepsilon j)^{(p-2)/2} + \varepsilon C_T^{p-2/p} (\varepsilon j)^{(p-2)/2} + \varepsilon^{p/2} \right) \\ &\leq C_* C_T^{(p-1)/p} (T^{1/2} + 1) \varepsilon^{p/2} \sum_{j=1}^{k} \left(j^{(p-2)/2} + 1 \right) \end{aligned}$$

(recall that we assumed $C_T \geq 1$). It is elementary to prove that for some $C_p > 0$

$$\sum_{j=1}^{k} \left(j^{(p-2)/2} + 1 \right) \leq C_p (k+1)^{p/2}.$$

If C_T was chosen

$$C_T \geq (C_* C_p (T^{1/2} + 1))^{1/p},$$

then eq. (6.3.7) holds true for $k + 1$, as well. This completes the entire proof. □

To formulate the weak convergence result, we introduce more notation and assumptions. First, we assume that $\mathbf{H}_0$ holds true and the following decomposition is available:

$$\gamma^{\varepsilon}_{k,L_\varepsilon} = A(\zeta^{\varepsilon}_k) + \chi^{\varepsilon}_k, \quad k \geq 0. \tag{6.3.11}$$

Here function A is interpreted as an "asymptotic drift coefficient" of the family $\{\zeta^{\varepsilon}\}$, and χ^{ε}_k is interpreted as a "residual term" which will be assumed to be negligible, in a sense.

Next, we assume the following.

$\mathbf{H}_1$. For each $j \geq 0$,

$$\mathsf{E}\left[\xi^{\varepsilon}_{k+L^{\varepsilon}} \xi^{\varepsilon}_{k+L^{\varepsilon}+j} \middle| \mathscr{F}^{\varepsilon}_k \right] = \varepsilon B_j(\zeta^{\varepsilon}_k) + \varepsilon \varkappa^{\varepsilon}_{k,j}, \quad k \geq 0. \tag{6.3.12}$$

The functions B_j in the above condition should be understood as "asymptotic covariance coefficients" of the family $\{\zeta^{\varepsilon}\}$. The function

$$B(\zeta) = B_0(\zeta) + 2\sum_{j=1}^{\infty} B_j(\zeta), \tag{6.3.13}$$

which will be assumed well defined, will play the role of the "asymptotic diffusion coefficient". The "residual terms" $\varkappa_{k,j}^{\varepsilon}$ will be assumed to be negligible, in a sense.

Theorem 6.3.5. *Let conditions of Theorem 6.3.3 hold true and eqs.* (6.3.11) *and* (6.3.12) *are valid. Assume in addition the following.*

1. *(Conditions on the coefficients). The function A is bounded and continuous. The series* (6.3.13) *converge point-wise; the sequence of partial sums*

$$B^m(\zeta) = B_0(\zeta) + 2\sum_{j=1}^{m} B_j(\zeta)$$

   is uniformly bounded, and each B^m, $m \geq 1$ *is continuous.*
2. *(Negligibility of the residual terms).*

$$\sup_k E\left|\chi_k^{\varepsilon}\right| \to 0, \tag{6.3.14}$$

   and for each $j \geq 0$

$$\sup_k E\left|\varkappa_{k,j}^{\varepsilon}\right| \to 0, \quad \varepsilon \to 0. \tag{6.3.15}$$

Then each weak limit point Y *of the family* $\{Y^{\varepsilon}\}$ *is a solution to the martingale problem* $(\mathscr{L}, \mathscr{D})$ *with*

$$\mathscr{L}\phi = A\phi' + \frac{1}{2}B\phi'', \quad \phi \in \mathscr{D} = C_b^3(\mathbb{R}).$$

In particular, if the above martingale problem is well posed, then the family Y^{ε} *weakly converges in* $\mathbb{D}(0, T)$ *to the diffusion process uniquely defined by the martingale problem* $(\mathscr{L}, \mathscr{D})$*. Furthermore, the family*

$$\widetilde{Y}^{\varepsilon}(t) = Y^{\varepsilon}(t) + \left(\varepsilon[\varepsilon^{-1}t] + \varepsilon - t\right)\xi^{\varepsilon}_{[\varepsilon^{-1}t]+1}$$

weakly converges in $\mathbb{C}(0, T)$ *to the same diffusion process.*

Remark 6.3.6. We will show in the proof that B is continuous. Hence for the martingale problem $(\mathscr{L}, \mathscr{D})$ to be well posed it is sufficient that

$$B(\zeta) > 0, \quad \zeta \in \mathbb{R}.$$

Proof of the theorem. Once the weak convergence of Y^{ε} is proved, the weak convergence of $\widetilde{Y}^{\varepsilon}$ in $\mathbb{C}(0, T)$ follows easily by Proposition 6.1.1; see the proof of

Theorem 6.2.1. In the rest of the proof, we fix a weak limit point Y for $\{Y^\varepsilon\}$ in $\mathbb{D}(0,T)$ and a function $f \in C_b^3(\mathbb{R})$. Our aim is to show that the process

$$\phi(Y(t)) - \int_0^t \mathscr{L}\phi(Y(s))\,ds$$

is a martingale w.r.t. the natural filtration for Y. To do that, it is enough to prove eq. (6.2.23). To simplify the notation, we will consider the case $s = 0$, only; the general case is completely analogous. Recall that $Y^\varepsilon(0) = y$ is a constant, hence our actual aim is to show that

$$\mathsf{E}\phi(Y(t)) - \mathsf{E}\phi(Y(0)) - \mathsf{E}\int_0^t \mathscr{L}\phi(Y(s))\,ds = 0.$$

To do that, it is clearly enough to show that

$$\mathsf{E}\phi(Y^\varepsilon(t)) - \mathsf{E}\phi(Y^\varepsilon(0)) - \mathsf{E}\int_0^t \mathscr{L}\phi(Y^\varepsilon(s))\,ds \to 0, \quad \varepsilon \to 0. \tag{6.3.16}$$

To simplify the notation, we assume that entire family Y^ε weakly converges to Y in $\mathbb{D}(0,T)$. Recall that we assumed $p \le 3$, hence ϕ'' is Hölder continuous with the index $p-2$, because it is bounded and has a bounded derivative. Then we have by the Taylor formula

$$\phi(\zeta_k^\varepsilon) = \phi(\zeta_{k-1}^\varepsilon) + \phi'(\zeta_{k-1}^\varepsilon)\xi_k^\varepsilon + \frac{1}{2}\phi''(\zeta_{k-1}^\varepsilon)(\xi_k^\varepsilon)^2 + \varsigma_k^\varepsilon$$

with

$$|\varsigma_k^\varepsilon| \le C|\xi_k^\varepsilon|^p.$$

To simplify the notation, we will systematically use the following convention. If families $\{f^\varepsilon\} \subset \mathbb{R}$, $\{g^\varepsilon\} \subset \mathbb{R}^+$ satisfy inequality $|f^\varepsilon| \le Cg^\varepsilon$, we write $f^\varepsilon \lesssim g^\varepsilon$; the same notation is also used for families of random variables if the above inequality holds true a.s. with a nonrandom C. The same notation will be used for families which depend on additional indices; the constant C in this case should not depend on the additional indices. We also write $f^\varepsilon \approx g^\varepsilon$ if $f^\varepsilon - g^\varepsilon \to 0, \varepsilon \to 0$.

By eq. (6.3.5), the above estimate yields $\mathsf{E}\varsigma_k^\varepsilon \lesssim \varepsilon^{p/2}$. Since $p/2 > 1$, this yields

$$\mathsf{E}\phi(Y^\varepsilon(t)) - \mathsf{E}\phi(Y^\varepsilon(0)) \approx \sum_{k=1}^{[t/\varepsilon]} \mathsf{E}\phi'(\zeta_{k-1}^\varepsilon)\xi_k^\varepsilon + \frac{1}{2}\sum_{k=1}^{[t/\varepsilon]} \mathsf{E}\phi''(\zeta_{k-1}^\varepsilon)(\xi_k^\varepsilon)^2. \tag{6.3.17}$$

The rest of the proof is essentially based on the "time delay" trick, which we had already used in the proof of Theorem 6.3.3. We would like to replace $\phi'(\zeta_{k-1}^\varepsilon)$ by

$\phi'(\zeta^\varepsilon_{(k-L^\varepsilon)_+})$ in the first sum in eq. (6.3.17). To do that, we apply the Taylor formula of the first order, and get

$$\phi'(\zeta^\varepsilon_{k-1}) = \phi'(\zeta^\varepsilon_{(k-L^\varepsilon)_+}) + \sum_{j=(k-L^\varepsilon)_+ +1}^{k-1} \left(\phi''(\zeta^\varepsilon_{j-1})\xi^\varepsilon_j + \iota^\varepsilon_j\right).$$

Here the residual terms satisfy $\iota^\varepsilon_k \lesssim |\xi^\varepsilon_k|^{p/2}$ because $p/2-1<1$ and thus ϕ'' can be treated as a Hölder continuous function with the Hölder index $p/2-1$. Then by eqs. (6.3.4) and (6.3.5), we have that, for some C,

$$\mathsf{E}|\iota^\varepsilon_j||\upsilon^\varepsilon_{j,k-j}| \le C\varepsilon^{p/2}, \quad \mathsf{E}|\iota^\varepsilon_j||\gamma^\varepsilon_{j,k-j}| \le C\varepsilon^{p/2}$$

for each k, j with $k-j \le L^\varepsilon$. By $\mathbf{H}_0$,

$$\left|\mathsf{E}\sum_{k=1}^{[t/\varepsilon]}\sum_{j=(k-L^\varepsilon)_+ +1}^{k-1} \iota^\varepsilon_j\xi^\varepsilon_k\right| \le \sum_{k=1}^{[t/\varepsilon]}\sum_{j=(k-L^\varepsilon)_+ +1}^{k-1} \mathsf{E}|\iota^\varepsilon_j||\varepsilon^{1/2}R_{k-j}\upsilon^\varepsilon_{j,k-j} + \varepsilon\gamma^\varepsilon_{j,k-j}|.$$

Recall that $\sum_l R_l < \infty$, hence the above estimates yield

$$\left|\mathsf{E}\sum_{k=1}^{[t/\varepsilon]}\sum_{j=(k-L^\varepsilon)_+ +1}^{k-1} \iota^\varepsilon_j\xi^\varepsilon_k\right| \le C\varepsilon^{p/2-1/2} + C\varepsilon^{p/2-1/2-\rho} \to 0, \quad \varepsilon \to 0.$$

Summarizing the above calculation, we get

$$\begin{aligned}\mathsf{E}\phi(Y^\varepsilon(t)) - \mathsf{E}\phi(Y^\varepsilon(0)) &\approx \sum_{k=1}^{[t/\varepsilon]} \mathsf{E}\phi'(\zeta^\varepsilon_{(k-L^\varepsilon)_+})\xi^\varepsilon_k \\ &+ \frac{1}{2}\sum_{k=1}^{[t/\varepsilon]}\left(\mathsf{E}\phi''(\zeta^\varepsilon_{k-1})(\xi^\varepsilon_k)^2 + \sum_{j=(k-L^\varepsilon)_+ +1}^{k-1} \mathsf{E}\phi''(\zeta^\varepsilon_{j-1})\xi^\varepsilon_j\xi^\varepsilon_k\right) \\ &=: \Sigma^{t,\varepsilon}_1 + \Sigma^{t,\varepsilon}_2.\end{aligned}$$

In the sum $\Sigma^{t,\varepsilon}_1$ the "time delay" is already performed, and we can complete easily the study of its limit behavior. The sum of the first $L^\varepsilon \ll \varepsilon^{-1/2}$ terms in $\Sigma^{t,\varepsilon}_1$ is negligible, because by eq. (6.3.5) each of these terms allows the bound $\lesssim \varepsilon^{1/2}$. Next, for $k \ge L^\varepsilon$, we have

$$\begin{aligned}\left|\mathsf{E}\phi'(\zeta^\varepsilon_{k-L^\varepsilon})(\xi^\varepsilon_k - \varepsilon A(\zeta_{k-L^\varepsilon}))\right| &\le C(\varepsilon^{1/2}\mathsf{E}|\upsilon^\varepsilon_{k-L^\varepsilon,L^\varepsilon}|R_{L^\varepsilon} + \varepsilon\mathsf{E}|\chi^\varepsilon_{k-L^\varepsilon}|) \\ &\le C\varepsilon\left(\varepsilon^{r\rho-1/2}\sup_{k,l}\mathsf{E}|\upsilon^\varepsilon_{k,l}| + \sup_k \mathsf{E}|\upsilon^\varepsilon_k|\right) \ll \varepsilon, \quad \varepsilon \to 0.\end{aligned}$$

Hence,

$$\Sigma_1^{t,\varepsilon} \approx \varepsilon \sum_{k=L^\varepsilon}^{[t/\varepsilon]} A(\zeta_{k-L^\varepsilon})\phi'(\zeta_{k-L^\varepsilon}^\varepsilon),$$

and it can be easily shown that

$$\Sigma_1^{t,\varepsilon} \to \mathsf{E}\int_0^t A(Y(s))\phi'(Y(s))\,ds, \quad \varepsilon \to 0. \tag{6.3.18}$$

Let us proceed with the second sum $\Sigma_2^{t,\varepsilon}$. A simple rearrangement gives

$$\Sigma_2^{t,\varepsilon} = \frac{1}{2}\sum_{j=1}^{[t/\varepsilon]} \mathsf{E}\phi''(\zeta_{j-1}^\varepsilon)\left((\xi_j^\varepsilon)^2 + 2\sum_{k=j+1}^{(j+L^\varepsilon-1)\wedge[t/\varepsilon]} \xi_j^\varepsilon\xi_k^\varepsilon\right).$$

We have by $\mathbf{H}_0$, eqs. (6.3.4) and (6.3.5)

$$\begin{aligned}\left|\mathsf{E}\phi''(\zeta_{j-1}^\varepsilon)\xi_j^\varepsilon \sum_{k=j+1}^{(j+L^\varepsilon-1)\wedge[t/\varepsilon]} \xi_k^\varepsilon\right| &\le C\varepsilon\left(\sum_l R_l\right)\left(\mathsf{E}\sup_{k,l}|\upsilon_{k,l}^\varepsilon|^2)\right)^{1/2} \\ &\quad + C\varepsilon^{3/2}L^\varepsilon\left(\sup_{k,l}\mathsf{E}|\gamma_{k,l}^\varepsilon|^2\right)^{1/2} \le C\varepsilon,\end{aligned}$$

hence the sums of the first L^ε terms and the last L^ε terms in $\Sigma_2^{t,\varepsilon}$ are negligible; that is,

$$\Sigma_2^{t,\varepsilon} \approx \frac{1}{2}\sum_{j=L^\varepsilon}^{[t/\varepsilon]-L^\varepsilon} \mathsf{E}\phi''(\zeta_{j-1}^\varepsilon)\left((\xi_j^\varepsilon)^2 + 2\xi_j^\varepsilon\Xi_j^\varepsilon\right),$$

where we denote

$$\Xi_j^\varepsilon = \sum_{k=j+1}^{j+L^\varepsilon-1} \xi_k^\varepsilon.$$

To manage the latter sum, we again make a "time delay trick"; however, now we simply change ζ_{j-1}^ε to $\zeta_{j-L^\varepsilon}^\varepsilon$; that is, in a sense we apply the Taylor formula of the order 0, instead of the Tayor formula of the order 1 applied in $\Sigma_1^{t,\varepsilon}$.

Since $p \in (2,3]$, we have $2(p-2)/p < 1$. Then ϕ'' is Hölder continuous with the index $2(p-2)/p$, and by eq. (6.3.5), we have

$$\mathsf{E}\left|\phi''(\zeta_{j-1}^\varepsilon) - \phi''(\zeta_{j-L^\varepsilon})^\varepsilon)\right|(\xi_j^\varepsilon)^2 \le C\left(\mathsf{E}(\zeta_{j-1} - \zeta_{j-L^\varepsilon}^\varepsilon))^2\right)^{(p-2)/p}\left(\mathsf{E}(\xi_j^\varepsilon)^p\right)^{2/p}.$$

By eq. (6.3.6) and the choice of L^ε,

$$\mathsf{E}(\zeta_{j-1} - \zeta^\varepsilon_{j-L^\varepsilon}))^2 \le C\varepsilon^{1-\rho}.$$

Hence,

$$\mathsf{E}\left|\phi''(\zeta^\varepsilon_{j-1}) - \phi''(\zeta^\varepsilon_{j-L^\varepsilon})\right|(\xi^\varepsilon_j)^2 \le C\varepsilon^{1+(1-\rho)(p-2)/p},$$

and the sum of these error terms is negligible. Next, by $\mathbf{H}_0$ we have

$$\begin{aligned}\mathsf{E}\Big(\phi''(\zeta^\varepsilon_{j-1}) - \phi''(\zeta^\varepsilon_{j-L^\varepsilon})\Big)\xi^\varepsilon_j \Xi^\varepsilon_j &= \varepsilon^{1/2}\sum_{k=j+1}^{k+L^\varepsilon} \mathsf{E}\Big(\phi''(\zeta^\varepsilon_{j-1}) - \phi''(\zeta^\varepsilon_{j-L^\varepsilon})\Big)\xi^\varepsilon_j R_{k-j} v^\varepsilon_{j,k-j} \\ &\quad + \varepsilon \sum_{k=j+1}^{k+L^\varepsilon} \mathsf{E}\Big(\phi''(\zeta^\varepsilon_{j-1}) - \phi''(\zeta^\varepsilon_{j-L^\varepsilon})\Big)\xi^\varepsilon_j \gamma^\varepsilon_{j,k-j}.\end{aligned} \tag{6.3.19}$$

By eq. (6.3.4) and (6.3.5), each term under the second sum can be simply estimated as $\lesssim \varepsilon^{1/2}$ because ϕ'' is bounded. Then the entire second term in the right-hand side of eq. (6.3.19) is

$$\lesssim \varepsilon^{3/2} L^\varepsilon \ll \varepsilon,$$

and the sum of $\sim t/\varepsilon$ of such terms is negligible. Finally, using that ϕ'' is Hölder continuous with the index $(p-2)/p$, we get

$$\begin{aligned}&\left|\mathsf{E}\Big(\phi''(\zeta^\varepsilon_{j-1}) - \phi''(\zeta^\varepsilon_{j-L^\varepsilon})\Big)\xi^\varepsilon_j v^\varepsilon_{j,k-j}\right| \\ &\qquad \le \left(\mathsf{E}(\zeta^\varepsilon_{j-1} - \zeta_{j-L^\varepsilon})^2\right)^{(p-2)/(2p)} \left(\mathsf{E}(\xi^\varepsilon_j)^p\right)^{1/p} \left(\mathsf{E}(v^\varepsilon_{j,k-j})^2\right)^{1/2}.\end{aligned}$$

Using eqs. (6.3.4), (6.3.5) and (6.3.6), and recalling that $\sum_l R_l < \infty$, we conclude that the entire first term in the right-hand side of eq. (6.3.19) is

$$\lesssim \varepsilon^{1+(1-\rho)(p-2)/(2p)},$$

and the sum of $\sim t/\varepsilon$ of such terms is negligible. Summarizing the above calculation, we get

$$\Sigma_2^{t,\varepsilon} \approx \frac{1}{2}\sum_{j=L^\varepsilon}^{[t/\varepsilon]-L^\varepsilon} \mathsf{E}\phi''(\zeta^\varepsilon_{j-L^\varepsilon})\left((\xi^\varepsilon_j)^2 + 2\xi^\varepsilon_j \Xi^\varepsilon_j\right).$$

Let us show that

$$\Sigma_2^{t,\varepsilon} \to \frac{1}{2}\mathsf{E}\int_0^t B(Y(s))\phi''(Y(s))\,ds, \quad \varepsilon \to 0. \tag{6.3.20}$$

Fix $m \geq 1$, and define

$$\Xi_j^{m,\varepsilon} = \sum_{k=j+1}^{j+m} \xi_k^\varepsilon.$$

Applying eq. (6.3.12), the negligibility assumption for the terms $\varkappa_{k,j}^\varepsilon$, and continuity of B^m, we easily obtain

$$\frac{1}{2} \sum_{j=L^\varepsilon}^{[t/\varepsilon]-L^\varepsilon} \mathsf{E}\phi''(\zeta_{j-L^\varepsilon}^\varepsilon)\left((\xi_j^\varepsilon)^2 + 2\xi_j^\varepsilon \Xi_j^{m,\varepsilon}\right) \to \frac{1}{2}\mathsf{E}\int_0^t B^m(Y(s))\phi''(Y(s))\,ds, \quad \varepsilon \to 0.$$

On the other hand, by $\mathbf{H}_0$, eqs. (6.3.4) and (6.3.5),

$$\begin{aligned}\mathsf{E}\phi''(\zeta_{j-L^\varepsilon}^\varepsilon)\xi_j^\varepsilon(\Xi_j^\varepsilon - \Xi_j^{m,\varepsilon}) &= \mathsf{E}\phi''(\zeta_{j-L^\varepsilon}^\varepsilon)\xi_j^\varepsilon \sum_{k=j+m+1}^{j+L^\varepsilon-1} \left(\varepsilon^{1/2} R_{k-j} v_{j,k-j}^\varepsilon + \varepsilon \gamma_{j,k-j}^\varepsilon\right) \\ &\lesssim C\left(\varepsilon \sum_{l=m+1}^{\infty} R_l + \varepsilon^{3/2} L^\varepsilon\right),\end{aligned}$$

and therefore,

$$\begin{aligned}\limsup_{\varepsilon\to 0} \Sigma_2^{t,\varepsilon} &\leq \frac{1}{2}\mathsf{E}\int_0^t B^m(Y(s))\phi''(Y(s))\,ds + C\sum_{l=m+1}^{\infty} R_l, \\ \liminf_{\varepsilon\to 0} \Sigma_2^{t,\varepsilon} &\geq \frac{1}{2}\mathsf{E}\int_0^t B^m(Y(s))\phi''(Y(s))\,ds - C\sum_{l=m+1}^{\infty} R_l.\end{aligned}$$

By the dominated convergence theorem,

$$\mathsf{E}\int_0^t B^m(Y(s))\phi''(Y(s))\,ds \to \mathsf{E}\int_0^t B(Y(s))\phi''(Y(s))\,ds, \quad m \to \infty,$$

hence we derive eq. (6.3.20) from the above relations passing to the limit as $m \to \infty$. Combining eqs. (6.3.18) and (6.3.20), we complete the proof. □

6.3.2 Averaging Principle and Diffusion Approximation for Systems with Fast Markov Perturbations

Here we apply the general statements obtained in Section 6.3.1 to the family of processes defined by eq. (6.1.2) with the coefficients given by eq. (6.2.2). That is, we consider a family of processes $\{Y^\varepsilon\}$ with

$$\begin{aligned} dY^\varepsilon(t) &= f\big(Y^\varepsilon(t), X^\varepsilon(t)\big)\,dt + \varepsilon^{-1/2} g\big(Y^\varepsilon(t), X^\varepsilon(t)\big)\,dt + h\big(Y^\varepsilon(t), X^\varepsilon(t)\big)\,dW_t, \\ Y^\varepsilon(0) &= y_0, \end{aligned} \tag{6.3.21}$$

where $X^\varepsilon(t) = X_{\varepsilon^{-1}t}$, X is a given Markov process, and W is a Wiener process independent of X. Process X^ε in eq. (6.3.21) plays the role of an external perturbation, which is plugged into the coefficients, and thus effects the process Y^ε. Since X^ε is obtained from X by the time change $t \mapsto \varepsilon^{-1}t$, which actually accelerates the time scale, eq. (6.3.21) is called a *system with fast Markov perturbation.*

To clarify the structure of the available results, and to improve readability of the proofs, we separate the exposition in two parts. First, we establish the *averaging principle* in the case where the "most dangerous" term with the multiplier $\varepsilon^{-1/2}$ is not actually presented in system (6.3.21).

Theorem 6.3.7 (The averaging principle). *Let the Markov process X satisfy eq.* (5.3.6), *and $g(\zeta, x) \equiv 0$. Assume the following.*

1. *For some $\gamma \in (0, 1]$ and function W, the functions*

$$f(\zeta, \cdot), \quad h(\zeta, \cdot)$$

belong to $H_{\gamma,W}(\mathbb{X}, d)$. In addition,

$$\sup_{\zeta\in\mathbb{R}} \big(\|f(\zeta, \cdot)\|_{d,\gamma,W} + \|h(\zeta, \cdot)\|_{d,\gamma,W}\big) < \infty,$$

and there exists $\delta > 0$ such that

$$\sup_{\zeta_1\neq\zeta_2} |\zeta_1 - \zeta_2|^{-\delta} \|f(\zeta_1, \cdot) - f(\zeta_2, \cdot)\|_{d,\gamma,W} < \infty,$$
$$\sup_{\zeta_1\neq\zeta_2} |\zeta_1 - \zeta_2|^{-\delta} \|h(\zeta_1, \cdot) - h(\zeta_2, \cdot)\|_{d,\gamma,W} < \infty.$$

2. *For every $\zeta \in \mathbb{R}$, the functions $f(\zeta, \cdot), h(\zeta, \cdot) \in L_1(\mathbb{X}, \mu)$ satisfy*

$$\sup_{\zeta\in\mathbb{R}} \left|\int_{\mathbb{X}} f(\zeta, x)\,\mu(dx)\right| < \infty, \quad \sup_{\zeta\in\mathbb{R}} \left|\int_{\mathbb{X}} h(\zeta, x)\,\mu(dx)\right| < \infty,$$

$$\sup_{\zeta_1\neq\zeta_2} |\zeta_1 - \zeta_2|^{-\delta} \left|\int_{\mathbb{X}} f(\zeta_1, x)\,\mu(dx) - \int_{\mathbb{X}} f(\zeta_2, x)\,\mu(dx)\right| < \infty,$$

$$\sup_{\zeta_1\neq\zeta_2} |\zeta_1 - \zeta_2|^{-\delta} \left|\int_{\mathbb{X}} h(\zeta_1, x)\,\mu(dx) - \int_{\mathbb{X}} h(\zeta_2, x)\,\mu(dx)\right| < \infty.$$

3.

$$\sup_t EV^{2+\delta}(X_t) < \infty, \quad \sup_t EW^{2+\delta}(X_t) < \infty.$$

Then the family $\{Y^\varepsilon\}$ *defined by eq.* (6.3.21) *is weakly relatively compact in* $\mathbb{D}(0,T)$, *and any weak limit point of* $Y^\varepsilon, \varepsilon \to 0$ *is a solution of the martingale problem* $(\mathscr{L}, \mathscr{D})$ *with*

$$\mathscr{L}\phi(\zeta) = A(\zeta)\phi'(\zeta) + \frac{1}{2}B(\zeta)\phi''(\zeta), \qquad \mathscr{D} = C_0^\infty(\mathbb{R}^d), \tag{6.3.22}$$

where

$$A(\zeta) = \int_{\mathbb{X}} f(\zeta, x)\,\mu(dx), \quad B(\zeta) = \int_{\mathbb{X}} h^2(\zeta, x)\,\mu(dx).$$

If the martingale problem (6.3.22) *is well posed, then the family* Y^ε *weakly converges in* $\mathbb{C}(0,T)$ *to a diffusion process* Y, *which is the unique solution of this martingale problem.*

Remark 6.3.8. Note that coefficients A, B of the limiting process are obtained by averaging of corresponding coefficients f, h in x w.r.t. the invariant measure μ of the Markov process X. This explains the name *averaging principle*, typically used for this type of results.

Remark 6.3.9. Conditions of Theorem 6.3.7 are well comparable with the assumptions of Theorem 6.2.1. The moment condition 3 has slightly different form than condition 4 of Theorem 6.2.1, because now we allow the Markov process X to be nonstationary. We note that in Theorem 6.3.7 we do not require any specific bounds on the ergodic rate r_t for X, assuming only

$$r_t \to 0, \quad t \to \infty.$$

This well corresponds to the heuristics that the averaging principle is an extension of LLN, which does not require the ergodic rates to be specified, see Section 5.4, comment **2.**

Before proceeding with the proof, let us formulate several inequalities, which will be repeatedly used in the sequel. Condition 3 and the Fatou lemma yield that $W \in L_{2+\delta}(\mathbb{X}, \mu)$. Without losing generality, from now on we assume that $W(x) \geq 1$. Then for any x,

$$\int_{\mathbb{X}} W(y)\mu(dy) \leq CW(x),$$

and thus by Corollary 5.1.5, we have

$$|f(\zeta, x)| \leq CV^\gamma(x)W^{1-\gamma}(x), \tag{6.3.23}$$

$$|f(\zeta_1, x) - f(\zeta_2, x)| \leq C|\zeta_1 - \zeta_2|^\delta V^\gamma(x)W^{1-\gamma}(x). \tag{6.3.24}$$

Next, denote

$$\mathbb{F}^{\varepsilon} = \left\{ \mathscr{F}_k^{\varepsilon} = \sigma\big(X(s), W_{\varepsilon s}, s \le k\big), k \ge 0 \right\}.$$

For any $k \ge 0$ and any $\mathscr{F}_k^{\varepsilon}$-measurable random variable Υ, we have by condition 1, Proposition 5.1.4, and eq. (5.3.6)

$$\begin{aligned}
&\left| \mathsf{E}\left[f(\Upsilon, X^{\varepsilon}(s)) \middle| \mathscr{F}_k^{\varepsilon} \right] - \int_{\mathbb{X}} f(\Upsilon, x)\mu(dx) \right| \\
&\quad = \left| \mathsf{E}\left[f(\zeta, X^{\varepsilon}(s)) \middle| \mathscr{F}_k^{\varepsilon} \right] - \int_{\mathbb{X}} f(x)\mu(dx) \right|_{\zeta=\Upsilon} \\
&\quad \le C r_{s-\varepsilon k}^{\gamma} V\big(X^{\varepsilon}(\varepsilon k)\big)^{\gamma} \mathsf{E}\left[W(X^{\varepsilon}(s)) \middle| \mathscr{F}_k^{\varepsilon} \right]^{1-\gamma}, \quad s \ge \varepsilon k.
\end{aligned} \tag{6.3.25}$$

Clearly, eqs. (6.3.23)–(6.3.25) hold true for function $h(\zeta, x)$, as well. We will also need an analogue of eq. (6.3.25) for the function $h^2(\zeta, x)$, but $h^2(\zeta, x)$ does not necessarily belong to $H_{\gamma,W}(\mathbb{X}, d)$. This motivates the following calculation. Without loss of generality, we will assume that $V(x) \ge 1$. We have

$$|h(\zeta, x) - h(\zeta, y)| \le C d^{\gamma}(x, y)\Big(V^{\gamma}(x) W^{1-\gamma}(x) + V^{\gamma}(y) W^{1-\gamma}(y) \Big)$$

(which follows by condition 1 and assumption $V(x) \ge 1$) and

$$|h(\zeta, x) - h(\zeta, y)| \le C\Big(V^{\gamma}(x) W^{1-\gamma}(x) + V^{\gamma}(y) W^{1-\gamma}(y) \Big)$$

(which follows by eq. (6.3.23)). Then for any $\gamma' \le \gamma$, we have

$$|h(\zeta, x) - h(\zeta, y)| \le C d^{\gamma'}(x, y)\Big(V^{\gamma}(x) W^{1-\gamma}(x) + V^{\gamma}(y) W^{1-\gamma}(y) \Big),$$

and thus

$$\begin{aligned}
|h^2(\zeta, x) - h^2(\zeta, y)| &= |h(\zeta, x) - h(\zeta, y)||h(\zeta, x) + h(\zeta, y)| \\
&\le C d^{\gamma'}(x, y)\Big(V^{\gamma}(x) W^{1-\gamma}(x) + V^{\gamma}(y) W^{1-\gamma}(y) \Big)^2.
\end{aligned}$$

The latter inequality, after simple re-arrangements, provides that

$$h^2(\zeta, \cdot) \in H_{\gamma', \widetilde{W}}(\mathbb{X}, d) \tag{6.3.26}$$

with

$$\widetilde{W}(x) = V^{2\gamma/(1-\gamma')(x)} W^{2(1-\gamma)/(1-\gamma')}(x),$$

and analogues of conditions 1 and 2 hold true for $h^2(\zeta, x)$ with γ, W changed to γ', $\widetilde{W}$. In what follows, we fix $\gamma' > 0$ small enough, so that

$$\frac{2}{1-\gamma'} < 2+\delta.$$

Then, $\widetilde{W}$ satisfies an analogue of condition 3: for

$$p' = \frac{(2+\delta)(1-\gamma')}{2} > 1,$$

we have

$$\sup_t \mathsf{E}\widetilde{W}^{p'}(X_t) \le \left(\sup_t \mathsf{E}V^{2+\delta}(X_t)\right)^{\gamma} \left(\sup_t \mathsf{E}W^{2+\delta}(X_t)\right)^{1-\gamma} < \infty. \tag{6.3.27}$$

Proof of the theorem. We will apply Theorems 6.3.3 and 6.3.5 to the AR family (5.2.1) with

$$\xi_k^\varepsilon = \int_{\varepsilon(k-1)}^{\varepsilon k} f(Y^\varepsilon(s), X^\varepsilon(s))\, ds + \int_{\varepsilon(k-1)}^{\varepsilon k} h(Y^\varepsilon(s), X^\varepsilon(s))\, dW_s =: \xi_k^{f,\varepsilon} + \xi_k^{h,\varepsilon}$$

and filtration $\mathbb{F}^\varepsilon$. For any $p \ge 1$, we have

$$\mathsf{E}\left|\xi_k^{f,\varepsilon}\right|^p = \mathsf{E}\left|\int_{\varepsilon(k-1)}^{\varepsilon k} f(Y^\varepsilon(s), X^\varepsilon(s))\, ds\right|^p \le \varepsilon^{p-1} \int_{\varepsilon(k-1)}^{\varepsilon k} \mathsf{E}\left|f(Y^\varepsilon(s), X^\varepsilon(s))\right|^p ds.$$

By eq. (6.3.23) and condition 3, we have for $p = 2+\delta$

$$\mathsf{E}\left|f(Y^\varepsilon(s), X^\varepsilon(s))\right|^p \le C\,(\mathsf{E}V^p(X_{\varepsilon^{-1}s}))^{\gamma}\,(\mathsf{E}W^p(X_{\varepsilon^{-1}s}))^{1-\gamma} \le C.$$

That is,

$$\mathsf{E}\left|\xi_k^{f,\varepsilon}\right|^p \le C\varepsilon^p. \tag{6.3.28}$$

Next, by the Burkholder–Davis–Gundy inequality (e.g., Ref. [69, Theorem 23.12]) we have

$$\mathsf{E}\left|\xi_k^{h,\varepsilon}\right|^p = C_p \mathsf{E}\left(\int_{\varepsilon(k-1)}^{\varepsilon k} h^2(Y^\varepsilon(s), X^\varepsilon(s))\, ds\right)^{p/2},$$

and then a similar calculation gives

$$\mathsf{E}\left|\xi_k^{h,\varepsilon}\right|^p \le C\varepsilon^{p/2}; \tag{6.3.29}$$

we omit the details. That is, condition (6.3.5) is satisfied. Next, the sequence $\{\xi_k^{h,\varepsilon}\}$ is a martingale difference w.r.t. $\mathbb{F}^\varepsilon$, hence assumption $\mathbf{H}_0$ is satisfied with

$$\gamma_{k,l}^\varepsilon = \mathsf{E}\left[\varepsilon^{-1}\xi_{k+l}^{f,\varepsilon}|\mathscr{F}_k^\varepsilon\right], \quad \upsilon_{k,l} = 0, \quad k \ge 0, \ge l \ge 1.$$

Then by eq. (6.3.28), we have

$$\mathsf{E}|\gamma_{k,l}^\varepsilon|^p \le C.$$

That is, all the assumptions of Theorem 6.3.3 are verified with arbitrary $\rho \in (0, 1/2]$ and $r > 1/(2\rho)$. We will apply this theorem, taking care of the slight abuse of notation: in the current setting, Y^ε denotes the solution to eq. (6.3.21), while in Theorem 6.3.3 this was a piece-wise constant process. To avoid misunderstanding, from now on we denote the latter process by $\overline{Y}^\varepsilon$. Note that

$$Y^\varepsilon(\varepsilon k) = \overline{Y}^\varepsilon(\varepsilon k), \quad k \ge 0,$$

hence using eq. (6.3.5) it is easy to show that the uniform (in $t \in [0, T]$) distance between Y^ε and $\overline{Y}^\varepsilon$ tends to 0 in probability. One can also easily derive eq. (6.3.6) for Y^ε; Theorem 6.3.3 gives this bound for $\overline{Y}^\varepsilon$.

Next, let us verify eqs. (6.3.11) and (6.3.14). Denote

$$\widetilde{\gamma}_{k,L_\varepsilon}^\varepsilon = \varepsilon^{-1}\mathsf{E}\left[\int_{\varepsilon(k+L_\varepsilon-1)}^{\varepsilon(k+L_\varepsilon)} f(Y^\varepsilon(\varepsilon k), X^\varepsilon(s))\,ds\Big|\mathscr{F}_k^\varepsilon\right].$$

By eq. (6.3.24),

$$\begin{aligned}
\mathsf{E}|\gamma_{k,L_\varepsilon}^\varepsilon - \widetilde{\gamma}_{k,L_\varepsilon}^\varepsilon| &\le \varepsilon^{-1}\mathsf{E}\int_{\varepsilon(k+L_\varepsilon-1)}^{\varepsilon(k+L_\varepsilon)} |f(Y^\varepsilon(s), X^\varepsilon(s)) - f(Y^\varepsilon(\varepsilon k), X^\varepsilon(s))|\,ds \\
&\le C\varepsilon^{-1}\int_{\varepsilon(k+L_\varepsilon-1)}^{\varepsilon(k+L_\varepsilon)} \left(\mathsf{E}|Y^\varepsilon(s) - Y^\varepsilon(\varepsilon k)|^p\right)^{\delta/p} \\
&\qquad \times \left(\mathsf{E}V^p(X_{\varepsilon^{-1}s})\right)^{\gamma/p}\left(\mathsf{E}W^p(X_{\varepsilon^{-1}s})\right)^{(1-\gamma)/p}\,ds \\
&\le C(\varepsilon L_\varepsilon)^{\delta/2} \to 0, \quad \varepsilon \to 0.
\end{aligned} \tag{6.3.30}$$

Next, we have

$$\tilde{\gamma}^{\varepsilon}_{k,L_\varepsilon} = \varepsilon^{-1} \int\limits_{\varepsilon(k+L_\varepsilon-1)}^{\varepsilon(k+L_\varepsilon)} \mathsf{E}\left[f(Y^\varepsilon(\varepsilon k), X^\varepsilon(s)) \middle| \mathcal{F}^\varepsilon_k \right] ds$$

and, since $Y^\varepsilon(\varepsilon k)$ is $\mathcal{F}^\varepsilon_k$-measurable, by eq. (6.3.25)

$$\left| \mathsf{E}\left[f(Y^\varepsilon(\varepsilon k), X^\varepsilon(s)) \middle| \mathcal{F}^\varepsilon_k \right] - A(Y^\varepsilon(\varepsilon k)) \right| \le C r^\gamma_{s-\varepsilon k} V\big(X^\varepsilon(\varepsilon k)\big)^\gamma \mathsf{E}\left[W\big(X^\varepsilon(s)\big) \middle| \mathcal{F}^\varepsilon_k \right]^{1-\gamma}.$$

Using condition 3 once again, we obtain

$$\mathsf{E}|\tilde{\gamma}^{\varepsilon}_{k,L_\varepsilon} - A(Y^\varepsilon(\varepsilon k))| \le C r^\gamma_{L_\varepsilon-1} \to 0, \quad \varepsilon \to 0,$$

which together with eq. (6.3.30) completes the proof of eqs. (6.3.11) and (6.3.14).

The proof of eqs. (6.3.12) and (6.3.15) is similar. Namely, it follows from eqs. (6.3.28) and (6.3.29) that for each $j \ge 0$

$$\left| \mathsf{E}\xi^{f,\varepsilon}_{k+L^\varepsilon} \xi^{f,\varepsilon}_{k+L^\varepsilon+j} \right| \le C\varepsilon^2, \quad \left| \mathsf{E}\xi^{f,\varepsilon}_{k+L^\varepsilon} \xi^{h,\varepsilon}_{k+L^\varepsilon+j} \right| \le C\varepsilon^{3/2}, \quad \left| \mathsf{E}\xi^{h,\varepsilon}_{k+L^\varepsilon} \xi^{f,\varepsilon}_{k+L^\varepsilon+j} \right| \le C\varepsilon^{3/2}.$$

That is,

$$\left| \mathsf{E}\left[\xi^\varepsilon_{k+L^\varepsilon} \xi^\varepsilon_{k+L^\varepsilon+j} \middle| \mathcal{F}^\varepsilon_k \right] - \mathsf{E}\left[\xi^{h,\varepsilon}_{k+L^\varepsilon} \xi^{h,\varepsilon}_{k+L^\varepsilon+j} \middle| \mathcal{F}^\varepsilon_k \right] \right| \le C\varepsilon^{3/2}. \tag{6.3.31}$$

Since the sequence $\{\xi^{h,\varepsilon}_k\}$ is a martingale difference w.r.t. $\mathbb{F}^\varepsilon$, we have

$$\mathsf{E}\left[\xi^{h,\varepsilon}_{k+L^\varepsilon} \xi^{h,\varepsilon}_{k+L^\varepsilon+j} \middle| \mathcal{F}^\varepsilon_k \right] = 0, \quad j \ge 1;$$

that is, for $j \ge 1$ eqs. (6.3.12) and (6.3.15) hold true with

$$B_j(\zeta) = 0.$$

Next, we have

$$\mathsf{E}\left[(\xi^{h,\varepsilon}_{k+L^\varepsilon})^2 \middle| \mathcal{F}^\varepsilon_k \right] = \mathsf{E}\left[\int\limits_{\varepsilon(k+L_\varepsilon-1)}^{\varepsilon(k+L_\varepsilon)} h^2(Y^\varepsilon(s), X^\varepsilon(s))\, ds \middle| \mathcal{F}^\varepsilon_k \right].$$

Since $h^2(\zeta, x)$ satisfies eq. (6.3.26) and corresponding analogues of conditions 1–3, we can repeat, with obvious changes, the calculation used above in the proof of eqs. (6.3.11) and (6.3.14), and get eqs. (6.3.12) and (6.3.15) for $j = 0$ with

$$B_0(\zeta) = B(\zeta).$$

Both functions $A(\zeta)$ and $B(\zeta)$ are obtained by averaging in x of the functions which are Hölder continuous in ζ, hence these functions are continuous. That is, all the assumptions of Theorem 6.3.5 are verified.

We complete the proof, applying Theorem 6.3.5: since the uniform distance between Y^ε and the piece-wise constant process $\overline{Y}^\varepsilon$, which appears in Theorem 6.3.5, tends to 0 in probability, required statement now follows by Proposition 6.1.1. □

Next, we establish the *diffusion approximation* type theorem for a system with a nontrivial term

$$\varepsilon^{-1/2} g\big(Y^\varepsilon(t), X^\varepsilon(t)\big). \tag{6.3.32}$$

Theorem 6.3.10. *(The diffusion approximation). Let the Markov process X satisfy* (5.3.6) *and conditions 1 – 3 of Theorem 6.3.7 be satisfied. Assume also the following.*

1. *There exist derivatives $\partial_\zeta g(\zeta,x), \partial^2_{\zeta\zeta} g(\zeta,x)$ and, for $\gamma \in (0,1]$ and function W same as in Theorem 6.3.7, the functions $g(\zeta,\cdot), \partial_\zeta g(\zeta,\cdot), \partial^2_{\zeta\zeta} g(\zeta,x)$ belong to $H_{\gamma,W}(\mathbb{X},d)$. In addition,*

$$\sup_{\zeta\in\mathbb{R}} \left(\|g(\zeta,\cdot)\|_{d,\gamma,W} + \|\partial_\zeta g(\zeta,\cdot)\|_{d,\gamma,W} + \|\partial^2_{\zeta\zeta} g(\zeta,x)\|_{d,\gamma,W} \right) < \infty,$$

and there exists $\delta > 0$ such that

$$\sup_{\zeta_1\neq\zeta_2} |\zeta_1-\zeta_2|^{-\delta} \left(\|g(\zeta_1,\cdot)-g(\zeta_2,\cdot)\|_{d,\gamma,W} + \|\partial_\zeta g(\zeta,\cdot)-\partial_\zeta g(\zeta_2,\cdot)\|_{d,\gamma,W} \right) < \infty,$$

$$\sup_{\zeta_1\neq\zeta_2} |\zeta_1-\zeta_2|^{-\delta} \|\partial^2_{\zeta\zeta} g(\zeta,\cdot)-\partial^2_{\zeta\zeta} g(\zeta_2,\cdot)\|_{d,\gamma,W} < \infty.$$

2. *For every $\zeta \in \mathbb{R}$, the functions $g(\zeta,\cdot), \partial_\zeta g(\zeta,\cdot), \partial^2_{\zeta\zeta} g(\zeta,\cdot) \in L_1(\mathbb{X},\mu)$ are centered; that is,*

$$\int_{\mathbb{X}} g(\zeta,x)\,\mu(dx) = 0, \quad \int_{\mathbb{X}} \partial_\zeta g(\zeta,x)\,\mu(dx) = 0, \quad \int_{\mathbb{X}} \partial^2_{\zeta\zeta} g(\zeta,x)\,\mu(dx) = 0.$$

3. *For some $\sigma > 1/\gamma$ and $C > 0$, the erodic rate r_t (see* (5.3.6)*) satisfies*

$$r_t \le C(1+t)^{-\sigma}, \quad t \ge 0.$$

4. *For some $C > 0$,*

$$|h(\zeta,x)| \le C.$$

Then the family $\{Y^\varepsilon\}$ defined by (6.3.21) is weakly compact in $\mathbb{D}(0,T)$, and any weak limit point of Y^ε, $\varepsilon \to 0$ is a solution of the martingale problem $(\mathscr{L}, \mathscr{D})$ with

$$\mathscr{L}\phi(\zeta) = A(\zeta)\phi'(\zeta) + \frac{1}{2}B(\zeta)\phi''(\zeta), \qquad \mathscr{D} = C_0^\infty(\mathbb{R}^d), \tag{6.3.33}$$

where

$$A(\zeta) = \int_{\mathbb{X}} f(\zeta,x)\,\mu(dx) + \int_0^\infty \iint_{\mathbb{X}^2} g(\zeta,x)\partial_\zeta g(\zeta,y)\,\mu(dx)P^t(x,dy)\,dt, \tag{6.3.34}$$

$$B(\zeta) = \int_{\mathbb{X}} h^2(\zeta,x)\,\mu(dx) + 2\int_0^\infty \iint_{\mathbb{X}^2} g(\zeta,x)g(\zeta,y)\,\mu(dx)P^t(x,dy)\,dt. \tag{6.3.35}$$

If the martingale problem (6.3.33) is well-posed, then the family Y^ε weakly converges in $\mathbb{C}(0,T)$ to a diffusion process Y, which is the unique solution of this martingale problem.

Proof. Likewise to the previous proof, we will apply Theorems 6.3.3 and 6.3.5. Let

$$\xi_k^\varepsilon = \xi_k^{f,\varepsilon} + \xi_k^{g,\varepsilon} + \xi_k^{h,\varepsilon},$$

where $\xi_k^{f,\varepsilon}, \xi_k^{h,\varepsilon}$ are the same as above, and

$$\xi_k^{g,\varepsilon} = \varepsilon^{-1/2} \int\limits_{\varepsilon(k-1)}^{\varepsilon k} g(Y^\varepsilon(s), X^\varepsilon(s))\,ds.$$

Repeating literally the proof of eq. (6.3.28), we get

$$\mathsf{E}\left|\xi_k^{g,\varepsilon}\right|^p \le C\varepsilon^{p/2}. \tag{6.3.36}$$

Combined with eqs. (6.3.28) and (6.3.29), this gives eq. (6.3.5).

Next, let us verify that the assumption $\mathbf{H}_0$ and bound (6.3.4) hold true for $p = 2+\delta$ and certain

$$\rho < \frac{1}{2}, \quad r > \frac{1}{2\rho}. \tag{6.3.37}$$

Since for the terms

$$\xi_k^{f,\varepsilon}, \quad \xi_k^{h,\varepsilon}$$

these statements were verified in the previous proof, here we focus on the term $\xi_k^{g,\varepsilon}$. Note that this term brings a substantially new difficulty to the current setting, because it is not a martingale difference, and has an L_p-bound of the order $\varepsilon^{1/2} \gg \varepsilon$, only.

We decompose

$$\xi^{g,\varepsilon}_{k+l} = \varepsilon^{-1/2} \int\limits_{\varepsilon(k+l-1)}^{\varepsilon(k+l)} \Gamma^{\varepsilon}_{k,l}(s)\, ds + \varepsilon^{-1/2} \int\limits_{\varepsilon(k+l-1)}^{\varepsilon(k+l)} \Delta^{\varepsilon}_{k,l}(s)\, ds =: \xi^{g,\varepsilon,\Gamma}_{k+l} + \xi^{g,\varepsilon,\Delta}_{k+l},$$

where

$$\Gamma^{\varepsilon}_{k,l}(s) = g(Y^{\varepsilon}(\varepsilon k), X^{\varepsilon}(s)), \quad \Delta^{\varepsilon}_{k,l}(s) = g(Y^{\varepsilon}(s), X^{\varepsilon}(s)) - g(Y^{\varepsilon}(\varepsilon k), X^{\varepsilon}(s)).$$

The term $\xi^{g,\varepsilon,\Gamma}_{k+l}$ can be treated using the stabilization property of the Markov process X. Namely, an analogue of eq. (6.3.25) holds true for $g(\zeta, x)$, which gives

$$\left| \mathsf{E}\left[\Gamma^{\varepsilon}_{k,l}(s) \middle| \mathscr{F}^{\varepsilon}_{k} \right] \right| \le C r^{\gamma}_{s-\varepsilon k} V\big(X^{\varepsilon}(\varepsilon k)\big)^{\gamma} \mathsf{E}\left[W\big(X^{\varepsilon}(s)\big) \middle| \mathscr{F}^{\varepsilon}_{k} \right]^{1-\gamma}$$

because $g(\zeta, x)$ is assumed to be centered. That is,

$$\left| \mathsf{E}\left[\xi^{g,\varepsilon,\Gamma}_{k+l} \middle| \mathscr{F}^{\varepsilon}_{k} \right] \right| \le \varepsilon^{1/2} \upsilon^{g,\varepsilon,\Gamma}_{k,l} R_l$$

with

$$R_l = r^{\gamma}_{l-1}, \quad \upsilon^{g,\varepsilon,\Gamma}_{k,l} = \frac{C}{\varepsilon} V\big(X^{\varepsilon}(\varepsilon k)\big)^{\gamma} \int\limits_{\varepsilon(k+l-1)}^{\varepsilon(k+l)} \mathsf{E}\left[W\big(X^{\varepsilon}(s)\big) \middle| \mathscr{F}^{\varepsilon}_{k} \right]^{1-\gamma} ds.$$

Take

$$r = \sigma\gamma > 1, \quad \rho \in \left(\frac{1}{2\sigma\gamma}, \frac{1}{2} \right),$$

then eq. (6.3.37) is satisfied. Using the Hölder inequality and condition 3, one easily obtains

$$\mathsf{E}|\upsilon^{g,\varepsilon,\Gamma}_{k,l}|^{p} \le C;$$

that is, $\xi^{g,\varepsilon,\Gamma}_{k+l}$ satisfies (analogues of) $\mathbf{H}_0$ and eq. (6.3.4).

Next, we decompose $\xi^{g,\varepsilon,\Delta}_{k+l}$ using the Itô formula. For that purpose, we introduce new filtration

$$\widetilde{\mathscr{F}}^{\varepsilon}_{k} = \sigma\big(X(v), W_{\varepsilon s}, s \le k, v \ge 0\big) \supset \mathscr{F}^{\varepsilon}_{k}, \quad k \ge 0.$$

Clearly, we have that, w.r.t. this enlarged filtration, W_t is a Wiener process and $Y^{\varepsilon}(t)$ is an Itô process. Since each $X^{\varepsilon}(s)$ is $\widetilde{\mathscr{F}}^{\varepsilon}_{0}$-measurable, we can apply the Itô formula and get

$$\begin{aligned}\Delta^{\varepsilon}_{k,l}(s) = &\int_{\varepsilon k}^{s} \partial_\zeta g(Y^\varepsilon(v), X^\varepsilon(s)) f(Y^\varepsilon(v), X^\varepsilon(v))\, \mathrm{d}v \\ &+ \varepsilon^{-1/2} \int_{\varepsilon k}^{s} \partial_\zeta g(Y^\varepsilon(v), X^\varepsilon(s)) g(Y^\varepsilon(v), X^\varepsilon(v))\, \mathrm{d}v \\ &+ \frac{1}{2}\int_{\varepsilon k}^{s} \partial^2_{\zeta\zeta} g(Y^\varepsilon(v), X^\varepsilon(s)) h^2(Y^\varepsilon(v), X^\varepsilon(v))\Big)\, \mathrm{d}v \\ &+ \int_{\varepsilon k}^{s} \partial_\zeta g(Y^\varepsilon(v), X^\varepsilon(s)) h(Y^\varepsilon(v), X^\varepsilon(v))\, \mathrm{d}W_v =: \sum_{i=1}^{4} \Delta^{\varepsilon,i}_{k,l}(s).\end{aligned}$$

We will analyze separately each of the terms

$$\xi^{g,\varepsilon,\Delta,i}_{k+l} = \varepsilon^{-1/2} \int_{\varepsilon(k+l-1)}^{\varepsilon(k+l)} \Delta^{\varepsilon,i}_{k,l}(s)\, \mathrm{d}s, \quad i = 1, \ldots, 4.$$

Clearly,

$$\mathsf{E}\left[\Delta^{\varepsilon,4}_{k,l}(s) \middle| \widetilde{\mathscr{F}}^{\varepsilon}_k\right] = 0.$$

Since $\widetilde{\mathscr{F}}^{\varepsilon}_k$ enlarges $\mathscr{F}^{\varepsilon}_k$, this means that

$$\mathsf{E}\left[\xi^{g,\varepsilon,\Delta,4}_{k+l} \middle| \mathscr{F}^{\varepsilon}_k\right] = 0;$$

that is, for $\xi^{g,\varepsilon,\Delta,4}_{k+l}$ (an analogue of) $\mathbf{H}_0$ holds true trivially.

Next, inequalities (6.3.23) hold true for $f(\zeta, x)$ and $\partial_\zeta g(\zeta, x)$. Then by the Hölder inequality and condition 3 of Theorem 6.3.7,

$$\|\Delta^{\varepsilon,1}_{k,l}(s)\|_{L_p} \le C(s - \varepsilon k).$$

Note that

$$s - \varepsilon k \le \varepsilon l \le \varepsilon L_\varepsilon,$$

and

$$L_\varepsilon = [\varepsilon^{-\rho}] + 1,$$

where $\rho < 1/2$ is yet to be specified. For any $\rho < 1/2$, we have

$$\left|\mathsf{E}\left[\xi^{g,\varepsilon,\Delta,1}_{k+l} \middle| \mathscr{F}^{\varepsilon}_k\right]\right| \le \varepsilon \gamma^{g,\varepsilon,\Delta,1}_{k,l}$$

with

$$\|\gamma_{k,l}^{g,\varepsilon,\Delta,1}\|_{L_p} \le C\varepsilon^{1/2}L_\varepsilon \to 0,$$

which means that (analogues of) $\mathbf{H}_0$ and eq. (6.3.4) hold true for the term $\xi_{k+l}^{g,\varepsilon,\Delta,1}$, as well.

Similarly, since $h(\zeta,x)$ is assumed to be bounded, we have

$$\left|\mathsf{E}\left[\xi_{k+l}^{g,\varepsilon,\Delta,3}\middle|\mathscr{F}_k^\varepsilon\right]\right| \le \varepsilon\gamma_{k,l}^{g,\varepsilon,\Delta,3}$$

with

$$\|\gamma_{k,l}^{g,\varepsilon,\Delta,3}\|_{L_p} \le C\varepsilon^{1/2}L_\varepsilon \to 0.$$

The estimates here are completely analogous, hence we omit the details.

Finally, we denote

$$Q^t(\zeta,x) = \mathsf{E}_x\partial_\zeta g(\zeta,X_t).$$

Then,

$$\mathsf{E}\left[\partial_\zeta g(Y^\varepsilon(v),X^\varepsilon(s))\middle|\mathscr{F}_v^\varepsilon\right] = Q^{(s-v)/\varepsilon}(Y^\varepsilon(v),X^\varepsilon(v))$$

and

$$\mathsf{E}\left[\xi_{k+l}^{g,\varepsilon,\Delta,2}\middle|\mathscr{F}_k^\varepsilon\right] = \varepsilon\gamma_{k,l}^{g,\varepsilon,\Delta,2},$$

where we denote

$$\gamma_{k,l}^{g,\varepsilon,\Delta,2} = \varepsilon^{-2}\int\limits_{\varepsilon(k+l-1)}^{\varepsilon(k+l)}\int\limits_{\varepsilon k}^{s}\mathsf{E}\left[g(Y^\varepsilon(v),X^\varepsilon(v))Q^{(s-v)/\varepsilon}(Y^\varepsilon(v),X^\varepsilon(v))\middle|\mathscr{F}_k^\varepsilon\right]\,dv\,ds.$$

Using eq. (6.3.23) for $g(\zeta,x)$, (6.3.25) for $\partial_\zeta g(\zeta,x)$, and condition 3 of Theorem 6.3.7, we get

$$\left\|g(Y^\varepsilon(v),X^\varepsilon(v))Q^{(s-v)/\varepsilon}(Y^\varepsilon(v),X^\varepsilon(v))\right\|_{L_p} \le Cr^\gamma_{(s-v)/\varepsilon}.$$

Then by condition 3

$$\|\gamma_{k,l}^{g,\varepsilon,\Delta,2}\|_{L_p} \le C,$$

which completes the proof of (analogues of) $\mathbf{H}_0$ and eq. (6.3.4) for the last term $\xi_{k+l}^{g,\varepsilon,\Delta,2}$ in the decomposition of $\xi_{k+l}^{g,\varepsilon}$. That is, all the assumptions of Theorem 6.3.3 are verified.

Next, we check the assumptions of Theorem 6.3.5. Denote

$$\gamma_{k,l}^{\varepsilon,f} = \mathsf{E}\left[\varepsilon^{-1}\xi_{k+l}^{f,\varepsilon}|\mathscr{F}_k^\varepsilon\right],$$

then, summarizing the previous calculations, we have that $\mathbf{H}_0$ holds true with

$$\sup_{k,l}\left\|\gamma_{k,l}^{\varepsilon} - \gamma_{k,l}^{\varepsilon,f} - \gamma_{k,l}^{g,\varepsilon,\Delta,2}\right\|_{L_p} \to 0, \quad \varepsilon \to 0.$$

Denote

$$A^f(\zeta) = \int_{\mathbb{X}} f(\zeta, x)\,\mu(dx),$$

then repeating literally the argument from the proof of Theorem 6.3.7, we get

$$\sup_k \mathsf{E}|\gamma_{k,L_\varepsilon}^{\varepsilon,f} - A^f(Y^\varepsilon(\varepsilon k))| \to 0, \quad \varepsilon \to 0. \tag{6.3.38}$$

Next, for $L \geq 0$ denote

$$\gamma_{k,l}^{g,\varepsilon,\Delta,2,L} = \varepsilon^{-2}\int_{\varepsilon(k+l-1)}^{\varepsilon(k+l)}\int_{s-\varepsilon L}^{s} \mathsf{E}\left[g(Y^\varepsilon(v), X^\varepsilon(v))Q^{(s-v)/\varepsilon}(Y^\varepsilon(v), X^\varepsilon(v))\middle|\mathscr{F}_k^\varepsilon\right] dv\, ds,$$

$$A^{g,L}(\zeta) = \int_0^L \iint_{\mathbb{X}^2} g(\zeta, x)\partial_\zeta g(\zeta, y)\,\mu(dx)P^t(x, dy)\,dt.$$

Similarly to the calculation which led to eq. (6.3.26), we can show that the family of functions

$$g(\zeta, x)Q^t(\zeta, x), \quad t \in [0, L]$$

uniformly satisfies corresponding analogues of conditions 1–3 of Theorem 6.3.7 with some $\gamma' \in (0, 1)$, $\widetilde{W}(x)$. Then, repeating the argument which led to eq. (6.3.38), we get

$$\sup_k \mathsf{E}|\gamma_{k,L_\varepsilon}^{g,\varepsilon,\Delta,2,L} - A^{g,L}(Y^\varepsilon(\varepsilon k))| \to 0, \quad \varepsilon \to 0. \tag{6.3.39}$$

On the other hand, using eq. (6.3.25) for $\partial_\zeta g(\zeta, x)$ and conditions 3 and 4, we get

$$\mathsf{E}|\gamma_{k,L_\varepsilon}^{g,\varepsilon,\Delta,2} - \gamma_{k,L_\varepsilon}^{g,\varepsilon,\Delta,2,L}| \leq CL^{1-\sigma\gamma}, \quad |A^g(\zeta) - A^{g,L}(\zeta)| \leq CL^{1-\sigma\gamma}. \tag{6.3.40}$$

That is, for each L we have

$$\limsup_{\varepsilon\to 0}\sup_k \mathsf{E}|\gamma_{k,L_\varepsilon} - A(Y^\varepsilon(\varepsilon k))| \le CL^{1-\sigma\gamma},$$

which completes the proof of eqs. (6.3.11) and (6.3.14).

The proof of eqs. (6.3.12) and (6.3.15) does not contain any essentially new ideas, hence we just sketch it, and omit the details. The terms in eq. (6.3.12), which contain the products

$$\xi^{f,\varepsilon}_{k+L^\varepsilon}\xi^{g,\varepsilon}_{k+L^\varepsilon+j}, \quad \xi^{g,\varepsilon}_{k+L^\varepsilon}\xi^{f,\varepsilon}_{k+L^\varepsilon+j},$$

are negligible, see the argument which led to eq. (6.3.31). The terms with the products

$$\xi^{h,\varepsilon}_{k+L^\varepsilon}\xi^{g,\varepsilon}_{k+L^\varepsilon+j}, \quad \xi^{g,\varepsilon}_{k+L^\varepsilon}\xi^{h,\varepsilon}_{k+L^\varepsilon+j}$$

are negligible, as well, though now one should use a slightly more precise argument, because eqs. (6.3.29) and (6.3.36) only provide an L_1-bound $\le C\varepsilon$ for these terms. However, using the Itô formula and repeating the argument which led to eq. (6.3.39), one can show the required negligibility; we omit the details. That is, for $j \ge 1$ the only nonnegligible term in the decomposition eq. (6.3.12) is

$$\mathsf{E}\left[\xi^{g,\varepsilon}_{k+L^\varepsilon}\xi^{g,\varepsilon}_{k+L^\varepsilon+j}\middle|\mathscr{F}^\varepsilon_k\right].$$

Denote

$$G^t(\zeta, x) = \mathsf{E}_x g(\zeta, X_t),$$

then, repeating the argument which led to eqs. (6.3.39), we get eq. (6.3.12) and (6.3.15) with

$$B_j(\zeta) = \int_j^{j+1}\int_0^1\int_{\mathbb{X}} g(\zeta, x)G^{t-s}(\zeta, x)\,\mu(\mathrm{d}x)\,\mathrm{d}s\,\mathrm{d}t.$$

For $j = 0$, the same argument gives eqs. (6.3.12) and (6.3.15) with

$$B_0(\zeta) = \int_{\mathbb{X}} h^2(\zeta, x)\mu(\mathrm{d}x) + \int_0^1\int_0^1\int_{\mathbb{X}} g(\zeta, x)G^{|t-s|}(\zeta, x)\,\mu(\mathrm{d}x)\,\mathrm{d}s\,\mathrm{d}t.$$

The functions $A(\zeta)$, $B_j(\zeta)$, $j \ge 0$ are continuous because they are obtained as averages of Hölder continuous functions. Similarly to eq. (6.3.40), we have

$$|B_j(\zeta)| \le C(1+j)^{-\sigma\gamma},$$

hence the series eq. (6.3.13) converges uniformly. That is, all the assumptions of Theorem 6.3.5 are verified. The end of the proof is literally the same as in the proof of Theorem 6.3.7. □

6.3.3 A Fully Coupled System with Dissipative Fast Component

Model (6.3.21) studied above has the strong structural restriction, that the Markov process X^ε, which performs the "fast perturbation", does not depend on the "slow" component Y^ε. This restriction simplifies the situation drastically, since it allows one to analyze separately ergodic properties of the "fast" component. In various more realistic models, one has to deal with systems, which still can be naturally separated in "fast" and "slow" components, but both these components effect each other, and none of them can be treated separately. Such systems are called *fully coupled* in the literature. In this section, we show that the general results obtained by the time delay method (Theorems 6.3.3 and 6.3.5) are well designed for a study of (at least some type of) the fully coupled systems.

In what follows, the pair $X^\varepsilon, Y^\varepsilon$ is assumed to satisfy

$$\begin{aligned}
dX^\varepsilon(t) &= \varepsilon^{-1}a\big(X^\varepsilon(t), Y^\varepsilon(t)\big)\,dt + \sigma\big(X^\varepsilon(t-)\big)\,dZ_{\varepsilon^{-1}t},\\
dY^\varepsilon(t) &= f\big(Y^\varepsilon(t), X^\varepsilon(t)\big)\,dt + \varepsilon^{-1/2}g\big(Y^\varepsilon(t), X^\varepsilon(t)\big)\,dt,\\
X^\varepsilon(0) &= x_0, \quad Y^\varepsilon(0) = y_0,
\end{aligned} \tag{6.3.41}$$

where Z is a Lévy process, and x_0, y_0 are given initial values; x_0 is allowed to be random, and then it should be independent on Z.

If coefficient a does not depend on Y^ε, then X^ε can be interpreted as a time-changed solution to an SDE driven by Z, and in this case eq. (6.3.41) is embedded into the previously studied model (6.3.21). The equation for Y^ε in eq. (6.3.41), when compared with eq. (6.3.21), lacks the stochastic term

$$h\big(Y^\varepsilon(t), X^\varepsilon(t)\big)\,dW_t.$$

This difference is not substantial, we remove this term just in order to simplify the presentation: since now X^ε depends on Y^ε, presence of this stochastic term will cause new technicalities required to apply the Itô formula. By the similar reason, we restrict ourselves by the case where only the drift coefficient a of the process X^ε depends on Y^ε.

We will consider the auxiliary family X^ε_ζ of Markov processes satisfying SDEs

$$dX^\varepsilon_\zeta(t) = \varepsilon^{-1}a\big(X^\varepsilon_\zeta(t), \zeta\big)\,dt + \sigma\big(X^\varepsilon_\zeta(t-)\big)\,dZ_{\varepsilon^{-1}t}. \tag{6.3.42}$$

These processes have a natural interpretation as approximations to the "fast" component, with the "slow" component being "frozen". We have

$$X^\varepsilon_\zeta(t) = X_\zeta(\varepsilon^{-1}t),$$

with X_ζ being defined by

$$\mathrm{d}X_\zeta(t) = a\big(X_\zeta(t), \zeta\big)\,\mathrm{d}t + \sigma\big(X_\zeta(t-)\big)\,\mathrm{d}Z_t. \tag{6.3.43}$$

We will assume the following.

A$_1$. The Lévy measure ν of the Lévy process Z is symmetric and satisfies

$$\int_{\mathbb{R}} u^4 \nu(\mathrm{d}u) < \infty.$$

The Lévy–Khinchin decomposition of Z has the form

$$L_t = B_t + \int_0^t \int_{\mathbb{R}} u\widetilde{N}(\mathrm{d}u, \mathrm{d}s).$$

Let $b \ge 0$ be such that $\mathsf{E}B_t^2 = bt$.

A$_2$. Functions $f(\zeta, x), g(\zeta, x), \sigma(x)$, and $a(0, \zeta)$ are bounded.

A$_3$. Functions $a(x, \zeta), f(\zeta, x), g(\zeta, x)$ have continuous derivatives in x, ζ, function σ has a continuous derivative, and all these derivatives except $\partial_\zeta a$ are bounded. For the latter derivative, there exist C such that

$$|\partial_\zeta a(x, \zeta)| \le C(1 + |x|).$$

Consequently, functions $\sigma(x), f(\zeta, x), g(\zeta, x)$ satisfy the Lipschitz condition w.r.t. x and x, ζ, respectively, and the functions $a(x, \cdot)$, $a(\cdot, \zeta)$ are Lipschitz continuous with the constants C and $C(1 + |x|)$, respectively.

A$_4$. There exist bounded continuous derivatives

$$\partial^2_{xx} a(x, \zeta), \quad \partial^2_{x\zeta} a(x, \zeta), \quad \partial^2_{xx} \sigma(x, \zeta), \quad \partial^2_{xx} g(\zeta, x), \quad \partial^2_{x\zeta} g(\zeta, x), \quad \partial^2_{\zeta\zeta} g(\zeta, x)$$

There exists also a continuous derivative $\partial^2_{\zeta\zeta} a(x, \zeta)$ which satisfies

$$\left|\partial^2_{\zeta\zeta} a(x, \zeta)\right| \le C\big(1 + |x|\big).$$

A$_5$. The following *drift condition* holds true: there exists $\theta > 0$ such that, for all $x \ne y, \zeta$,

$$\begin{aligned}\big(a(x, \zeta) - a(y, \zeta)\big)(x - y) &+ \frac{3(\sigma(x) - \sigma(y))^2}{2}\left(b + \int_{\mathbb{R}} u^2 \nu(\mathrm{d}u)\right) \\ &+ \frac{(\sigma(x) - \sigma(y))^4}{4(x - y)^2} \int_{\mathbb{R}} u^4 \nu(\mathrm{d}u) \le -\theta (x - y)^2.\end{aligned} \tag{6.3.44}$$

In the following proposition, we collect the properties of the auxiliary family $\{X_\zeta\}$ defined by eq. (6.3.43).

Proposition 6.3.11. *Let A_1–A_5 hold true. Then we have the following.*

1. *For two solutions $X_\zeta, \tilde{X}_\zeta$ to eq. (6.3.43) with the same Z and initial values $x, \tilde{x}$,*

$$E\big(X_\zeta(t) - \tilde{X}_\zeta(t)\big)^4 \le (x - \tilde{x})^4 e^{-4\theta t}. \tag{6.3.45}$$

2. *There exists C such that, for any x, ζ, corresponding solution to eq. (6.3.43) satisfies*

$$E\big(X_\zeta(t)\big)^4 \le C(1 + x^4), \quad t \ge 0.$$

3. *For X_ζ there exists a unique invariant measure μ^ζ, and the family $\{\mu^\zeta\}$ satisfies*

$$\int_{\mathbb{R}} x^4 \mu^\zeta(dx) \le C.$$

Transition probabilities

$$P^{t,\zeta}(x, \cdot) = P_x(X_\zeta(t) \in \cdot)$$

converge to μ^ζ, $t \to \infty$ exponentially fast w.r.t. the coupling distance, which corresponds to $d(x, y) = (x - y)^4$.

4. *Denote*

$$G^t(\zeta, x) = E_x g(\zeta, X_\zeta(t)), \quad \tilde{G}(\zeta) = \int_{\mathbb{R}} g(\zeta, x)\mu^\zeta(dx).$$

Then

$$|G^t(\zeta, x) - \tilde{G}(\zeta)| \le Ce^{-\theta t}\big(1 + |x|\big), \tag{6.3.46}$$

there exist derivatives $\partial_\zeta G^t(\zeta, x), \partial_\zeta \tilde{G}(\zeta)$, and

$$|\partial_\zeta G^t(\zeta, x) - \partial_\zeta \tilde{G}(\zeta)| \le Ce^{-\theta t}\big(1 + |x|\big). \tag{6.3.47}$$

Proof. 1. The proof of this statement is very similar to the proof of Proposition 4.2.1, with minor changes caused by new terms brought to the Itô formula by the jump part of the noise. Namely, applying the Itô formula to the C^2-function $F(x) = x^4$ and the Itô process $\Delta_\zeta(t) = X_\zeta(t) - \tilde{X}_\zeta(t)$, we get

$$\mathrm{d}\big(X_\zeta(t) - \tilde{X}_\zeta(t)\big)^4 = \eta_t \,\mathrm{d}t + \mathrm{d}M_t$$

with a local martingale M_t and

$$\eta_t = 4\big(\Delta_\zeta(t)\big)^3\big(a(X_\zeta(t),\zeta) - a(\widetilde{X}_\zeta(t),\zeta)\big)$$
$$+ 6\big(\Delta_\zeta(t)\big)^3\big(\sigma(X_\zeta(t),\zeta) - \sigma(\widetilde{X}_\zeta(t),\zeta)\big)^2 b \,\mathrm{d}t$$
$$+ \int_{\mathbb{R}} \Big[\Big(\Delta_\zeta(t) + \big(\sigma(X_\zeta(t),\zeta) - \sigma(\widetilde{X}_\zeta(t),\zeta)\big)u\Big)^4 - \big(\Delta_\zeta(t)\big)^4\Big]\, \nu(\mathrm{d}u)\,\mathrm{d}t.$$

By the drift condition, we have

$$\eta_t \le -4\theta\big(\Delta_\zeta(t)\big)^4.$$

Hence the process

$$e^{4\theta t}\big(\Delta_\zeta(t)\big)^4$$

is a local supermartingale. Repeating the argument from the proof of Proposition 4.2.1, based on time localization and the Fatou lemma, we get eq. (6.3.45).

2. Taking $y = 0$ in (6.3.44) and using that $a(0,\zeta)$ is bounded, we deduce that

$$a(x,\zeta)\operatorname{sgn} x \le -\theta|x| + C.$$

That is, functions $a(\cdot,\zeta)$ satisfy the drift condition eq. (3.3.4) uniformly in ζ. Since σ is bounded, we can apply Proposition 3.4.4 (actually, its uniform version) and get that the function $V(x) = x^4$ belongs to the domain of the extended generator $\mathscr{L}^\zeta$ of each of the processes X_ζ, and for each $\theta' < \theta$ there exists C such that for each x, ζ

$$\mathscr{L}^\zeta V(x) \le -4\theta' V(x) + C.$$

Applying the Itô formula, we get that an analogue of eq. (3.2.2) holds true uniformly in ζ. Then the required statement follows by eq. (3.2.3).

3. This statement follows directly from two previous statements, and Theorem 4.5.1.

4. The bound (6.3.46) follows from eq. (6.3.45): the argument here is similar to the one used in Proposition 5.1.4, which we repeatedly used before. Let $X_\zeta(0) = x$ and $\widetilde{X}_\zeta$ be the stationary solution to eq. (6.3.43) with the initial condition, independent on Z, and such that $\mathrm{Law}(\widetilde{X}_\zeta(0)) = \mu^\zeta$. Then by eq. (6.3.45) and statement 2, we have

$$\mathsf{E}\big(X_\zeta(t) - \widetilde{X}_\zeta(t)\big)^4 \le C\big(1 + x^4\big)e^{-4\theta t}. \tag{6.3.48}$$

Since

$$\Big|G^t(\zeta,x) - \widetilde{G}(\zeta)\Big| = \Big|\mathsf{E}g(X_\zeta(t),\zeta) - \mathsf{E}g(\widetilde{X}_\zeta(t),\zeta)\Big| \le C\mathsf{E}\Big|X_\zeta(t) - \widetilde{X}_\zeta(t)\Big|,$$

this completes the proof of eq. (6.3.46).

Next, it is standard (e.g., Ref. [45, Chapter 2, §8]) to show that, under the given conditions, the solution $X_\zeta(t)$ to eq. (6.3.43) is L_4-differentiable w.r.t. ζ, and its derivative

$$\mathcal{H}^\zeta(t) = \partial_\zeta X_\zeta(t)$$

satisfies the linear SDE

$$\begin{aligned} d\mathcal{H}_\zeta(t) &= (\partial_x a)\big(X_\zeta(t), \zeta\big)\mathcal{H}_\zeta(t)\, dt + (\partial_x \sigma)\big(X_\zeta(t-), \zeta\big)\mathcal{H}_\zeta(t-)\, dZ_t \\ &\quad + (\partial_\zeta a)\big(X_\zeta(t), \zeta\big)\, dt, \qquad (6.3.49) \\ \mathcal{H}_\zeta(0) &= 0. \end{aligned}$$

Let $X_\zeta(0) = x$, then we have

$$\partial_\zeta G^t(\zeta, x) = \partial_\zeta \mathsf{E} g\big(X_\zeta(t), \zeta\big) = \mathsf{E}\partial_\zeta g\big(X_\zeta(t), \zeta\big) + \mathsf{E}\partial_x g\big(X_\zeta(t), \zeta\big)\mathcal{H}_\zeta(t).$$

On the other hand, if $\tilde{X}_\zeta(t)$ is a stationary solution to eq. (6.3.43) and $\tilde{\mathcal{H}}_\zeta(t)$ is the corresponding solution to (the analogue of) eq. (6.3.49), then we have

$$\partial_\zeta \tilde{G}(\zeta) = \partial_\zeta \mathsf{E} g\big(\tilde{X}_\zeta(t), \zeta\big) = \mathsf{E}\partial_\zeta g\big(\tilde{X}_\zeta(t), \zeta\big) + \mathsf{E}\partial_x g\big(\tilde{X}_\zeta(t), \zeta\big)\tilde{\mathcal{H}}_\zeta(t).$$

It follows from eq. (6.3.44) that

$$\partial_x a(x, \zeta) + \frac{3}{2}\big(\partial_x \sigma(x)\big)^2 \left(b + \int_{\mathbb{R}} u^2 \nu(du) \right) + \frac{1}{4}\big(\partial_x \sigma(x)\big)^4 \int_{\mathbb{R}} u^4 \nu(du) \le \theta. \qquad (6.3.50)$$

Then repeating, with straightforward changes, the proof of statement 1, we easily derive that

$$\mathsf{E}\big(\mathcal{H}_\zeta(t) - \tilde{\mathcal{H}}_\zeta(t)\big)^4 \le C\big(1 + x^4\big)e^{-4\theta t}.$$

Combining this bound with eq. (6.3.48) and using that $\partial_\zeta g(\zeta, x)$, $\partial_x g(\zeta, x)$ are Lipschitz continuous in x, we easily get eq. (6.3.47). □

Theorem 6.3.12. *Assume $\mathbf{A}_1$–$\mathbf{A}_5$. In addition, assume the following.*
$\mathbf{C}_1$. *For each ζ, the function $g(\zeta, \cdot)$ is centered w.r.t. μ^ζ.*
$\mathbf{C}_2$.

$$E\big(X^\varepsilon(0)\big)^4 \le C.$$

Then the family $\{Y^\varepsilon\}$, defined by eq. (6.3.41), is weakly relative compact in $\mathbb{D}(0, T)$, and any weak limit point of Y^ε, $\varepsilon \to 0$ is a solution of the martingale problem (6.3.22) with

$$A(\zeta) = \int_{\mathbb{R}} f(\zeta, x)\,\mu^{\zeta}(dx) + \int_0^{\infty}\int_{\mathbb{R}} g(\zeta, x)\partial_{\zeta} G^t(\zeta, x)\,\mu^{\zeta}(dx)\,dt,$$

$$B(\zeta) = 2\int_0^{\infty}\int_{\mathbb{R}} g(\zeta, x)G^t(\zeta, x)\,\mu^{\zeta}(dx)\,dt.$$

If the martingale problem (6.3.22) is well posed, then the family Y^{ε} weakly converges in $\mathbb{C}(0, T)$ to a diffusion process Y, which is the unique solution of this martingale problem.

Before proving the theorem, we give some auxiliary constructions and results. For given x, ζ, we define iteratively

$$\begin{aligned}
&Y^{\varepsilon,0}(t) \equiv \zeta\\
&dX^{\varepsilon,0}(t) = \varepsilon^{-1} a\big(X^{\varepsilon,0}(t), Y^{\varepsilon,0}(t)\big)\,dt + \sigma\big(X^{\varepsilon,0}(t-)\big)\,dZ_{\varepsilon^{-1}t},\\
&dY^{\varepsilon,1}(t) = f\big(Y^{\varepsilon,0}(t), X^{\varepsilon,0}(t)\big)\,dt + \varepsilon^{-1/2} g\big(Y^{\varepsilon,0}(t), X^{\varepsilon,0}(t)\big)\,dt,\\
&dX^{\varepsilon,1}(t) = \varepsilon^{-1} a\big(X^{\varepsilon,1}(t), Y^{\varepsilon,1}(t)\big)\,dt + \sigma\big(X^{\varepsilon,1}(t-)\big)\,dZ_{\varepsilon^{-1}t}
\end{aligned} \tag{6.3.51}$$

with

$$X^{\varepsilon,0}(0) = X^{\varepsilon,1}(0) = x, \quad Y^{\varepsilon,1}(0) = \zeta.$$

Clearly, $X^{\varepsilon,0}$ is just another notation for X^{ε}_{ζ}, the solution to eq. (6.3.42).

We fix $\rho \in (0, 1/2)$, and denote

$$\tau_{\varepsilon} = \varepsilon^{1-\rho}.$$

Since $f(\zeta, x), g(\zeta, x)$ are bounded, we have

$$\big|Y^{\varepsilon}(t) - Y^{\varepsilon,0}(t)\big| = \big|Y^{\varepsilon}(t) - \zeta\big| \le Ct\varepsilon^{-1/2} \le C\varepsilon^{1/2-\rho}, \quad t \le \tau_{\varepsilon}. \tag{6.3.52}$$

Proposition 6.3.13. *There exists C such that for $t \le \tau_{\varepsilon}$*

$$\|X^{\varepsilon}(t) - X^{\varepsilon,0}(t)\|_{L_4} \le C\big(1 + |x|\big)\varepsilon^{1/2-\rho}, \tag{6.3.53}$$

$$\|Y^{\varepsilon}(t) - Y^{\varepsilon,1}(t)\|_{L_4} \le C\big(1 + |x|\big)\varepsilon^{1-2\rho}, \tag{6.3.54}$$

$$\|X^{\varepsilon}(t) - X^{\varepsilon,1}(t)\|_{L_2} \le C\big(1 + |x|\big)\varepsilon^{1-2\rho}. \tag{6.3.55}$$

Proof. We have

$$\begin{aligned}
d\big(X^{\varepsilon}(t) - X^{\varepsilon,0}(t)\big) &= \varepsilon^{-1}\Big(a\big(X^{\varepsilon}(t), Y^{\varepsilon}(t)\big) - a\big(X^{\varepsilon,0}(t), Y^{\varepsilon}(t)\big)\Big)\,dt\\
&\quad + \Big(\sigma\big(X^{\varepsilon}(t-)\big) - \sigma\big(X^{\varepsilon,0}(t-)\big)\Big)\,dZ_{\varepsilon^{-1}t} + \varepsilon^{-1}\Delta_a(t)\,dt
\end{aligned}$$

Here

$$\Delta_a(t) = \left(a\left(X^{\varepsilon,0}(t), Y^{\varepsilon}(t)\right) - a\left(X^{\varepsilon,0}(t), Y^{\varepsilon,0}(t)\right)\right)$$
$$= \left(a\left(X^{\varepsilon,0}(t), Y^{\varepsilon}(t)\right) - a\left(X^{\varepsilon,0}(t), \zeta\right)\right),$$

and we have by eq. (6.3.52) and assumption $\mathbf{A}_3$ that

$$|\Delta_a(t)| \le C\left(1 + \left|X^{\varepsilon,0}(t)\right|\right)\varepsilon^{1/2-\rho}, \quad t \le \tau_\varepsilon. \tag{6.3.56}$$

We repeat, with proper changes, the argument from the proof of statement 1 in Proposition 6.3.11. Namely, we apply the Itô formula and get that

$$\mathrm{d}\left(X^{\varepsilon}(t) - X^{\varepsilon,0}(t)\right)^4 = \varepsilon^{-1}\eta_t^{\varepsilon}\,\mathrm{d}t + 4\varepsilon^{-1}\left(X^{\varepsilon}(t) - X^{\varepsilon,0}(t)\right)^3\Delta_a(t)\,\mathrm{d}t + \mathrm{d}M^{\varepsilon}(t)$$

with a local martingale M^{ε} and

$$\eta_t^{\varepsilon} \le -4\theta\left(X^{\varepsilon}(t) - X^{\varepsilon,0}(t)\right)^4,$$

where θ is the constant from the drift condition $\mathbf{A}_5$. Fix ϑ such that

$$1 \le \vartheta u^{-3} + \frac{\theta}{2}u, \quad u > 0.$$

Then,

$$\left|\left(X^{\varepsilon}(t) - X^{\varepsilon,0}(t)\right)^3\Delta_a(t)\right| \le \vartheta\left(\Delta_a(t)\right)^4 + \frac{\theta}{2}\left(X^{\varepsilon}(t) - X^{\varepsilon,0}(t)\right)^4.$$

Applying the Itô formula once again, and using the same argument as in the proof of Proposition 5.1.4, based on time localization and the Fatou lemma, we get

$$e^{2\theta t/\varepsilon}\mathsf{E}\left(X^{\varepsilon}(t) - X^{\varepsilon,0}(t)\right)^4 \le C\varepsilon^{-1}\int_0^t e^{2\theta s/\varepsilon}\mathsf{E}\left(\Delta_a(t)\right)^4\,\mathrm{d}s.$$

By eq. (6.3.56) and Proposition 6.3.11, statement 2, we have

$$\mathsf{E}\left(\Delta_a(t)\right)^4 \le C\left(\varepsilon^{1/2-\rho}\right)^4,$$

hence

$$\mathsf{E}\left(X^{\varepsilon}(t) - X^{\varepsilon,0}(t)\right)^4 \le C\left(\varepsilon^{1/2-\rho}\right)^4\varepsilon^{-1}\int_0^t e^{-2\theta(t-s)/\varepsilon}\,\mathrm{d}s,$$

which completes the proof of eq. (6.3.53).

Inequality (6.3.54) follows from eqs. (6.3.52) and (6.3.53) by Lipschitz continuity of $f(\zeta, x), g(\zeta, x)$. Namely, we have

$$Y^\varepsilon(t) - Y^{\varepsilon,1}(t) = \int_0^t \Big(f\big(Y^\varepsilon(s), X^\varepsilon(s)\big) - f\big(Y^{\varepsilon,0}(s), X^{\varepsilon,0}(s)\big)\Big)\, ds + \varepsilon^{-1/2}\int_0^t \Big(g\big(Y^\varepsilon(s), X^\varepsilon(s)\big) - g\big(Y^{\varepsilon,0}(s), X^{\varepsilon,0}(s)\big)\Big)\, ds.$$

Since

$$\left\| \int_0^t \Big(f\big(Y^\varepsilon(s), X^\varepsilon(s)\big) - f\big(Y^{\varepsilon,0}(s), X^{\varepsilon,0}(s)\big)\Big)\, ds \right\|_{L_4} \le Ct\varepsilon^{1/2-\rho} \le C\varepsilon^{3/2-2\rho},$$

$$\left\| \varepsilon^{-1/2}\int_0^t \Big(f\big(Y^\varepsilon(s), X^\varepsilon(s)\big) - f\big(Y^{\varepsilon,0}(s), X^{\varepsilon,0}(s)\big)\Big)\, ds \right\|_{L_4} \le C\varepsilon^{1-2\rho},$$

we get eq. (6.3.54).

Next, we note that, repeating literally the argument used above in order to get eq. (6.3.53) from eq. (6.3.52), one can prove also

$$\|X^{\varepsilon,1}(t) - X^{\varepsilon,0}(t)\|_{L_4} \le C\big(1 + |x|\big)\varepsilon^{1/2-\rho}. \tag{6.3.57}$$

The same argument can be used also to derive inequality eq. (6.3.55) from eq. (6.3.54). The minor difference here is that eq. (6.3.52) is a uniform bound, while eq. (6.3.54) is a (weaker) L_4-bound. Hence instead of eq. (6.3.56) we have

$$\|a\big(X^{\varepsilon,1}(t), Y^\varepsilon(t)\big) - a\big(X^{\varepsilon,1}(t), Y^{\varepsilon,1}(t)\big)\|_{L_2} \le C\Big\|1 + \big|X^{\varepsilon,1}(t)\big|\Big\|_{L_4} \|Y^\varepsilon(t) - Y^{\varepsilon,1}(t)\|_{L_4}.$$

Then it follows from Proposition 6.3.11 (statement 2), eqs. (6.3.57) and (6.3.54) that

$$\|a\big(X^{\varepsilon,1}(t), Y^\varepsilon(t)\big) - a\big(X^{\varepsilon,1}(t), Y^{\varepsilon,1}(t)\big)\|_{L_2} \le C\varepsilon^{1-2\rho}.$$

The rest of the proof remains the same: applying the Itô formula we get

$$\mathsf{E}\big(X^\varepsilon(t) - X^{\varepsilon,1}(t)\big)^2 \le C\Big(\varepsilon^{1-2\rho}\Big)^2 \varepsilon^{-1}\int_0^t e^{-2\theta(t-s)/\varepsilon}\, ds. \qquad \square$$

We deduce from eqs. (6.3.52) and (6.3.53) the following

Corollary 6.3.14.

$$Ef\big(Y^\varepsilon(t), X^\varepsilon(t)\big) = Ef\big(\zeta, X^\varepsilon_\zeta(t)\big) + Q^{t,\varepsilon,f}(\zeta, x),$$

where

$$|Q^{t,\varepsilon,f}(\zeta, x)| \le C\big(1 + |x|\big)\varepsilon^{1/2-\rho}, \quad t \le \tau_\varepsilon.$$

Clearly, the same statement holds true for $g(\zeta, x)$, as well, but it is not strong enough for our further needs because the corresponding term in SDE (6.3.21), which defines Y^ε, contains an extra $\varepsilon^{-1/2}$. To handle this term, we perform further analysis. Denote

$$\mathscr{U}^\varepsilon(t) = \varepsilon^{-1/2} \int_0^t g(X^{\varepsilon,0}(s), \zeta)\, ds,$$

and consider linear SDE

$$\begin{aligned} d\mathscr{V}^\varepsilon(t) &= \varepsilon^{-1}(\partial_x a)\big(X^{\varepsilon,0}(t), \zeta\big)\mathscr{V}^\varepsilon(t)\, dt + (\partial_x \sigma)\big(X^{\varepsilon,0}(t), \zeta\big)\mathscr{V}^\varepsilon(t)\, dZ_{\varepsilon^{-1}t} \\ &\quad + \varepsilon^{-1}(\partial_\zeta a)\big(X^{\varepsilon,0}(t), \zeta\big)\mathscr{U}^\varepsilon(t)\, dt, \\ \mathscr{V}^\varepsilon(0) &= 0. \end{aligned} \tag{6.3.58}$$

Proposition 6.3.15. *There exists C such that for $t \le \tau_\varepsilon$*

$$\big\|X^\varepsilon(t) - X^{\varepsilon,0}(t) - \mathscr{V}^\varepsilon(t)\big\|_{L_2} \le C\big(1 + |x|\big)\varepsilon^{1-2\rho}. \tag{6.3.59}$$

Proof. We decompose

$$\begin{aligned} &\Big(a\big(X^{\varepsilon,1}(t), Y^{\varepsilon,1}(t)\big) - a\big(X^{\varepsilon,0}(t), Y^{\varepsilon,0}(t)\big)\Big) \\ &\qquad = \Big(a\big(X^{\varepsilon,1}(t), Y^{\varepsilon,1}(t)\big) - a\big(X^{\varepsilon,1}(t), Y^{\varepsilon,0}(t)\big)\Big) \\ &\qquad + \Big(a\big(X^{\varepsilon,1}(t), Y^{\varepsilon,0}(t)\big) - a\big(X^{\varepsilon,0}(t), Y^{\varepsilon,0}(t)\big)\Big). \end{aligned} \tag{6.3.60}$$

Using the Taylor formula and condition $\mathbf{A}_4$, we get

$$\begin{aligned} &\Big(a\big(X^{\varepsilon,1}(t), Y^{\varepsilon,1}(t)\big) - a\big(X^{\varepsilon,1}(t), Y^{\varepsilon,0}(t)\big)\Big) \\ &\qquad = (\partial_\zeta a)\big(X^{\varepsilon,1}(t), Y^{\varepsilon,0}(t)\big)\big(Y^{\varepsilon,1}(t) - Y^{\varepsilon,0}(t)\big) + \Theta^{\varepsilon,1}(t), \\ &\Big(a\big(X^{\varepsilon,1}(t), Y^{\varepsilon,0}(t)\big) - a\big(X^{\varepsilon,0}(t), Y^{\varepsilon,0}(t)\big)\Big) \\ &\qquad = (\partial_x a)\big(X^{\varepsilon,0}(t), Y^{\varepsilon,0}(t)\big)\big(X^{\varepsilon,1}(t) - X^{\varepsilon,0}(t)\big) + \Theta^{\varepsilon,2}(t), \end{aligned}$$

where

$$|\Theta^{\varepsilon,1}(t)| \le C\Big(1 + |X^{\varepsilon,1}(t)|\Big)C\big(Y^{\varepsilon,1}(t) - Y^{\varepsilon,0}(t)\big)^2,$$
$$|\Theta^{\varepsilon,2}(t)| \le C\big(X^{\varepsilon,1}(t) - X^{\varepsilon,1}(t)\big)^2.$$

We have

$$\Big|Y^{\varepsilon,1}(t) - Y^{\varepsilon,0}(t)\Big| \le C\varepsilon^{1/2-\rho}, \quad t \le \tau_\varepsilon,$$

hence by Proposition 6.3.11 (statement 2) and eq. (6.3.57) we have

$$\Big\|\Theta^{\varepsilon,1}(t)\Big\|_{L_2} \le C\big(1 + |x|\big)\varepsilon^{1-2\rho}.$$

By eq. (6.3.57), we also have

$$\Big\|\Theta^{\varepsilon,2}(t)\Big\|_{L_2} \le C\big(1 + |x|\big)\varepsilon^{1-2\rho}.$$

Similarly, one can show that

$$\begin{aligned}(\partial_\varsigma a)\big(X^{\varepsilon,1}(t), Y^{\varepsilon,0}(t)\big)\Big)\big(Y^{\varepsilon,1}(t) - Y^{\varepsilon,0}(t)\big)\\ = (\partial_\varsigma a)\big(X^{\varepsilon,0}(t), Y^{\varepsilon,0}(t)\big)\Big)\big(Y^{\varepsilon,1}(t) - Y^{\varepsilon,0}(t)\big) + \Theta^{\varepsilon,3}(t)\end{aligned}$$

with

$$\Big\|\Theta^{\varepsilon,3}(t)\Big\|_{L_2} \le C\big(1 + |x|\big)\varepsilon^{1-2\rho},$$

we leave details for the reader.

Note that

$$|(Y^{\varepsilon,1}(t) - Y^{\varepsilon,0}(t) - \mathscr{U}^\varepsilon(t)| = \left|\int_0^t f\big(X^{\varepsilon,0}(s), Y^{\varepsilon,0}(s)\big)\,\mathrm{d}s\right| \le C\varepsilon^{1-\rho}, \quad t \le \tau_\varepsilon. \tag{6.3.61}$$

Performing a similar analysis for the term

$$\Big(\sigma\big(X^{\varepsilon,1}(t-)\big) - \sigma\big(X^{\varepsilon,0}(t-)\big)\Big),$$

we finally conclude that

$$\mathscr{W}^\varepsilon(t) = X^\varepsilon(t) - X^{\varepsilon,0}(t) - \mathscr{V}^\varepsilon(t)$$

satisfies

$$d\mathscr{W}^{\varepsilon}(t) = \varepsilon^{-1}(\partial_x a)\big(X^{\varepsilon,0}(t),\zeta\big)\mathscr{W}^{\varepsilon}(t)\,dt + (\partial_x\sigma)\big(X^{\varepsilon,0}(t),\zeta\big)\mathscr{V}^{\varepsilon}(t)\,dZ_{\varepsilon^{-1}t} + \varepsilon^{-1}\Theta^{\varepsilon}(t)\,dt + \Sigma^{\varepsilon}(t-)\,dZ_{\varepsilon^{-1}t},$$

where

$$\|\Theta^{\varepsilon}(t)\|_{L_2} \le C\varepsilon^{1-2\rho}, \quad \|\Sigma^{\varepsilon}(t)\|_{L_2} \le C\varepsilon^{1-2\rho}.$$

The rest of the proof is similar to the proofs of two previous propositions, hence we just outline it. Namely, we have eq. (6.3.50), hence for $\mathscr{W}^{\varepsilon}(t)$ the argument, used in the proofs of eqs. (6.3.53) and (6.3.55), can be repeated with obvious changes, which gives

$$\|\mathscr{W}^{\varepsilon}(t)\|_{L_2} = \left\|X^{\varepsilon}(t) - X^{\varepsilon,0}(t) - \mathscr{V}^{\varepsilon}(t)\right\|_{L_2} \le C\big(1+|x|\big)\varepsilon^{1-2\rho}.$$

Combined with eq. (6.3.55), this completes the entire proof. □

Using the Taylor formula and condition $\mathbf{A}_4$ once again, we directly get the following

Corollary 6.3.16.

$$Eg\big(Y^{\varepsilon}(t), X^{\varepsilon}(t)\big) = Eg\big(\zeta, X^{\varepsilon}_{\zeta}(t)\big) + E(\partial_{\zeta}g)\big(\zeta, X^{\varepsilon}_{\zeta}(t)\big)\mathscr{U}^{\varepsilon}(t) + E(\partial_x g)\big(\zeta, X^{\varepsilon}_{\zeta}(t)\big)\mathscr{V}^{\varepsilon}(t) + Q^{t,\varepsilon,g}(\zeta, x),$$

where

$$|Q^{t,\varepsilon,g}(\zeta,x)| \le C\big(1+|x|\big)\varepsilon^{1-2\rho}, \quad t \le \tau_{\varepsilon}.$$

Since $\mathscr{U}^{\varepsilon}(t)$ is expressed directly and $\mathscr{V}^{\varepsilon}(t)$ solves a linear equation, the latter formula can be rearranged into a more explicit form. Namely, consider the family

$$\mathscr{E}^{\varepsilon}_{\zeta}(s;t) \quad t \ge s$$

defined by the linear SDE

$$\begin{aligned} d\mathscr{E}^{\varepsilon}_{\zeta}(s;t) &= \varepsilon^{-1}(\partial_x a)\big(X^{\varepsilon}_{\zeta}(t),\zeta\big)\mathscr{E}^{\varepsilon}_{\zeta}(s;t)\,dt + (\partial_x\sigma)\big(X^{\varepsilon}_{\zeta}(s;t-)\big)\mathscr{E}^{\varepsilon}_{\zeta}(s;t-)\,dZ_{\varepsilon^{-1}t}, \\ \mathscr{E}^{\varepsilon}_{\zeta}(s;s) &= 1. \end{aligned} \tag{6.3.62}$$

Then the solution $\mathscr{V}^\varepsilon(t)$ to eq. (6.3.58) can be written as follows:

$$\begin{aligned}\mathscr{V}^\varepsilon(t) &= \varepsilon^{-1}\int_0^t \mathscr{E}_\zeta^\varepsilon(s;t)(\partial_\zeta a)\big(X_\zeta^\varepsilon(s),\zeta\big)\mathscr{U}^\varepsilon(s)\,\mathrm{d}s \\ &= \varepsilon^{-3/2}\int_0^t\int_0^s \mathscr{E}_\zeta^\varepsilon(s;t)(\partial_\zeta a)\big(X_\zeta^\varepsilon(s),\zeta\big)g\big(X_\zeta^\varepsilon(v),\zeta\big)\,\mathrm{d}v\,\mathrm{d}s.\end{aligned}$$

Consider also the family

$$\mathscr{H}_\zeta^\varepsilon(s;t),\quad t\geq s,$$

given by

$$\begin{aligned}\mathrm{d}\mathscr{H}_\zeta^\varepsilon(s;t) &= \varepsilon^{-1}(\partial_x a)\big(X_\zeta^\varepsilon(t),\zeta\big)\mathscr{H}_\zeta^\varepsilon(s;t)\,\mathrm{d}t + (\partial_x\sigma)\big(X_\zeta^\varepsilon(s;t-)\big)\mathscr{E}_\zeta^\varepsilon(s;t-)\,\mathrm{d}Z_{\varepsilon^{-1}t}, \\ &\quad + \varepsilon^{-1}(\partial_\zeta a)\big(X_\zeta^\varepsilon(t),\zeta\big)\,\mathrm{d}t, \\ \mathscr{H}_\zeta^\varepsilon(s;s) &= 0.\end{aligned}\tag{6.3.63}$$

Then,

$$\mathscr{H}_\zeta^\varepsilon(s;t) = \varepsilon^{-1}\int_s^t \mathscr{E}_\zeta^\varepsilon(v;t)(\partial_\zeta a)\big(X_\zeta^\varepsilon(v),\zeta\big)\,\mathrm{d}v,$$

and we get finally

$$\mathscr{V}^\varepsilon(t) = \varepsilon^{-1/2}\int_0^t \mathscr{H}_\zeta^\varepsilon(s;t)g\big(X_\zeta^\varepsilon(s),\zeta\big)\,\mathrm{d}s.$$

Comparing eq. (6.3.63) with (6.3.58), we observe the following interpretation of the family $\{\mathscr{H}_\zeta^\varepsilon(s;t)\}$. Consider the family

$$X_\zeta^\varepsilon(\chi;s;t),\quad t\geq s$$

of solutions to eq. (6.3.42) with the initial conditions

$$X_\zeta^\varepsilon(\chi;s;s) = \chi.$$

Then,

$$\mathscr{H}_\zeta^\varepsilon(s;t) = \partial_\zeta X_\zeta^\varepsilon(\chi;s;t)\Big|_{\chi=X_\zeta^\varepsilon(s)}.$$

Summarizing all the above, we finally get the following

Corollary 6.3.17.

$$\begin{aligned} Eg\big(Y^\varepsilon(t), X^\varepsilon(t)\big) &= Eg\big(\zeta, X_\zeta^\varepsilon(t)\big) \\ &\quad + \varepsilon^{-1/2}\int_0^t E\Big(\partial_\zeta G^{(t-s)/\varepsilon}\Big)\big(\zeta, X_\zeta^\varepsilon(s)\big)g\big(\zeta, X_\zeta^\varepsilon(s)\big)\, ds + Q^{t,\varepsilon,g}(\zeta, x), \end{aligned}$$

with the same $Q^{t,\varepsilon,g}(\zeta, x)$ as in Corollary 6.3.16.

Proof. We have

$$\begin{aligned} &\mathsf{E}(\partial_\zeta g)\big(\zeta, X_\zeta^\varepsilon(t)\big)\mathscr{U}^\varepsilon(t) + \mathsf{E}(\partial_x g)\big(\zeta, X_\zeta^\varepsilon(t)\big)\mathscr{V}^\varepsilon(t) \\ &\quad = \varepsilon^{-1/2}\int_0^t \mathsf{E}\Big[(\partial_\zeta g)\big(\zeta, X_\zeta^\varepsilon(t)\big) + (\partial_\zeta g)\big(\zeta, X_\zeta^\varepsilon(t)\big)\mathscr{H}_\zeta^\varepsilon(s;t)\Big]g\big(\zeta, X_\zeta^\varepsilon(s)\big)\, \mathrm{ds} \\ &\quad = \varepsilon^{-1/2}\int_0^t \Big[\partial_\zeta \mathsf{E}g(\zeta, X_\zeta^\varepsilon(\chi; s; t))\Big]_{\chi = X_\zeta^\varepsilon(s)} g\big(\zeta, X_\zeta^\varepsilon(s)\big)\, \mathrm{ds} \\ &\quad = \varepsilon^{-1/2}\int_0^t \Big(\partial_\zeta G^{(t-s)/\varepsilon}\Big)\big(\zeta, X_\zeta^\varepsilon(s)\big)g\big(\zeta, X_\zeta^\varepsilon(s)\big)\, \mathrm{ds}. \end{aligned}$$

□

Proof of Theorem 6.3.12. We will apply Theorems 6.3.3 and 6.3.5 to the AR family (5.2.1) with

$$\xi_k^\varepsilon = \int_{\varepsilon(k-1)}^{\varepsilon k} f(Y^\varepsilon(s), X^\varepsilon(t))\, \mathrm{d}t + \varepsilon^{-1/2}\int_{\varepsilon(k-1)}^{\varepsilon k} g(Y^\varepsilon(t), X^\varepsilon(t))\, \mathrm{d}t =: \xi_k^{f,\varepsilon} + \xi_k^{g,\varepsilon}$$

and filtration $\mathbb{F}^\varepsilon = \{\mathscr{F}_k^\varepsilon\}$,

$$\mathscr{F}_k^\varepsilon = \sigma(X^\varepsilon(0), Z_s, s \le k).$$

Since $f(\zeta, x), g(\zeta, x)$ are bounded, we easily have eq. (6.3.5) with any $p > 2$. The most difficult part of the proof will be to show that $\mathbf{H}_0$ holds true with $L^\varepsilon = [\varepsilon^{-\rho}]+1$ with properly chosen ρ and the terms $\gamma_{k,l}^\varepsilon, \upsilon_{k,l}^\varepsilon, R_l$ having the required set of properties; namely, eqs. (6.3.3), (6.3.4), (6.3.11) and (6.3.14).

Note that the conditional law of

$$\Big\{\big(X^\varepsilon(t - \varepsilon k), Y^\varepsilon(t - \varepsilon k)\big), t \ge \varepsilon k\Big\}$$

w.r.t. $\mathscr{F}_k^\varepsilon$ equals to the law of the solution to eq. (6.3.41) with initial values

$$x_0 = X^\varepsilon(\varepsilon k), \quad y_0 = Y^\varepsilon(\varepsilon k)$$

and another (say, independent) noise Z. Denote

$$F^t(\zeta, x) = \mathsf{E}_x f(\zeta, X_\zeta(t)),$$

then it follows from Corollary 6.3.14 that

$$\mathsf{E}\Big[f\big(Y^\varepsilon(t), X^\varepsilon(t)\big)\Big|\mathscr{F}_k^\varepsilon\Big] = F^{t/\varepsilon-k}\big(Y^\varepsilon(\varepsilon k), X^\varepsilon(\varepsilon k)\big) + \delta_k^{\varepsilon,f}(t),$$

where

$$\delta_k^{\varepsilon,f}(t) \le C\varepsilon^{1/2-\rho}\big(1+|X^\varepsilon(\varepsilon k)|\big), \quad t \in [\varepsilon k, \varepsilon(k+l)], \quad l \le L_\varepsilon.$$

Clearly, for $F^t(\zeta, x)$ the analogue of eq. (6.3.46) holds true:

$$|F^t(\zeta, x) - \tilde{F}(\zeta)| \le Ce^{-\theta t}\big(1+|x|\big), \quad \tilde{F}(\zeta) = \int_{\mathbb{R}} g(\zeta, x)\mu^\zeta(dx).$$

On the other hand, one can show that, for some C,

$$\mathsf{E}\big(X^\varepsilon(t)\big)^4 \le C, \quad t \ge 0, \tag{6.3.64}$$

the proof here is standard, for example, statement 2 of Proposition 6.3.11, and we leave it for the reader. We can conclude the above calculations as follows:

$$\mathsf{E}\Big[\xi_{k+l}^{\varepsilon,f}|\mathscr{F}_k^\varepsilon\Big] = \varepsilon\tilde{F}(\zeta_k^\varepsilon) + \Delta_{k,l}^{\varepsilon,f},$$

where

$$\big\|\Delta_{k,l}^{\varepsilon,f}\big\|_{L_4} \le C\varepsilon^{3/2-\rho} + Ce^{-\theta l}.$$

Next, by Corollary 6.3.17 and eq. (6.3.64) we have

$$\mathsf{E}\Big[\xi_{k+l}^{\varepsilon,g}|\mathscr{F}_k^\varepsilon\Big] = \varepsilon^{1/2}\Upsilon_{k,l}^\varepsilon + \varepsilon\Gamma_{k,l}^\varepsilon + \Lambda_{k,l}^{\varepsilon,g}$$

with

$$\Upsilon_{k,l}^\varepsilon = \varepsilon^{-1}\int_{\varepsilon(k+l-1)}^{\varepsilon(k+l)} G^{t/\varepsilon-k}\big(Y^\varepsilon(\varepsilon k), X^\varepsilon(\varepsilon k)\big)\, dt,$$

$$\Gamma_{k,l}^\varepsilon = \varepsilon^{-2}\int_{\varepsilon(k+l-1)}^{\varepsilon(k+l)}\int_{\varepsilon k}^{t} \mathsf{E}_x\Big(\partial_\zeta G^{t/\varepsilon-s/\varepsilon-k}\Big)\big(\zeta, X_\zeta^\varepsilon(s)\big) \times g\big(\zeta, X_\zeta^\varepsilon(s)\big)\, ds\, dt\Big|_{x=X^\varepsilon(\varepsilon k), \zeta=Y^\varepsilon(\varepsilon k)},$$

and

$$\left\|\Lambda_{k,l}^{\varepsilon,g}\right\|_{L_4} \le C\varepsilon^{3/2-2\rho}.$$

By the centering condition $\mathbf{C}_1$, "stabilization" bound (6.3.46), and the moment inequality (6.3.64), we have

$$\left\|\Gamma_{k,l}^{\varepsilon}\right\|_{L_4} \le Ce^{-\theta l}.$$

Similarly, by eq. (6.3.47) we have

$$\left\|\Gamma_{k,l}^{\varepsilon} - \tilde{\Gamma}\big(Y^{\varepsilon}(\varepsilon k)\big)\right\|_{L_4} \le Ce^{-\theta l}$$

with

$$\tilde{\Gamma}(\zeta) = \int_0^{\infty}\int_{\mathbb{R}} (\partial_\zeta G^t)(\zeta, x) g(\zeta, x)\, \mu^{\zeta}(\mathrm{d}x)\,\mathrm{d}t.$$

Summarizing the above calculation, we conclude that, for

$$\rho < \frac{1}{4},$$

condition $\mathbf{H}_0$ holds true with

$$L^{\varepsilon} = [\varepsilon^{-\rho}] + 1, \quad R_l = e^{-\theta l},$$

and the terms $\gamma_{k,l}^{\varepsilon}, \upsilon_{k,l}^{\varepsilon}$ satisfying eq. (6.3.4) with $p = 4$, eq. (6.3.11) with the coefficient $A(\zeta)$ specified in the formulation of the theorem, and eq. (6.3.14).

In a similar and simpler way, we can verify assumptions eqs. (6.3.12) and (6.3.15). The terms $\xi_k^{\varepsilon,f}$ now are negligible, and for the calculation of

$$\mathsf{E}\left[\xi_{k+L^{\varepsilon}}^{g,\varepsilon}\xi_{k+L^{\varepsilon}+j}^{g,\varepsilon}\middle|\mathscr{F}_k^{\varepsilon}\right]$$

we can use just eqs. (6.3.52) and (6.3.53), which, combined with the moment bound (6.3.64) give

$$\left\|\mathsf{E}\left[\xi_{k+L^{\varepsilon}}^{g,\varepsilon}\xi_{k+L^{\varepsilon}+j}^{g,\varepsilon}\middle|\mathscr{F}_k^{\varepsilon}\right] - B_j\big(Y^{\varepsilon}(\varepsilon k)\big)\right\|_{L_4} \le C\varepsilon^{1/2-\rho} + C\varepsilon^{-1}e^{-L_{\varepsilon}},$$

where

$$B_j(\zeta) = \int_j^{j+1}\int_0^1\int_{\mathbb{X}} g(\zeta, x)G^{t-s}(\zeta, x)\,\mu(\mathrm{d}x)\,\mathrm{d}s\,\mathrm{d}t.$$

The functions $A(\zeta), B_j(\zeta), j \geq 0$ are continuous because they are obtained as averages of Hölder continuous functions. We have

$$|B_j(\zeta)| \leq Ce^{-\theta j},$$

hence the series (6.3.13) converges uniformly. That is, all the assumptions of Theorem 6.3.5 are verified. The end of the proof is literally the same as in the proof of Theorem 6.3.7. □

We finish this section, giving one typical example where Theorem 6.3.12 is well applicable.

Example 6.3.18. Let

$$a(x,\zeta) = -a(\zeta)x + \alpha(\zeta), \quad \sigma(x) = \sigma \neq 0;$$

that is, X^ε satisfies the SDE similar to the one for the general Lévy-driven Ornstein-Uhlenbeck process, see eq. (6.3.65), with the coefficients a, α depending on the "slow" component. If a, α, σ and f, g belong to C_b^2, then conditions $\mathbf{A}_2$–$\mathbf{A}_4$ are satisfied. If

$$a(\zeta) \geq a_0 > 0,$$

then the drift condition (6.3.44) is satisfied as well. Assume also conditions $\mathbf{A}_1$ and $\mathbf{C}_2$ to hold true, which are just independent conditions on the Lévy process Z and the initial value of the "fast" component. Then, in order to apply Theorem 6.3.12 we only need to verify $\mathbf{C}_1$.

For the SDE

$$\mathrm{d}U(t) = \Big(-aU(t) + \alpha \Big)\,\mathrm{d}t + \sigma\,\mathrm{d}Z_t, \quad U(0) = x, \tag{6.3.65}$$

the solution can be given in the integral form:

$$U(t) = e^{-at}x + \alpha \int_0^t e^{-a(t-s)}\,\mathrm{d}s + \sigma \int_0^t e^{-a(t-s)}\,\mathrm{d}Z_s,$$

and its law is completely (though a bit implicitly) specified by its characteristic function

$$\phi_t^{a,\alpha,\sigma}(x,\xi) = \mathsf{E}e^{i\xi U(t)} = \exp\left(i\xi\left(e^{-at}x + \alpha\int_0^t e^{-as}\,\mathrm{d}s \right) + \int_0^t \psi\big(\sigma e^{-as}\xi\big)\,\mathrm{d}s \right),$$

where $\xi \in \mathbb{R}$ and we denote

$$\psi(\xi) = -\frac{b}{2}\xi^2 + \int_{\mathbb{R}} \left(e^{i\xi u} - 1 - i\xi u\right) \nu(du),$$

the characteristic exponent for the process Z. Corresponding IPM has the characteristic function

$$\phi^{a,\alpha,\sigma}(\xi) = \exp\left(i\xi m + \int_0^\infty \psi\left(\sigma e^{-as}\xi\right) ds \right), \quad m = \frac{\alpha}{a},$$

and since the Lévy measure ν of Z is assumed to be symmetric, this distribution is symmetric w.r.t. the point m. Consequently, for a function $g(\zeta, x)$ to be centered w.r.t. μ^ζ, it is sufficient that it is anti-symmetric w.r.t. the point

$$m(\zeta) = \frac{\alpha(\zeta)}{a(\zeta)};$$

that is,

$$g\left(\zeta + m(\zeta), x\right) = -g\left(- \zeta + m(\zeta), x\right). \tag{6.3.66}$$

That is, assuming eq. (6.3.66), Theorem 6.3.12 can be applied. Note that for the corresponding "frozen" process the transition probabilities $P^{t,\zeta}(x, dy)$ and the IPM μ^ζ are specified by their characteristic functions

$$\phi^{t,\zeta}(x,\xi) = \exp\left(i\xi \left(e^{-a(\zeta)t}x + \alpha(\zeta)\int_0^t e^{-a(\zeta)s}\, ds \right) + \int_0^t \psi\left(\sigma e^{-a(\zeta)s}\xi\right) ds \right),$$

$$\phi^\zeta(\xi) = \exp\left(i\xi m(\zeta) + \int_0^t \psi\left(\sigma e^{-a(\zeta)s}\xi\right) ds \right).$$

6.4 Comments

1. For the early history and the physical background of the averaging principle and diffusion approximation, we refer to the survey in Ref. [42, Chapter 7]. For studying the models with fast Markov perturbations, one natural method is provided by the semigroup approach, see Ref. [38, Chapter 12] and the survey in Ref. [70]. This method appears to be well applicable in various settings where the semigroup associated with the Markov component exhibits stabilization as $t \to \infty$ in a sufficiently strong sense, see Ref. [19] for diffusion approximation for models with boundary

conditions, and Ref. [81] for the asymptotic analysis of the models with fast semi-Markov switching. The complexity of the semigroup methods increases drastically, once uniform (in a sense) stabilization condition is replaced by locally uniform one; one can clearly see this difference comparing Refs. [38, Theorem 12.2.4] and [38, Theorem 12.3.1], formulated for the driving noise valued in a *compact state space* and a *locally compact state space*, respectively. Such a difference is not so crucial within the martingale methods, which date back to Refs. [126] and [110], see Ref. [112]. The aim of Section 6.2 is to illustrate these methods by giving a sample theorem, where several limitations, hidden in the previous versions of the method (see the discussion at the end of Section 6.2), are mainly removed.

2. The time delay method, developed in Section 6.3.1, apparently is new. Since it does not require construction of the corrector term, it is well adapted to the asymptotic study of the "fully coupled" systems, where the fast component depends on the slow one, and hence the study of correspondent potentials are complicated. Another, more hidden, motivation for this method, is given by paper Ref. [108], where the accuracy bound for CLT, given by Ref. [85, Theorem 3.13], is established. The accuracy bound therein, up to a slowly varying term, has the form $n^{-1/4}$, which is better than the general bound $n^{-1/8}$ in the martingale CLT by Ref. [96], but still is far from the classical Berry-Esseen bound $n^{-1/2}$. The latter bound is not genuine for the i.i.d. setting, it can be achieved for Markov models with L_2-spectral gaps, see Ref. [41], and (up to slowly varying terms) for stable models with fast mixing, see Ref. [127]. A thorough check of the proofs in Ref. [108] shows that the error, brought to the overall estimate by the corrector term itself, has the order $n^{-1/4}$. The time delay method is free from such a limitation, and it is plausible that the error bound within this method can be made close to the Berry-Esseen bound.

3. The approximation principle for a "fully coupled" system is not specific: taking $g \equiv 0$ in Theorem 6.3.12 we obtain a statement close to the one from Theorem 6.3.7, with the fast Markov component dependent just on the "frozen" slow one. The main difficulty in the "fully coupled" setting is brought by diffusion approximation type theorems, where the dependence of the fast Markov component on the slow one cannot be ignored. To our best knowledge, a "fully coupled" diffusion approximation was systematically studied in the diffusion setting, only, see Ref. [113]. We separately mention closely related *homogenization* type results for solutions of stochastic equations with oscillating coefficients. The way to treat such equations, commonly adopted nowadays, dates back to [6], and is actually a particular form of the martingale corrector term method. In this approach, the fast component is treated as a diffusion process on a torus, which allows one to prove C^2-smoothness of the auxiliary solutions to the Poisson equations. For further applications and extensions, see [67, 99, 111, 114]. We also shortly note another type of asymptotic results, which dates back to Ref. [72], and provides a kind of the CLT in the averaging principle. For the detailed study of that type of results in the "fully coupled" setting and related physical models we refer to Refs. [2] and [74]; see also references therein.

4. The framework, adopted in Section 6.3.3, in the particular diffusion setting is rather restrictive, when compared with the one studied in Ref. [113]. This comparison shows that, in general, it should be possible to avoid using the dissipativity condition $\mathbf{A}_4$, which is a strong structural restriction on the model. In Section 6.3.3, thanks to such an assumption, we were able to design the asymptotic expansion of the pair $(X^\varepsilon, Y^\varepsilon)$ with the error bound $o(\varepsilon^{1/2})$; see eqs. (6.3.59) and (6.3.61). This calculation is comparatively simple and transparent, and can be modified to various settings where the dissipativity condition is available, for example, for SDDEs, see Section 4.2.2. One can expect that, without the dissipativity assumption, the required asymptotic expansion still holds true for the semigroup defined by the pair $(X^\varepsilon, Y^\varepsilon)$, rather than for this pair itself. In the diffusion setting, such an expansion was developed in Ref. [113] by means of the classical PDE theory. In many other cases (for Lévy-driven SDEs, SDDEs, etc.) these analytic tools are not available, but we expect that methods from Part I, after a proper extension, will be effective enough to provide the desired expansions; this is a wide and promising field for further research.

Bibliography

[1] Abourashchi, N., Veretennikov, A.Yu.: On stochastic averaging and mixing, Theory Stoch. Process. 16, no. 1, 111–129 (2010).

[2] Bakhtin, V., Kifer, Yu.: Diffusion approximation for slow motion in fully coupled averaging, Prob. Th. Related Fields, 129, 157–181 (2004).

[3] Bakry, D., Cattiaux, P., and Guillin A.: Rate of convergence for ergodic continuous Markov processes: Lyapunov versus Poincaré, J. Funct. Anal. 254, no. 3, 727–759 (2008).

[4] Bakry, D., Gentil, I., and Ledoux, M.: Analysis and Geometry of Markov Diffusion Operators, volume 348 of Grundlehren der mathematischen Wissenschaften, Springer (2014).

[5] Bally, V., Clément, E.: Integration by parts formula and applications to equations with jumps, Probab. Theory Relat. Fields, 151, no. 3–4, 613–657 (2011).

[6] Bensoussan, A. and Lions, J.-L. and Papanicolaou, G.: Studies in Mathematics and its Applications: Asymptotic analysis for periodic structures, North-Holland. 5, (1978).

[7] Bernstein, S.N.: Sur l'extension du théoréme limite du calcul des probabilités aux sommes de quantités dépendantes, Math. Ann., 97, 1–59 (1926).

[8] Bhattacharya, R.N.: On the functional central limit theorem and the law of the iterated logarithm for Markov processes. Z. Wahr. verw. Gebiete, 60, 185–201 (1982).

[9] Bichteler, K., Gravereaux, J.-B., Jacod, J.: Malliavin Calculus for Processes with Jumps. Gordon and Breach, New York (1987).

[10] Billingsley, P.: Convergence of Probability Measures (2nd edition), Wiley, New York (1999).

[11] Bodnarchuk, S., Kulik, A.: Conditions for existence and smoothness of the distribution density of an Ornstein-Uhlenbeck process with Lévy noise, Prob. Theory and Math. Stat., 79, 20–33 (2008).

[12] Bradley, R.: On some results of MI Gordin: a clarification of a misunderstanding, J. Theor. Probab., no. 2, 115–119 (1988).

[13] Butkovsky, O.: Subgeometric rates of convergence of Markov processes in the Wasserstein metric, Ann. Appl. Probab., 24, no. 2, 526–552 (2014).

[14] Butkovsky, O., Scheutzow, M.: Invariant measures for stochastic functional differential equations, arXiv:1703.05120.

[15] Butkovsky, O., Veretennikov, A.: On asymptotics for Vaserstein coupling of Markov chains, Stochastic Processes Appl., 123, 3518–3541 (2013).

[16] Cattiaux, P., Chafaï, D., and Guillin, A.: Central limit theorems for additive functionals of ergodic Markov diffusions processes, ALEA, Lat. Am. J. Probab. Math. Stat., 9, no. 2, 337–382 (2012).

[17] Cattiaux, P., Guillin, A.: Hitting times, functional inequalities, Lyapunov conditions and uniform ergodicity, J. Funct. Anal., 272, no. 6, 2361–2391 (2017).

[18] Cattiaux, P., Kulik, A., Fradon, M., and Roelly, S.: Long time behavior of stochastic hard ball systems, Bernoulli 22, no. 2, 681–710 (2016).

[19] Costantini, C., Kurtz, T.G.: Diffusion approximations of transport processes with general reflecting boundary conditions, Math. Models Methods Appl. Sci., 16, 717–762 (2006).

[20] Da Prato, G., Zabczyk, Z.: Stochastic Equations in Infinite Dimensions, Cambridge University Press, Cambridge (1992).

[21] Da Prato, G., Zabczyk, Z.: Ergodicity for Infinite-Dimensional Systems, Cambridge University Press, Cambridge (1996).

[22] Davies, E.B.: Heat Kernels and Spectral Theory, Cambridge University Press, Cambridge (1989).

[23] Derriennic, Y., Lin, M.: The central limit theorem for Markov chains with normal transitions, started at a point, Probab. Theory Relat. Fields, 119, 508–528 (2001).

DOI 10.1515/9783110458930-008

[24] Derriennic, Y., Lin, M.: The central limit theorem for Markov chains started at a point, Probab. Theory Relat. Fields, 125, 73–76 (2003).
[25] Dobrushin, R.L.: Central limit theorem for nonstationary Markov chains, Theor. Prob. Appl., 1, 65–80; 329–383 (1956).
[26] Doeblin, W.: Sur les propriétés asymptotiques de mouvements régis par certains types de chaînes simples, Thése de doctoratés Sciences Mathematiques, Paris (1938).
[27] Doeblin, W.: Eléments dùne théorie générale des chaînes simples constantes de Markoff, Ann. Sci. E.N.S., 57, 61–111 (1940).
[28] Doob, J.L.: Stochastic Processes. Wiley, New York (1953).
[29] Douc, R., Fort, G., Moulines, E., and Soulier, P.: Practical drift conditions for subgeometric rates of convergence. Ann. Appl. Probab., 14, no. 3, 1353–1377 (2004).
[30] Douc, R., Fort, G., Guillin, A.: Subgeometric rates of convergence of f-ergodic strong Markov processes, Stoch. Proc. Appl., 119, no. 2, 897–923 (2009).
[31] Dudley, R.M.: Real Analysis and Probability. Cambridge University Press, Cambridge (2002).
[32] Durmus, A., Fort, G., and Moulines, E.: Subgeometric rates of convergence in Wasserstein distance for Markov chains, Ann. Inst. H. Poincaré Probab. Statist., 52, no. 4, 1799–1822 (2016).
[33] Dynkin, E.B.: Markov Processes, vols. 1 and 2, Academic Press, New York (1965).
[34] Dynkin, E.B., Yushkevich, A.A.: Controlled Markov Processes, Springer, Berlin (1979).
[35] E, W., Mattingly, J.C., and Sinai, Y.: Gibbsian dynamics and ergodicity for the stochastically forced Navier-Stokes equation, Comm. Math. Phys., 224, no. 1, 83–106 (2001).
[36] Eidelman, S.D., Ivasyshen, S.D., and Kochubei, A.N.: Analytic Methods in the Theory of Differential and Pseudo-differential Equations of Parabolic Type, Birkhäuser, Basel (2004).
[37] Eberle, A., Guillin, A., and Zimmer, R.: Quantitative Harris type theorems for diffusions and McKean-Vlasov processes, arXiv:1606.06012.
[38] Ethier, S.N., Kurtz, T.G.: Markov Processes. Characterization and Convergence, Wiley, New York (1986).
[39] Feller, W.: Introduction to Probability Theory, vol. 1, Wiley, New York (1967).
[40] Feng, J., Kurtz, T.G.: Large Deviations for Stochastic Processes, Mathematical Surveys and Monographs, vol. 131, AMS, Providence (2006).
[41] Ferré, D., Hervé, L., and Ledoux, J.: Limit theorems for stationary Markov processes with L2-spectral gap, Annales de lÎnstitut Henri Poincaré Probabilités et Statistiques 48, no. 2, 396–423 (2012).
[42] Freidlin, M. I., Wentzell, A.D.: Random Perturbations of Dynamical Systems (2nd ed.), Springer-Verlag, New York (1998).
[43] Friedman, A.: Partial Differential Equations of Parabolic Type, Prentice-Hall, New York (1964).
[44] Gikhman, I.I., Skorokhod, A.V.: Introduction to the Theory of Random Processes, W. B. Saunders (1969).
[45] Gikhman, I.I., Skorokhod, A.V.: Stochastic Differential Equations, Springer, New York (1972).
[46] Glatt-Holtz, N.E., Mattingly, J.C., and Richards, G.: On unique ergodicity in nonlinear stochastic partial differential equations, J. Stat. Phys., 166, 618–649 (2017).
[47] Gordin, M.I.: On the central limit theorem for stationary processes, Sov. Math. Dokl., 10, 1174–1176 (1969).
[48] Gordin, M.I., Lifšic, B.A.: Central limit theorem for stationary Markov processes, Dokl. Akad. Nauk SSSR 239, no. 4, 766–767 (1978).
[49] Griffeath, D.: Uniform coupling of non-homogeneous Markov chains, J. Appl. Probab., 12, no. 4, 753–762 (1975).
[50] Hairer, M.: Exponential mixing properties of stochastic PDEs through asymptotic coupling. Probab. Theory Related Fields, 124, no. 3, 345–380 (2002).

[51] Hairer, M., Mattingly, J.: Ergodicity of the 2D Navier-Stokes equations with degenerate stochastic forcing, Ann. Math., 164, 993–1032 (2006).
[52] Hairer, M.: Ergodic properties of Markov processes, Lecture Notes (2006), http://www.hairer.org/notes/Markov.pdf
[53] Hairer, M.: Convergence of Markov Processes, Lecture Notes (2016), http://www.hairer.org/notes/Convergence.pdf
[54] Hairer, M., Mattingly, J.: Yet another look at Harris' ergodic theorem for Markov chains. In Seminar on Stochastic Analysis, Random Fields and Applications VI, Progr. Probab. Birkhauser, Basel (2010).
[55] Hairer, M., Mattingly, J., and Scheutzow, M.: Asymptotic coupling and a general form of Harris' theorem with applications to stochastic delay equations, Prob. Theory Rel. Fields, 149, no. 1–2, 223–259 (2011).
[56] Hall, P., Heyde, C.C.: Martingale Limit Theory and its Application, Academic, New York (1980).
[57] Harris, T.E.: The existence of stationary measures for certain Markov processes, In: Proceedings of the Third Berkeley Symposium on Mathematical Statistics and Probability, 1954–1955, vol. II, 113–124. University of California Press, Berkeley (1956).
[58] Helland, I.S.: Central limit theorem for martingales with discrete or continuous time., Scand. J. Statist., 9, 79–94 (1982).
[59] Ibragimov, I.A.: A central limit theorem for one class of dependent random variables, Teor. Veroyatnost. i Primenen., 8, no. 1, 89–94, (1963).
[60] Ibragimov, I.A., Linnik, Yu.V.: Independent and Stationary Sequences of Random Variables, Wolters-Noordhoff, Groningen (1971).
[61] Ikeda, N., Watanabe, S.: Stochastic differential equations and diffusion processes, North-Holland, Amsterdam (1981).
[62] Isaacson, D.L., Madsen R.W.: Markov Chains: Theory and Applications, Wiley, New York, London, Sydney, Toronto (1976).
[63] Ishikawa, Y., Kunita, H.: Existence of density for canonical differential equations with jumps, Stoch. Process. Appl., 116, no. 12, 1743–1769 (2006).
[64] Jacod, J., Shiryaev, A.: Limit Theorems for Stochastic Processes, Grundlehren der Mathematischen Wissenschaften, vol. 288, SpringerVerlag, Berlin, Heidelberg, New York (1987).
[65] Jarner, S., Roberts, G.: Polynomial convergence rates of Markov chains. Ann. Appl. Probab., 12, 224–247 (2002).
[66] Jeganathan, P., Le Cam, L.: On Lévy's Martingale central limit theorem, Sankhyā: Indian J. Stat., Ser A (1961–2002), 47, no. 2, 141–155 (1985).
[67] Jikov, V. V. and Kozlov, S. M. and Oleinik, O. A.: Homogenization of Differential Operators and Integral Functionals, Springer, (1994).
[68] Jones, G.L.: On the Markov chain central limit theorem, Probab. Surv., 1, 299–320 (2004).
[69] Kallenberg, O.: Foundations of Modern Probability, Probability and its Applications, Springer-Verlag, New York (1997).
[70] Kang, H.-W., Kurtz, T.G., and Popovic, L.: Central limit theorems and diffusion approximations for multiscale Markov chain models, Ann. App. Prob., 24, no. 2, 721–759 (2014).
[71] Kartashov, N.V.: Strong Stable Markov Chains, VSP, Utrecht (1996).
[72] Khas'minskii, R.Z.: On stochastic processes defined by differential equations with a small parameter, Theory Probab. Appl., 11, no. 2, 211–228 (1966).
[73] Khas'minskii, R.Z.: Stochastic Stability of Differential Equations, Sijthoff & Noordhoff, Rockville (1980) 2nd edition.: Springer, Berlin Heidelberg (2011).
[74] Kifer, Yu.: Large deviations and adiabatic transitions for dynamical systems and Markov processes in fully coupled averaging, Memoirs Am. Math. Soc. 944, AMS, Providence (2009).

[75] Kipnis, C., Varadhan, S.R.S.: Central limit theorem for additive functionals of reversible Markov processes and applications to simple exclusion, Comm. Math. Phys., 104, 1–19 (1986).
[76] Klokov S. A., Veretennikov, A.Yu.: Sub-exponential mixing rate for a class of Markov chains, Math. Comm., 9, 9–26 (2004).
[77] Klokov S. A., Veretennikov, A.Yu.: On subexponential mixing rate for Markov processes, Theory Probab. Appl., 49, no. 1, 110–122 (2005).
[78] Komorowski T., Walczuk, A.: Central limit theorem for Markov processes with spectral gap in the Wasserstein metric, Stoch. Proc. Appl., 122, 2155–218 (2012).
[79] Komorowski, T., Peszat, Sz., Szarek, T.: On ergodicity of some Markov processes, Ann. Prob., 38, 1401–1443 (2010).
[80] Kontoyiannis, I., Meyn, S.P.: Spectral theory and limit theorems for geometrically ergodic Markov processes., Ann. Appl. Prob., 13, no. 1, 304–362 (2003).
[81] Koroliuk, V.S., Limnios, N.: Stochastic Systems in Merging Phase Space, World Scientific, New Jersey (2005).
[82] Kulik, A.M.: Exponential ergodicity of the solutions to SDE's with a jump noise, Stoch. Proc. Appl., 119, No. 2, 602–632 (2009).
[83] Kulik, A.M.: Asymptotic and spectral properties of exponentially ϕ-ergodic Markov processes, Stoch. Proc. and Appl., 121, no. 5, 1044–1075 (2011).
[84] Kulik, A.M.: Absolute continuity and convergence in variation for distributions of functionals of Poisson point measure, J. Theor. Probab., 24, 1–38 (2011).
[85] Kulik, A.M.: Introduction to Ergodic Rates for Markov Chains and Processes, with Applications to Limit Theorems, Potsdam University Press, Potsdam (2015).
[86] Kulik, A.M., Scheutzow, M.: A coupling approach to Doob's theorem, Atti Accad. Naz. Lincei Rend. Lincei Mat. Appl., 26, 83–92 (2015).
[87] Kulik, A.M., Scheutzow, M.: Generalized couplings and convergence of transition probabilities, Probab. Theory Relat. Fields, DOI: 10.1007/s00440-017-0779-8, to appear.
[88] Kulik, A.M., Veretennikov, A.Yu.: On extended Poisson equation for weakly ergodic Markov processes, Prob. Theory Math. Stat., 85, 22–38 (2011).
[89] Kulik, A.M., Veretennikov, A.Yu.: Diffusion approximation for systems with weakly ergodic Markov perturbations I, II, Prob. Theory Math. Stat., 87, 1–16 (2012); 88, 1–16 (2013).
[90] Kunita, H.: Absolute continuity of Markov processes and generators, Nagoya Math. J., 36, 1–26 (1969).
[91] Kunita, H.: Stochastic Flows and Stochastic Differential Equations, Cambridge University Press, Cambridge (1997).
[92] Lévy, P.: Propriétés asymptotiques des sommes de variables al?atoires enchainees. C. R. Acad. Sei. Paris, 199, 627–629 (1934).
[93] Lévy, P.: Propriétés asymptotiques des sommes de variables aléatoires indpendantes ou enchainees. J. Math. Pures et Appliqués, 34, 347–402 (1935).
[94] Lévy, P.: Complément á un théoréme sur la loi de Gauss. Bull, des Sciences Mathématiques, 61, 115 (1937).
[95] Lindvall T.: Lectures on the Coupling Method, John Wiley & Sons, New York (1992).
[96] Liptser R.S., Shiryayev, A.N.: On the rate of convergence in the central limit theorem for cemimartingales, Theory Probab. Appl., 27, no. 1, 1–13 (1982).
[97] Liptser R.S., Shiryayev, A.N.: Martingales and limit theorems for stochastic processes. In book: Anulova S.V. et al., Stochastic calculus (in Russian), VINITI, Moscow (1989).
[98] Liptser, R.S., Shiryaev, A.N.: Statistics of Random Processes: I. General (2nd ed.), Springer, Berlin (2001).
[99] Makhno, S. Ya.: Stochastic Equations. Limit Theorems (in Russian), Naukova Dumka (2012).

[100] Markov, A.A.: Rasprostranenie zakona bolshikh chisel na velichiny, zavisyashchie drug ot druga (The extension of the law of large numbers on mutually dependent variables). Isvestia Fiz.-mat. obsch. Kazan univ., 1906, 15, no. 4, 135–156. See also 100, 339–361.
[101] Markov, A.A.: Rasprostranenie predel'nyh teorem ischisleniya veroyatnostej na summu velichin svyazannyh v cep' (Extension of the limit theorems of probability theory to a sum of variables connected in a chain), Zapiski Akademii Nauk po Fiziko-matematicheskomu otdeleniyu, VIII seriya 22, no. 9 (1908). Translated into English. In: Dynamic Probabilities Systems (R.A. Howard ed.), vol. 1, Wiley, New York, 1971, 552–576. See also 100, 364–397.
[102] Markov, A.A.: Issledovanie obschego sluchaya ispytanij svyazannyh v cep' (Studies of a general case of trials connected in a chain), Zapiski Akademii Nauk po Fiziko-matematicheskomu otdeleniyu, VIII seriya 25, no. 3 (1910). See also 100, 465–508.
[103] Markov, A.A.: Izbrannye trudy. Teoria chisel. Teoria veroyatnostei (Selected Works. Number Theory. Probability Theory) (Yu.V. Linnik ed.), Moscow-Leningrad (1951).
[104] Mattingly, J.C.: Exponential convergence for the stochastically forced Navier-Stokes equations and other partially dissipative dynamics, Comm. Math. Phys., 230, no. 3, 421–462 (2002).
[105] Maxwell, M., Woodroofe, M.: Central limit theorem for additive functionals of Markov chains., Ann. Prob., 28, 713–724 (2000).
[106] Meyn, S.P., Tweedie, R.L.: Markov Chains and Stochastic Stability, Springer, London (1993).
[107] Mohammed, S.E.A.: Stochastic functional differential equations, Research Notes in Mathematics, 99, Pitman, Boston (1984).
[108] Molyboga, G.: An analogue of the Berry-Esseen theorem for functions of weakly ergodic Markov processes, Theory Stochastic Process., 37, no. 1, 53–63 (2016).
[109] Numellin, E.: General Irreducible Markov Chains and Non-negative Operators, 2nd edition, Cambridge University Press, Cambridge (2004).
[110] Papanicolaou, G.C., Stroock, D.W., and Varadhan, S.R.S.: Martingale approach to some limit theorems, In: Conference on Statistical Mechanics, Dynamical Systems and Turbulence (M. Reed ed.), Duke University Press, Durham (1977).
[111] Pardoux, E.: Homogenization of linear and semilinear second order parabolic PDEs with periodic coefficients: a probabilistic approach, Journal of Functional Analysis., 167, no. 2, 498–520 (1999).
[112] Pardoux, E., Veretennikov A.Yu.: On Poisson equation and diffusion approximation 1, Ann. Prob., 29, 1061–1085 (2001).
[113] Pardoux, E., Veretennikov A.Yu.: On Poisson equation and diffusion approximation 2, Ann. Prob., 31, 1166–1192 (2003).
[114] Pavliotis, G. and Stuart, A.: Multiscale methods: averaging and homogenization, Springer (2008)
[115] Peszat, S., Zabczyk, J.: Stochastic Partial Differential Equations with Lévy Noise (an Evolution Equation Approach). Cambridge University Press, Cambridge (2007).
[116] Picard, J.: On the existence of smooth densities for jump processes, Probab. Theory Rel. Fields, 105, 481–511 (1996).
[117] Rachev, S.T., Rüschendorf, L.: Mass Transportation Problems: Volume I: Theory, Springer, Berlin (1998).
[118] von Renesse, M., Scheutzow, M.: Existence and uniqueness of solutions of stochastic functional differential equations, Random Oper. Stochastic Equ., 18, no. 3, 267–284 (2010).
[119] Revuz, D., Yor, M.L.: Continuous Martingales and Brownian Motion, 3rd ed. Grundlehren der Mathematischen Wissenschaften [Fundamental Principles of Mathematical Sciences], vol. 293, Springer-Verlag, Berlin (1999).
[120] Samorodnitsky, G., Taqqu, M.S.: Stable Non-Gaussian Random Processes: Stochastic Models with Infinite Variance, Chapman and Hall, New York (1994).

[121] Sato, K.: Lévy Processes and Infinitely Divisible Distributions, 2nd edition. Cambridge University Press (2013).
[122] Sato, K., Yamazato, M.: Operator-selfdecomposable distributions as limit distributions of processes of Ornstein-Uhlenbeck type, Stoch. Proc. Appl., 17, no. 1, 73–100 (1984).
[123] Scheutzow, M.: Exponential growth rates for stochastic delay differential equations, Stoch. Dyn., 5, 163–174 (2005).
[124] Skorokhod, A.V.: Asymptotic methods in the theory of Stochastic differential equations, Trans. Math. Monographs, AMS, Rhode Island, 2008.
[125] Szarek, T.: The Uniqueness of Invariant Measures for Markov Operators. Studia Math., 189, 225–233 (2008).
[126] Stroock, D.W., Varadhan, S.R.S.: Multidimensional Diffusion Processes, Springer, Berlin (1979).
[127] Tihomirov, A.N.: Convergence rate in the central limit theorem for weakly dependent random variables, Theory Probab. Appl., 25, no. 4, 790–809 (1980).
[128] Thorisson, H.: Coupling, Stationarity, and Regeneration. Springer, New York (2000).
[129] Tuominen, P., Tweedie, R.: Subgeometric rates of convergence of f-ergodic Markov chains. Adv. Appl. Probab., 26, 775–798 (1994).
[130] Tsybakov, A.: Introduction to Nonparametric Estimation, Springer, New York (2009).
[131] Veretennikov, A.Yu.: Bounds for the mixing rates in the theory of stochastic equations, Theory Probab. Appl., 32, 273–281 (1987).
[132] Veretennikov, A.Yu.: On polynomial mixing bounds for stochastic differential equations, Stoch. Proc. Appl., 70, no. 1, 115–127 (1997).
[133] Veretennikov, A.Yu.: Ergodic Markov Processes and Poisson Equations, Lecture Notes, arXiv:1610.09661.
[134] Wagner, D.H.: Survey of measurable selection theorems: An update, Lect. Notes Math., 794, 176–219 (1980).
[135] Wang, F.Y.: Functional Inequalities, Markov Processes and Spectral Theory, Science Press, Beijing (2005).
[136] Zolotarev, V.M.: Sums of Independent Random Variables: A modern Approach, Wiley, New York (1995).

Index

DOI 10.1515/97831104589301-009

De Gruyter Studies in Mathematics

Volume 66
Igor V. Nikolaev
Noncommutative Geometry. A Functorial Approach, 2017
ISBN 978-3-11-054317-9, e-ISBN 978-3-11-054525-8, Set-ISBN 978-3-11-054526-5

Volume 65
Günter Mayer
Interval Analysis. And Automatic Result Verification, 2017
ISBN 978-3-11-050063-9, e-ISBN 978-3-11-049946-9, Set-ISBN 978-3-11-049947-6

Volume 64
Dorina Mitrea, Irina Mitrea, Marius Mitrea, Michael Taylor
The Hodge-Laplacian. Boundary Value Problems on Riemannian Manifolds, 2016
ISBN 978-3-11-048266-9, e-ISBN 978-3-11-048438-0, Set-ISBN 978-3-11-048439-7

Volume 63
Evguenii A. Rakhmanov
Orthogonal Polynomials, 2016
ISBN 978-3-11-031385-7, e-ISBN 978-3-11-031386-4, Set-ISBN 978-3-11-038325-6

Volume 62
Ulrich Krause
Positive Dynamical Systems in Discrete Time, 2015
ISBN 978-3-11-036975-5, e-ISBN 978-3-11-036569-6, Set-ISBN 978-3-11-036571-9

Volume 61
Francesco Altomare, Mirella Cappelletti Montano, Vita Leonessa, Ioan Rasa
Markov Operators, Positive Semigroups and Approximation Processes, 2015
ISBN 978-3-11-037274-8, e-ISBN 978-3-11-036697-6, Set-ISBN 978-3-11-036698-3

Volume 60
Vladimir A. Mikhailets, Alexandr A. Murach, Peter V. Malyshev
Hörmander Spaces, Interpolation, and Elliptic Problems, 2014
ISBN 978-3-11-029685-3, e-ISBN 978-3-11-029689-1, Set-ISBN 978-3-11-029690-7

Volume 59
Jan de Vries
Topological Dynamical Systems, 2014
ISBN 978-3-11-034073-0, e-ISBN 978-3-11-034240-6, Set-ISBN 978-3-11-034241-3

www.ingramcontent.com/pod-product-compliance
Ingram Content Group UK Ltd.
Pitfield, Milton Keynes, MK11 3LW, UK
UKHW050953290926
15470UKWH00006B/1802

9 783110 458701